Global Warming

Stephen H. Schneider

Global Warming

Are We Entering the Greenhouse Century?

The Lutterworth Press
Cambridge

The Lutterworth Press
P.O. Box 60
Cambridge CB1 2NT

British Library Cataloguing in Publication Data
Schneider, Stephen H. (Stephen Henry)
Global warming.
1. Climate. Effects of carbon dioxide
I. Title
551.6

ISBN 0-7188-2815-1

First published in Great Britain by
The Lutterworth Press 1990

Printed in Great Britain by
St Edmundsbury Press Ltd
Bury St Edmunds, Suffolk

To Adam and Rebecca,

in the hope that their generation will be

more creative in adapting to the greenhouse century

than mine has been in preventing it.

Contents

Preface

Some Thoughts on the Greenhouse Century

In 1988 the environment was as big a story as politics, AIDS, or baseball. Heat, drought, air pollution, and forest fires filled the front pages of newspapers, newsweeklies, and TV news cover stories for months. The greenhouse effect and global warming had emerged from academia and government offices to mingle with popular culture. Unknown scientists appeared on national television or in the pages of *Time* magazine. New bills to control the climatic threat were introduced into Congress, international leaders called for global action, and some spoke of reviving the environmental ethic that was obscured during the 1980s. But others said it was premature to act and folly to over-react. Battle lines were being drawn for what promises to be one of the most important political debates of this—and the next—century: what can or should we do to avert the possibility of an unprecedented threat to the global environment, global warming?

This book is about our common future and how we might deal with the prospect of rapid climatic change caused by human economic and population growth that now looms dramatically before us. We'll start by examining the greenhouse effect, greenhouse gases and their control on global temperature. We'll also look at the earth's climate history, especially the extraordinary events of the last million years or so—the ice ages. Understanding them will help tremendously in understanding how climate may change in the future. We'll look at the case for climate stability to see if somehow nature might save us from ourselves. We'll examine the ways climate models—our cloudy crystal balls—work, how the media treats (and mistreats) climate news, and—perhaps most importantly—what the impact might be if we are entering the greenhouse century—a century of increasing global temperature. For example, if global temperatures rise 2° to 6°C (3.6° to 10.8°F), what might happen to agriculture, public health, sea levels, inland water resources, air quality, forests, the economy, and international security?

If the public is to exercise its right to balance environmental, economic, and social values, then it must give informed advice to its political leaders. But how can any citizen send signals to politicians on complex issues neither understands very well? More generally, how can any nonspecialist deal with controversial scientific questions? Just because an issue is complicated and involves technical debate does not mean the public is unqualified to participate in the *policymaking* process. On the contrary, if people learn what kinds of questions to ask experts, then not only can they participate more effectively in the democratic process (for example, to help choose strategies to deal with climatic change issues), but they also will be better prepared to solve other technically complicated problems with societal implications (for example, arms control, nuclear waste disposal, AIDS, budget deficits, or taxes and tariffs). But how to deal with experts on complex matters takes techniques that are all too often excluded from our education. We'll deal with these techniques extensively later on.

I believe that scientists have the obligation to explain to those who pay our salaries what we have learned about complex issues: what we think we know well, what we don't know, what we might be able to learn, when we might learn it, how much it might cost to learn, and, most importantly, what it might mean for civilization. (These are, incidentally, an outline of the kinds of questions I was referring to above.) This means explaining issues in plain language, using appropriate but familiar metaphors, and holding back nothing—that is, letting the public decide whether a certain amount of uncertainty or knowledge is sufficient for action. This scientific equivalent of full disclosure is preferable to nondemocratic choices made by an élite group of experts. In the 1912 presidential campaign, Woodrow Wilson recognized the risks of government by specialists when he said, "What I fear is a government of experts . . . What are we for if we are to be scientifically taken care of by a small number of gentlemen who are the only men who understand the job? Because if we do not understand the job, then we are not a free people."

However, it is difficult enough for most scientists to communicate complicated information to each other, let alone to the public or politicians. Although we scientists do not have a formal Hippocratic oath like our medical counterparts, all scientists are ethically bound to search for truth, no matter where it leads or what cherished beliefs new knowledge may overturn. At the same time, scientists are human beings: most want to leave the world a better place then they found it. This creates what I see as a double ethical bind. On the one hand, loyalty to the scientific method requires "truth, the whole truth, and nothing but the truth," which in the communication of science means including *all* the uncertainties and caveats. On the other hand, the scientist-as-citizen wants to make the world a better place by advancing his or her own worldview. This has two ethical requirements: first, that scientists acknowledge their value systems so that their worldviews are transparent; and second, that they effectively capture the public's attention, if not its imagination, in order to create positive change. Often this means simple and dramatic statements—the easiest way to get media coverage. Familiar and direct metaphors or other well-known expressions are usually necessary to get one's views into the media—as the cliché says: no media, no message. The double ethical bind for communicating science to the public, then, is for the scientist to find an appropriate balance between being an effective agent for change and being honest about the limitations of the state of knowledge.

There is no simple formula for resolving the dilemma of balancing effectiveness against full disclosure, for one scientist's clear simplification could well be another's irresponsible oversimplification. Each tries to find the best path across this treacherous ethical ground. In the end, I hope my solution to the double ethical bind presented here is neither too opaque for most nonspecialists nor too oversimplified for most scientists. I try to resist the natural temptation to take refuge in the scientific beauty of technical details—such detail is not usually the best way to present complicated and controversial points to the public. Instead, I emphasize what lies beneath the climate change debate: the vying for attention of opposing advocate groups, such as environmentalists and industrialists and their political allies, whose statements make up much of the media accounts of global warming. At the same time, I weave in the often conflicting positions of scientists, some of whom are more sophisticated than others in the art of communication and political advocacy. It is my overall goal to make less baffling for the nonspecialist the confusing set of conflicting statements reported from various members of these groups.

Despite all the confusion, we know a lot about how the climate works and might change; nevertheless, much remains uncertain. Whether our present knowledge is sufficient for major political action or the uncertainties are enough to delay action is the value choice that society must make. That choice can only effectively be made if the public understands the context in which global warming is reported in the media. Sometimes in the following pages, in order to make the climate debates more transparent, I interrupt the logical presentation of scientific ideas to digress to actual accounts of behind-the-scenes debates among scientists, environmentalists, politicians, business leaders, and journalists, many of whom have made headlines. I feel justified in minimizing logical and detailed scientific presentations here partly because an earlier book of mine, *The Coevolution of Climate and Life* (Sierra Club Books, 1984), extensively covers the science. I hope readers who are particularly curious about the detailed science I omit here (or interleave with private stories or the background to publicly visible debates) will refer to *Coevolution*.

My principal purpose with *Global Warming* is to make the basic issues and policy implications of global atmospheric change more accessible to the public, which means focusing on what is most visible and putting it in the context of scientific knowledge and public policy values. If the only consequence of my effort is to spark a wider debate over the appropriateness of immediate actions to avert change or make adaptation more effective in the greenhouse century, then I shall be satisfied. Incidentally, to answer responsibly the question posed in the subtitle of this book takes eight chapters. But the one line response is simple: I believe we are already a decade or two into the greenhouse century; a decade or two more should prove that beyond a reasonable doubt.

Acknowledgements

Writing a book in four months at such short notice that it is too late to alter an already packed schedule is a taxing venture for the author and everyone around him, both at home and in the office. Against that background, I wish to express deep appreciation to my office staff, who had to deal not only with the hundreds of additional inquiries that followed in the wake of the heat and drought of the summer of '88, but also with the added stress of daily doses of manuscript to be typed or edited, references checked and looming deadlines to be met. In particular, I wish to thank Mary Rickel for effectively co-ordinating this administrative nightmare, Susan Mikkleson for the bulk of the word processing and for a superb job of tracking down scores of obscure references given only scraps from my memory. I would also like to thank Lynda Faulds for efficient typing of early drafts of a few chapters. In addition, I wish to thank A. Brewster Rickel for producing an unusually thorough and cross-referenced index and Lee Fortier for efficient drafting of most of the figures.

No interdisciplinary work with as broad a scope as *Global Warming* should be published without the help of knowledgeable specialists to review early draft manuscripts. In particular, I thank Stephen G. Warren, David Pollard, and Linda O. Mearns for taking the time to read early draft manuscripts line by line. Their sharp eyes and good comments substantially improved the accuracy and clarity of the final manuscript. I want to thank my Sierra Club Books editor Jim Cohee for his part in initiating and seeing through to completion the publication process, especially in view of the compressed time frame. I also thank his staff, especially several copy editors whose manuscript reworkings improved the readability of the final product. Lastly, but never least, are the major victims of an author who writes in his spare time: his family members. My wife and children put up with a man whose mind was frequently in the next century when his body was at the dinner table. I also thank Cheryl for helpful comments on the organization and presentational style of the manuscript. For Rebecca and Adam I hope that whatever impact *Global Warming* may have on marginally easing the stress on their future will prove enough to compensate for the temporary loss of a full-time father in their growing years.

I also acknowledge the support of my institution, the National Center for Atmospheric Research, which is sponsored by the National Science Foundation. Of course, any opinions and findings expressed herein are my own.

Global Warming

1

Shadows of the Climate Future

No one can know the future, at least not in detail. But enough is known to allow us to fashion plausible scenarios of events that could well take place if current trends and present understanding are even partly true. This chapter, while obviously fictional, is meant to provide a feeling for what a year in the greenhouse century might have in store for us if nothing is done to deal with the growing problem of global warming. Just how plausible these shadows of future climate might be will be explored in subsequent chapters.

Old timers said they hadn't seen anything like it for decades. It reminded them of the protests in the 1980s when South African teams had tried to compete in Great Britain. After apartheid ended in the early 1990s, such political demonstrations at sporting events passed into the history books. Today, the Australian Rugby Union side were playing a crucial test match with an England XV, but the competitive suspense was secondary to the political tension.

"Free the refugees!", "Inhumane treatment for the boat people", "Homes not hovels". These and other placards were thrust in the faces of the rugby followers trying to sneak past the shouts and signs through the turnstyles. Some Australian fans reacted angrily as they shoved past the protestors: "We didn't make this problem!

We didn't ask for it; and you wouldn't do any better if it happened to you."

What had happened to Australia that led to such a display of animosity from old friends? The Australian supporters were right, the problem indeed was not primarily of their creation. Furthermore, it's not clear what Britain, France, the United States or anybody else would have done differently. Fortunately for those countries, they had yet to face an identical situation.

In the first few decades of the twenty-first century population growth did finally slow dramatically in South Asia, but not until the region's population had nearly doubled from the late twentieth century level. In the meantime, pressure for improved diets had caused intensive agricultural development, including the cutting of coastal forests in Indonesia and the extensive farming of marginal lowlands there and in Bangladesh. These developments proceeded rapidly, despite warnings in the 1990s that such farming might not be sustainable in the face of warming global temperatures, rising sea levels and an increased probability of more violent hurricanes and associated floods from storm surges. As the twenty-first century unfolded, population pressures, the need for space and the desire for better living standards continued to drive the transmigration of people from the overpopulated Indonesian island of Java onto the coastal plains of the outer islands like Sumatra and Kalimantan. In Bangladesh, the same kinds of population pressures forced people to continue to rely on every parcel of the Ganges Delta for agricultural sustenance. But, in the past five years three devastating hurricanes struck. Two-hundred-and-fifty thousand people were killed in Bangladesh in the first storm, and another 2.7 million were made homeless. Because of the intense competition for land and jobs away from the lowlands, there simply was little room inland for most of these people to move. So, some took to boats, as flotillas of displaced homeless began to criss-cross Asian seas. These "environmental refugees", as they were dubbed in the 1980s by Sir Crispin Tickell, the then British Ambassador to the United Nations, were refused admission in many ports. Boats were lost at sea, and horrific stories of death, disease and desparation trickled into the media. One small colony settled undetected in a remote part of Western Australia. The compassionate but concerned Australians debated and allowed them to stay. However, after the super-cyclone last year, thousands of people from Indonesia joined a growing stream from Bangladesh, climbing into anything that would float over the Indian Ocean or the Banda Sea. Merchant ships reported that they often couldn't sail a day without seeing debris, floating suitcases, bodies and makeshift conveyances providing evidence of the armada of environmental refugees. This year, the tropical storms were more moderate, but striking the battered coastal lowlands for the third time in five years was a trigger. Millions took to the seas, each with different sets of languages and little cultural familiarity with Australia. The Australian authorities were soon faced with the prospect of refusing entry and sinking the ships, setting up refugee camps, or with wide-scale integration of these unskilled homeless into their society. Millions of dollars were spent and a High Commission on Refugees was established. But the thrust of the effort was to contain the unwelcomed "colonists" temporarily while more suitable homes were sought—hopefully in Asia. Medical supplies, food, sanitation, and cultural amenities that citizens of the world are entitled to—and Australians, in particular, believed all people deserved—simply weren't provided at a rate or amount satisfactory to many outside critics, such as those picketing the Australian team.

Ironically, Australia was one of the nations that strongly advocated international global greenhouse gases emissions cuts in the 1990s. She now complained at the United Nations that her efforts decades earlier were watered-down by such nations as the United States, Britain, the U.S.S.R. and Japan, and thus these nations had the obligation, since they helped create the current problem by blocking substantial emissions cuts, to help Australia out both financially and by taking some refugees off their hands. Although the problem of rising seas and the prospect of stronger storms[1] was considered a plausible scenario in the early 1990s, traditional economists successfully convinced the major developed countries that the costs of cutting carbon dioxide emissions to slow the rate of global warming were in the trillions of dollars, too high given that sea walls to protect many coastlines against rising waters cost less. What these "rational" cost/benefit studies neglected was the human, political and security costs of a world destabilized by millions of environmental refugees.

Two days after the controversial rugby match, a special session of the United Nations was convened to deal with the growing problem of environmental refugees, and to urgently find solutions to the substantial human suffering and security threats to the region that the situation had created.

In New York City, water restrictions had been in effect for almost a month, lawns were browning, swimming pools had been covered, and the evening news was filled with stories of stiff fines and public reprimands for families caught violating the ordinance. For twelve of the fourteen days preceding that Sunday game, New York City had experienced temperatures above 35° C (95° F—by then the United States had finally gone metric), substantially increasing the demand for water. Torrents gushed onto city streets from illegally opened fire hydrants. The health department issued smog alerts announcing that low-level ozone had reached historic highs. The elderly and people with asthma or lung disease were cautioned to stay indoors on Monday, when normal traffic and industrial activity would once again dump pollutants into the still-hot air.

The evening news reported that already six hundred people—mostly the sick, elderly, or poor without protection of air-conditioning—had died that month of causes directly attributable to heat stress. Nevertheless, this was a substantial reduction from the heat-death numbers a few decades earlier. Beginning in the 2010s, New York had attempted to air-condition all buildings to reduce the physiological distress from ever-increasing heat waves. This city policy was not without opponents, however, for a number of critics pointed out that humans can acclimatize their bodies to hot weather, and that although air-conditioning would clearly protect the weak and vulnerable segments of the population, continuous air-conditioning on a massive scale would also reduce people's ability to acclimatize. Then, should a superhot week descend on the city, with an excessive demand for air-conditioning resulting in power cuts, even more people would be vulnerable to heat stress.

Later that week the national news media carried a story of growing conflict between New York City water authorities and those in charge of the reservoirs along the Delaware River. It seemed absurd to most New Yorkers that at this time of heat stress, Delaware River water was being sent downriver into the ocean rather than to New York simply because of an old legal requirement that mandated upstream releases of fresh water to protect Philadelphia, as well as downstream fisheries, from encroaching salinity. When the river flow was slow, as it was this drought year, ocean water backed up into the Delaware estuary, threatening fisheries and fragile ecosystems, as well as groundwater storage. The Hudson

River was also having salinity problems. The increased salinity of both rivers was due not only to drought and reduced fresh-water flows but also to the rising sea level—more than a centimetre (nearly half an inch) per year for the past several decades. The reservoir shortfall at upstate New York, combined with the reduced transfer from the Delaware Basin, was becoming so critical that the city looked to its emergency backup station, located nearly 150 kilometres (about 90 miles) north on the Hudson River at Hyde Park. Unfortunately, the river flow was so low this year, and sea level had crept so far, that salt water intrusion had rendered Hudson River water at Hyde Park essentially undrinkable.

One enterprising reporter dug up a quote from an old 1987 report by the mayor's Intergovernment Task Force on New York Water Supply that had warned of the likelihood of the current crisis:

> The greenhouse effect could have profound implications for the region's water supply with impacts on the salt front in the Hudson and Delaware rivers, the groundwater resources of Long Island, and rainfall patterns in the region as a whole. Planning for the city's water supply future must take these possible impacts into account, and planners should keep informed about continuing work on this important area.

Capping off the reporter's story were the now-fateful replies of city water planners to such a threat. When the prospect of the greenhouse century had gained professional attention in the late 1980s, the American Association for the Advancement of Science (AAAS) had sponsored studies on the implications of projected climatic changes on water supplies throughout the century. In a 1989 study by Clark University water engineer Harry E. Schwarz and his doctoral student Lee A. Dillard, dozens of district water planners and managers were questioned about what the then very uncertain projections of climatic change might mean to their planning efforts. New York City officials had boasted that "the system is robust" and probably could withstand considerable climatic change. They stressed that before making costly adjustments to anticipate the possibility of climatic change, superiors would need "new unanimity among scientific and professional bodies," since "prediction of change in fifty years or so would only affect thinking and not action." The officials concluded that "everyone has to start to consider change, but New York City is not going to be the first to act."[2]

The political aftermath of this heat combined with the refugee situation was swift and sure, even if late: the United States began a push to rapidly establish a much more vigorous "law of the atmosphere" to replace the weak and symbolic agreement grudgingly passed by the United Nations at the end of the twentieth century. The U.S., Japan, and the U.K. joined together to keep emission cuts to a minimum on the grounds that such cuts were expensive and science was too uncertain. The old agreement had simply called for nations to use energy more efficiently, develop renewable energy resources more vigorously, and reforest where economically feasible. It had set up a small fund to help bring these changes about and had resulted in an estimated 10% reduction in the rate of buildup of carbon dioxide, methane and other trace greenhouse gases. Nevertheless, these gases had more than doubled from their preindustrial values.

But now the United States was calling for substantial emission limits from each nation and a crash programme to power the world with hydrogen produced from electricity. The electricity would be produced at a number of desert solar collector sites as well as at a few remotely located power parks containing a large concentration of small-scale nuclear reactors with specially designed fuel cycles and cooling mechanisms to prevent meltdown even in the event of operator incompetence. It had been recognized for decades that the seriousness of fossil fuel pollution required building of more of these power plants as their costs also slowly decreased relative to coal-powered plants. But their implementation was too slow to prevent a rapid buildup of greenhouse gases, a buildup that could have been cut substantially if taxes on fossil fuels hadn't been rejected by Congress in the 1990s.

It had taken the enormity of a Hurricane George for politicians finally to realize the immediacy and urgency of the problem of climate warming and to take legislative steps to control the industrial and agricultural sources of the problem. Too bad so much time had been wasted, in which much of the current problem could have been abated.

Chicagoans were growing weary of this hot summer. What happened one Sunday afternoon in Chicago actually originated over a thousand miles away in Saskatchewan, where a very intense three-week heat wave had come on the heels of a dry spring. Relative humidities had been abnormally low for the previous week.

Then a weak cold front came through, triggering many relatively rainless thundershowers. From the lightning, fires broke out simultaneously over several thousand square miles. Within days, an area nearly the size of New Jersey had burned, and the drifting smoke was so thick over the Midwest that the late afternoon looked more like twilight for several days,[3] In downstate Illinois, in what used to be one of the most productive agricultural areas in the country—a spot that had provided a major shot in the arm to the U.S. economy through its export of corn—frequent drought was producing an economic collapse of crisis proportions. Large corporate farms that were able to purchase water rights and implement efficient irrigation systems had bought out failing family farms at low prices and were now growing winter wheat instead of corn in the cooler, moister weather of autumn and early spring. Most smaller farmers or family farmers had either gone out of the farm business or joined the steady flight of immigrants to northern Michigan, Wisconsin and Minnesota, which due to the 3° C (5° F) climate warming that had taken place so far in this greenhouse century were now warm enough for growing corn, yet because of their more northerly latitude were less afflicted with intense heat.

The evening news in Minneapolis was punctuated with Minnesota conservation officials' complaints about the rapid rate of deforestation of the state's northern areas accompanying the northward migration of the corn belt. At the same time as land was being cleared for new agriculture, Minnesota was experiencing a rapid rate of forest dieback brought about by the warmer winters and increasingly frequent intense heat waves of the last few decades. While forest managers were fairly certain that maples and other hardwoods would eventually move in to replace the dying firs, they were concerned that it would take a dedicated planting effort to fill the several-hundred-year gap between the death of the fir forest and the natural succession to a mature, robust hardwood forest. Officials therefore also called for new reserves to help keep the influx of farmers away from some of the shrinking forest areas. Also problematic was the extreme fire danger posed by the dead and dying trees, especially during major hot spells like this summer.

Chicagoans had become used to depressing farm news in many small towns in the former corn belt, but they were becoming increasingly alarmed as the shores of Lake Michigan kept receding, with far-reaching economic and health effects. To begin with, ten years ago a number of developers had built dock facilities and highrise condos out into the lake; with increasing drought, the lake level dropped several feet, leaving docks and condos high and dry. Then a freak wet period of five years caused lake levels to rise above the level at which the condos had been built, nearly destroying those speculative developments. Now, a series of dry years with warm summers had caused lake levels to drop well below the level of a decade ago, and politicians argued constantly about whether developers should be allowed to follow the receding lake shore. Who would insure against the loss if lake levels rose again?

Who also should pay for the rebuilding of port facilities and the dredging of channels and locks connecting the Chicago River with the receding lake? And who should fund the hundreds of millions of dollars needed to relocate water intake and outflow pipes? Should the cost be borne by local communities or by a multistate bureaucracy under consideration by the lake states of Wisconsin, Illinois, Indiana, and Michigan? Already an international agency had been set up between Michigan, Ohio, Pennsylvania, New York, and Ontario, Canada, to deal with Lake Erie problems. As Lake Erie water levels had dropped, ships had had to go with lighter loads to prevent hitting bottom, except in areas where dredging was done. After a number of decades of costly losses in shipping tonnage, the international commission had eventually undertaken massive dredging, and shipping through Lake Erie had already shown signs of rebound. Now Lake Michigan's neighbours were facing the same problems.

Another serious problem facing the four Lake Michigan states was the nearly one hundred identified "hot spots" where fifty years' worth of toxic contaminants had accumulated in shore sediments. With the lower lake level a wide variety of these toxic sediments had Begun to emerge and through erosion were threatening the water quality of the lake. Should the four states create a mini superfund to deal with this dramatic problem or could federal aid be brought in? Already the federal government was facing hundreds of billions of dollars of extra bills to accelerate renewable energy systems, shore up coastlines, build irrigation and water control facilities, and deal with the social problems of the outmigration from lower midwestern farms and the collapse of coal mining as coal increasingly became an internationally unacceptable fuel.

Although not every year had been as dramatic as this, the increasing frequency of droughts, severe flooding episodes, air pollution crises, heat-stress days, power cuts, and forest fires was already stretching the national treasury. Fortunately, the urgent need for environmentally sustainable development had become an international priority in recent decades, and the emergence of a politically united Europe (East and West) along with a nonexpansionist Soviet Union had allowed a halving of military budgets on both sides. Resources were finally freed from military expenditure to deal with the growing environmental pressures of the greenhouse century and the development demands of still-overpopulated, underdeveloped, and overpolluted Third World countries.

Californians had felt relatively fortunate for the past decade or so, having largely escaped the intense heat waves and damaging storm surges of the East Coast. To be sure, there were a number of intensely unpleasant days, especially in Los Angeles, with a week or so each year of temperatures above 38° C (100° F) and serious air pollution crises, but by and large Californians had come through unscathed.

The one group that had suffered from the climatic warming trends of recent decades was the Central Valley farmers, who were heavily dependent upon irrigation. Agricultural water use had been heavily subsidized in the twentieth century, thus allowing rich development of these naturally fertile but hot and dry lands. However, as urban populations in southern California and the Bay Area had increased, competition for water had tripled its price to agriculture—a price now more accurately reflecting the actual cost of water engineering projects and their operating and maintenance expenses.

Although California had not been as stricken by drought as the Midwest or the East, its water supply was being seriously threatened from out of state. That is, the upper-Colorado-River-basin states of Colorado, Utah, and Wyoming were now suing the lower-basin states of California, Nevada, and Arizona to reduce the amount of water allocated to the latter by the Colorado River Compact of 1922. The upper-basin states complained that the compact erroneously forced them to provide a fixed amount of water to the lower-basin states, an allocation naively based on a two-decade experience with Colorado River water flow rates, which then averaged more than 17 million acre-feet per year at Lee's Ferry in northern Arizona. It had seemed equitable at the time for the upper-basin states to keep 7.5 million acre-feet and for the lower-basin states to receive 7.5 million acre-feet, with the remaining 2 million acre-feet reserved for Mexico and for maintenance of the water quality. For some reason, the fixed number of 7.5 million acre-feet was set as the annual obligation of the upper-basin states to the lower-basin states, regardless of what the actual flow rates might be in the future.

Soon after 1922 came the infamous Dust Bowl years of the 1930s, and yearly Colorado River flows dropped by more than half. It also became clear from long-term analysis that the average annual flow rate for the Colorado when calculated over centuries was only something like 13 million acre-feet, and therefore the 1922 compact should have allocated only 5 or 6 million acre-feet to the lower-basin states to begin with.[4]

A law is a law, California had always argued, and that combined with the fact that throughout the twentieth century the upper-basin states had not fully developed their own allotment had kept the shaky compact temporarily free from legal challenge. However, with hotter summers and shorter snow seasons of late, annual river flows had averaged only about 10 million acre-feet, and the upper-basin states had been left with only 2.5 million acre-feet, well below their own needs. Water-threatened Californians and Arizonans were not taking the court challenge lightly.

This looked like a particularly difficult year. Projections made in the late 1980s for population increases and water demands in southern California were turning out to be underestimates, perhaps because planners had not anticipated the outmigration from the Northeast and Midwest due to the collapse of family farms and the growing intensity of summer heat waves. Furthermore, Dallas, which in the 1980s already had a one-in-three chance of five or more days of 38° C (100° F) in a row, had warmed to the point where that kind of heat wave routinely occurred three out of every four years.[5] Thousands of Texans chose to flee the heat and more frequent coastal flooding and move to southern California despite the increasing price of land, the overcrowding, the water shortages, and the air pollution. Although California had the potential for increased water storage—even the ability to cope with reduced rainfall and snowfall—through the damming of wilderness rivers and creation of lakes, only a small fraction of that potential had been tapped because of opposition from northern California conservation groups dedicated to protecting wilderness areas and those arguing it was fiscally irresponsible to spend billions to water more lawns in Los Angeles when it was already overpopulated.

What made this year particularly worrying was the fact that the ongoing rise in sea level had begun to push salt water further and further into San Francisco Bay and the Sacramento Delta. In anticipation of this increasing water level, California had built a large number of levees at a cost of over $4 billion, but this year, because of the reduced river flow into the Sacramento Delta, water officials had been legally required to release more fresh water into the delta to prevent loss of habitat critical to fisheries, the shrimp industry, and many species of birds. This regulation was meeting with intense hostility from water users in the south, who saw these northern water releases as being wasted on wildlife instead of being put to "productive" use in the Southland. California had already passed an emergency rule prohibiting watering of lawns and golf courses and the filling of swimming pools, the latter being a particularly hard blow to the influential, rich, and famous in this relatively hot summer.

Northern Californians were also experiencing a decrease in water availability since the storm track didn't migrate down into their area until later in midwinter and then left earlier in the spring. Nevertheless, water was not northern California's biggest concern, but rather the frequent episodes of smoky skies that were a constant reminder of the growing intensity and scale of bush fires. Extra-heavy storms in the winter, though less frequent now, generally nurtured good winter and spring growth in forest and chaparral areas. However, the early dry seasons and hot summers of late had increased the fire potential. Moreover, pressure from homeowners, ranchers, and the timber industry had kept firefighting activities at a maximum over the past several decades, thus allowing a critically high buildup of undergrowth and dead branches. This dry summer needed only a few irresponsible campers or some normal lightning activity to cover the northern half of the state—and much of the West—in smoke. That California would escape this fate for yet another year was by no means assured as the September fire-danger season began.

The scenarios just sketched for sometime in the greenhouse century are contrived, of course, but as we will explore in some detail in Chapter 6, they are based upon extensive studies. In fact, they could happen today, but at a much lower probability than in a warmer world. Many national and international scientific assessments of the environmental and societal impacts of plausible climatic changes caused by greenhouse gases have been undertaken in the 1980s. From these plausible scenarios, such as those just invented, we can examine the all-important question: So what if the climate changes?

Recently, I was challenged by a U.S. government undersecretary to justify my advocacy of active policy responses to the prospect of global warming given that "there could well be some strong buffering effect in nature that climatologists hadn't yet discovered that would negate most of the large change predictions." "I agree that such a buffer may exist," I said (and he seemed pleased with the concession), "but can you tell me what the probability is that it actually exists?" He quickly changed to a blank expression. "One in ten, one in five, or perhaps I'll even make it as large as one in two?" I went on. "OK, let's say one in two," he responded. "Fine," I responded. "Since uncertainty is as likely to make change greater as lesser, then there is also a one-in-two chance that change will be as large or *larger* than we currently project. In other words, we are flipping a coin with unprecedented environmental change on one of the faces. With those kinds of odds and potentially dramatic impacts, I feel quite justified in advocating policy consideration for prudent actions that can slow down the rate of change, buy time to forecast the outcome and at the same time offer multiple additional benefits to society." "I agree that those kinds of actions may be warranted," he conceded, "but until you scientists can offer very confident, less speculative scenarios, I doubt governments will take dramatic or costly actions."

How solid a scientific case, then, can we make to back up concern over global warming?

2

The Greenhouse Effect

The "greenhouse effect" is a phrase popularly used to describe the increased warming of the earth's surface and lower atmosphere due to increased levels of carbon dioxide and other atmospheric gases that, like the glass panels of a greenhouse, let heat in but prevent some of it from going back out. If it weren't for the greenhouse effect, temperatures at the earth's surface today would be some 33° C (60° F) colder than they are, and life as we know it could not exist. Scientists are now debating whether the amount of these "greenhouse gases" will soon be increased by human actions to levels harmful to life on earth. Feared is a rise in temperature of about 5° C (9° F) as early as the middle of the next century—a rate of climate change tens of times faster than the average rate of natural change. How much and how fast temperatures will change and how these changes will alter accustomed patterns of rainfall, drought, growing seasons, sea level, and so on is controversial. But the greenhouse effect as a scientific proposition is as widely accepted a theory as there is in the earth sciences today. How, then, does the greenhouse effect work?

The sun is the prime mover of the earth's climate and the source of its life. The sun's hot surface temperatures of about 6,000° C (10,800° F) give off immense quantities of radiant energy at very short wavelengths. About half this solar energy reaches the earth's surface. Particles and gases in the earth's atmosphere absorb about 25% of this energy, and another 25% is reflected back to space by the atmosphere, mostly from clouds. In addition, 5% or so of the incoming solar radiation is reflected back to space from the surface of the earth, largely from brighter areas, such as deserts and ice fields.

At the distance from the sun of the earth's orbit, if we took a 1-square-metre piece of plywood (39 inches by 39 inches), painted it black, put it above the atmosphere, and pointed it squarely at the sun, it would collect about 1,370 watts of radiant power. This is what scientists call the solar constant (even though it isn't constant but varies by some tenths of a percent over eleven-year-long sunspot cycles). However, the sun is not shining all the time on every square metre of the earth, since the earth is spinning around on its axis. The area of the surface of the earth is four times greater than the area of a disk with the earth's radius (which is the area in space that intercepts the sunlight reaching the earth). Thus, in order to find out how much solar energy reaches the top of the earth's atmosphere over twenty-four hours, we need to divide the 1,370 watts per square metre by four. However, as seen in Figure 1, about 30% of the incoming solar energy is reflected back to space and thus is not available to heat the earth or to be used by life.[1] (On Figure 1, the total amount of incoming energy—about 340 watts per square metre—is considered to be 100%, and the numbers in parentheses are actually percentages of the total incoming energy, not the number of watts per square metre.)

The sun isn't the only object to give off radiant energy. In fact, all bodies with temperature greater than absolute zero (the temperature at which molecular motion stops) give off radiant energy. The colder the body, the less energy it gives off per unit area, and the longer the wavelength of that radiation. The maximum amount of energy a body can give off at a given temperature is

Figure 1. The earth's radiation energy balance, which controls the way the greenhouse effect works, can be seen graphically here. The numbers in parentheses represent energy as a percentage of the average solar constant—about 340 watts per square metre—at the top of the atmosphere. Note that nearly half the incoming solar radiation penetrates the clouds and greenhouse gases to the earth's surface. These gases and clouds re-radiate most (i.e., 88 units) of the absorbed energy back down toward the surface. This is the mechanism of the greenhouse effect. [Source: Schneider, S. H. 1987. Climate Modeling, Scientific American *256:5, 72–80.]*

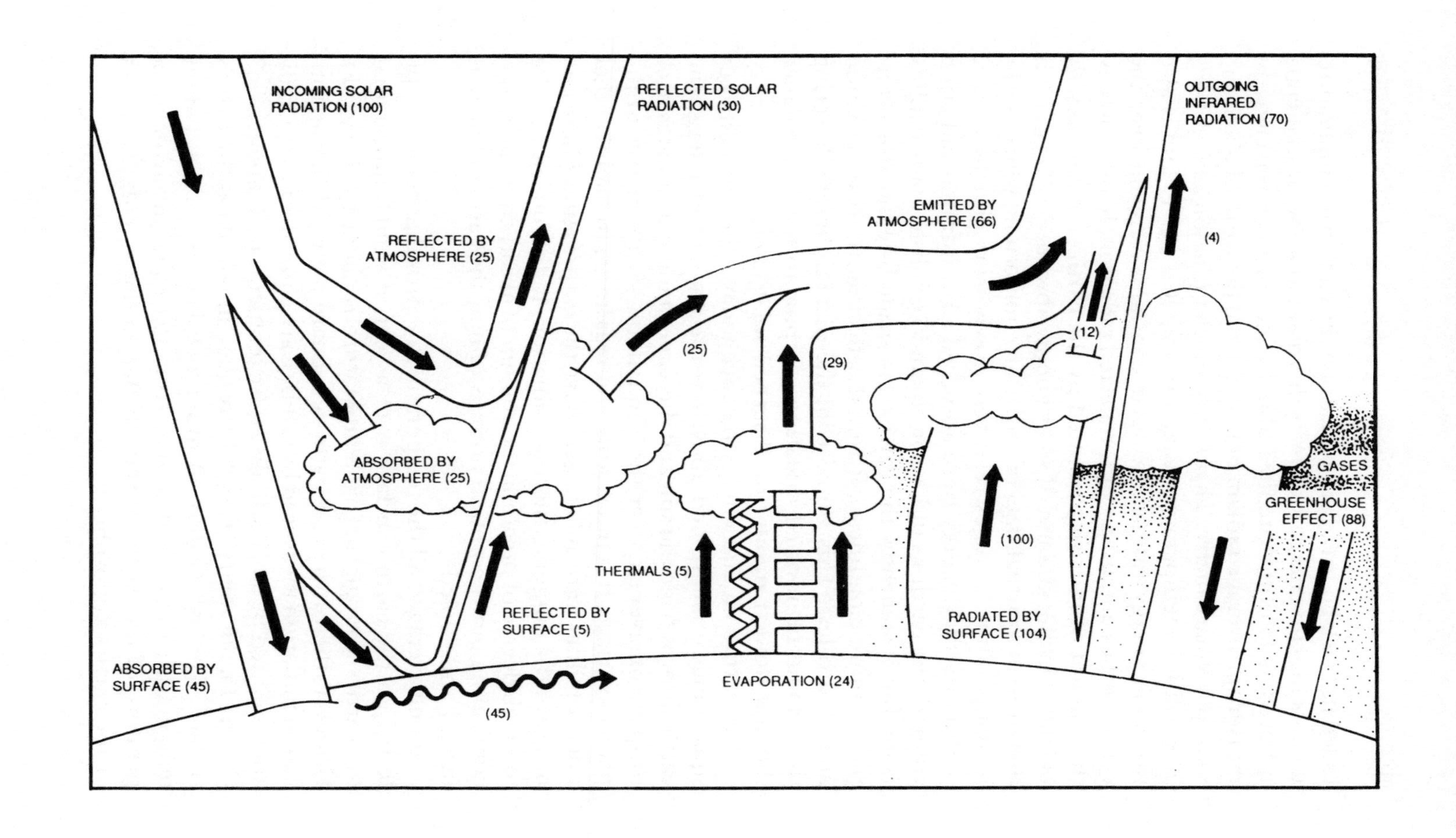

INCOMING SOLAR RADIATION (100)
REFLECTED SOLAR RADIATION (30)
OUTGOING INFRARED RADIATION (70)
REFLECTED BY ATMOSPHERE (25)
EMITTED BY ATMOSPHERE (66)
(4)
(12)
(25)
(29)
ABSORBED BY ATMOSPHERE (25)
GASES
GREENHOUSE EFFECT (88)
(100)
THERMALS (5)
REFLECTED BY SURFACE (5)
RADIATED BY SURFACE (104)
ABSORBED BY SURFACE (45)
EVAPORATION (24)
(45)

called its black body radiation. Because the earth has temperature, it emits radiant energy known as thermal radiation or planetary infrared radiation. Satellites able to see infrared wavelengths measure an average radiant emission from the earth of about 240 watts per square metre. That is how much radiation a black body gives off if its temperature is about −19° C (−3° F) and is also the same energy rate as the solar constant averaged over the earth's surface minus the 30% reflected radiation. In other words, the amount of radiation emitted by the earth is almost exactly balanced by the amount of solar energy absorbed.[2] The earth is therefore in a virtual state of radiation equilibrium, which is why its temperature changes relatively slowly from year to year.

If more solar energy were absorbed than infrared radiation emitted, the earth would warm up and establish a new equilibrium. But if somehow the earth were made brighter (for example, by more cloudiness), it would reflect more solar radiation and absorb less. This would cool the planet, thereby lowering the amount of infrared radiation escaping to space to balance the new (lower) amount of absorbed solar energy.

The earth's radiant energy balance today—the 240 watts per square metre mentioned above—is the amount of energy the earth absorbs from the sun and the same amount it radiates back to space on average over the 500 trillion square metres of the planet's surface. It is interesting and reassuring to scientists that when satellites were able to get above the earth's atmosphere and measure the outgoing thermal radiation, it did indeed balance to a high degree of precision the amount of absorbed solar energy, confirming our theoretical notions of how the earth's climate is heated and cooled.

If one averaged all the temperature records that could be collected on the earth's surface over a year, it would appear that the earth's average surface temperature is about 14° C (57° F). However, the earth's 240 watts per square metre of thermal infrared radiation measured by satellite are equivalent to the radiation emitted by a black body whose temperature is about −19° C (−3° F), not the 14° C (57° F) average measured at the earth's surface. The 33° C (60° F) difference between the apparent temperature of the earth as seen from space and the actual temperature of the earth's surface is due to the greenhouse effect of the earth's atmosphere.

The solar heat absorbed by the atmosphere and the earth's surface is given back to the atmosphere through the evaporation of water, the direct heating of air in contact with a warm surface (thermals), or the upward emission of 104 units of energy (see Figure 1). A few trace gases in the earth's atmosphere accounting for only a few percent of its composition have an important physical property that has the potential to make our planet habitable: they absorb radiant energy at infrared wavelengths much more efficiently than they absorb radiant energy at solar wavelengths and thus trap most of the radiant heat emitted from the earth's surface before it escapes from the planet. Not only do these gases—mainly water vapour and carbon dioxide (CO_2)—and particles (mostly in the form of water droplets in clouds) absorb infrared energy, but they also give it off. This reradiated infrared radiation is emitted both up to space (cooling the planet and maintaining the balance with incoming sunlight) and back toward the earth's surface (creating the greenhouse effect). Almost 88 units of infrared energy are emitted downward from the atmosphere to the earth's surface because of the trace greenhouse gases and clouds on earth. That downward reradiation is what warms the earth's surface and makes it 33° C (60° F) warmer than the effective radiating temperature of the earth as seen from a satellite. The earth looks so much colder to the infrared eye of a satellite because that instrument is actually looking primarily at the colder temperatures of the middle of the atmosphere and the tops of clouds; it sees only a little bit of the warmer earth's surface, mostly over arid, cloudless deserts or open ocean.

This brings us back to our greenhouse analogy. The gases and clouds in the earth's atmosphere let a larger amount of the sun's shorter-wavelength radiation in than they allow the longer-wavelength planetary infrared radiation to escape to space. This theory has been well established through millions of measurements in the atmosphere, in space, and in laboratories.

How fragile is the climate to changes in the earth's greenhouse properties? Is our planetary life support system the delicately balanced, fine-tuned mechanism some fear or rather a resilient web of internal control systems that have and will continue to sustain life comfortably as they have for 4 billion years? To investigate whether the earth's climate can maintain reasonable stability despite all the human and natural forces of change at work will require a broad look at the ancient past, at the role of life, and even at our planetary neighbours. The largest climatic changes in modern geological times are the ice ages. Why they come and go is a question that can have much relevance to predicting our future climate. How life responds to the physical environment, and at the same time alters that environment, is another question that has a direct bearing on the all-important concern about the resilience of the climate. Perhaps the best way to approach these critical questions is first to look for answers elsewhere in the solar system: on our nearest neighbours, Mars and Venus.

One of Mars' famous dust storms was raging when the Mars-orbiting spacecraft Mariner 9 first approached the intriguing red planet in 1971. The storm obscured most of the surface and for awhile delayed very close-up investigation. After a number of weeks, however, the atmosphere cleared, and the orbiting planetary probe revealed startling scenes of a crater-pocked surface, gigantic volcanic mountains, deep canyons, and frozen masses of ice at the poles.

While it is fascinating to speculate on why Mars at some ancient time owned a climate in which liquid water apparently flowed freely—thereby creating the giant canyons that the spacecraft so dramatically revealed—Mars today is a cold, dry planet that is almost certainly lifeless. In fact, at the poles it is so cold (about −120° C or −184° F) that the ice there is composed largely of frozen carbon dioxide—what is popularly called dry ice. Substantial supplies of frozen water appear to be underneath the ice, perhaps from a bygone era when the canyons were carved out. In addition, the atmosphere of Mars is made up almost entirely of carbon dioxide, but at a pressure less than 1% of the total pressure of the atmosphere on earth. The earth's atmosphere, on the other hand, is made up roughly of 78% nitrogen, 20% oxygen, and a percent or so of water vapour and other trace gases like carbon dioxide, methane, ammonia, and so on.

Venus is a study in contrast. Although its surface is obscured, too—continuously hidden from our eyes by a thick, planetary-scale layer of caustic sulphuric acid clouds drifting around in a largely carbon dioxide atmosphere—its atmospheric pressure is nearly 100 times more than that on earth. Surface temperatures are hotter than a baker's oven (450° C, 840° F), and almost certainly no life that we could imagine exists on this inhospitable planet.

Earth, of course, is radically different from its planetary sisters. A beautiful, shining blue-and-white ball, it contains large amounts of liquid water and water vapour, making life possible. Its total atmospheric pressure is midway between Mars and Venus, and the atmosphere itself contains a small amount of carbon dioxide and several unusual, chemically reactive gases like oxygen, methane, ammonia, and ozone. A number of planetary climatologists have noted that Venus is too hot, Mars is too cold, while Earth is just right for life—what they dubbed the planetary Goldilocks phenomenon.[3] But why does this phenomenon exist, and what does it have to do with the urgent issues of managing the greenhouse effect on earth?

The explanation of the Martian deep freeze, the Venusian oven, and the habitable Earth is as well-accepted a theory as exists in the atmospheric sciences. Although Venus is closest to the sun and Mars farthest away, their positions are not nearly enough to explain the contrasts in planetary climates. Instead the differences are largely a result of the gaseous compositions of these planets. Especially pertinent to our discussion is Venus's atmosphere, largely made up of carbon dioxide. As we've seen, carbon dioxide is one of the gases that permits more solar energy to reach a planet's surface than it allows radiant heat to escape. On Venus, the atmosphere forms a very effective heat trap: while only a small portion of the sun's energy actually permeates the thick clouds, thereby slowly heating the planet's surface, even less of the heat radiated from that surface escapes into space.

It is the trapping of infrared radiation in a planet's surface layers that makes the planetary greenhouse work. There is no doubt among climatologists and planetary astronomers that any change in the chemical composition of a planet's atmosphere that alters its greenhouse properties almost certainly will lead to a change in its climate. By looking at the earth's past climates we can gain a better understanding of the greenhouse phenomenon.

The earth has not always had today's climate, nor has it necessarily had today's greenhouse effect. Carbon dioxide, the second most important greenhouse gas after water vapour, has been of significance to the earth's climate from our planetary beginnings.

Our galactic inheritance some 4.6 billion years ago included heat from a sun that was perhaps 25% to 30% less energetic than it is today, and a collection of elements, including carbon and oxygen, that condensed to form the earth. As the earth's planetary crust cooled and hardened, hot gases from the interior were ejected, and carbon dioxide surely was one of the important ones. Its content in the primordial atmosphere has been estimated by some to have been many, many times greater than today—indeed this is one of the principal explanations of how the earth's climate could have been warm enough for liquid water to flow and life to evolve from the organic soup that brewed some 4 billion years ago.[4] This "super greenhouse effect" may have helped preserve the early habitability of the planet we have inherited today. Then, as life evolved, solar output increased and photosynthetic organisms used the carbon dioxide to create their tissues.

CO_2 is intimately connected with the weathering cycles that maintain mineral abundance on earth. Fossil fuels were laid down over a several-hundred-million-year period during the recent Phanerozoic era of abundant life—the past 600 million years or so. The richest fossil fuel deposits occurred mostly at times when proxy evidence suggests the earth was considerably warmer and possibly much richer in CO_2 than it is today. Over the past 2 million years, as the era of permanent polar ice descended, some evidence suggests that CO_2 levels dropped by severalfold relative to those in the times of the dinosaurs some 65 to 200 million years ago.[5]

Gas bubbles trapped in ancient ice in great glaciers on Greenland and Antarctica suggest that during the end of the last great ice age between 10,000 and 20,000 years ago, CO_2 levels were less than 60% what they are today, near the lowest limit for successful photosynthesis in many green plants. The same reduction was found for the preceding ice age, some 150,000 years ago.[6] After the last ice age—5,000 to 9,000 years ago, during the early to mid-Holocene (an era, known as the climate optimum, with some 2°C [3.6° F] warmer summers than today)—CO_2 concentrations grew to preindustrial levels. Since the beginning of the industrial era, human activities have accounted for a 25% increase in CO_2. As humanity burns the organic matter from past geologic periods (or the forests of today) to power the engines and economies of modern society, we are reinjecting our fossil carbon legacy into the atmosphere at an incredibly accelerated pace. CO_2 is dumped into the atmosphere at a much faster rate than it can be withdrawn or absorbed by the oceans or living things in the biosphere. CO_2 buildup in the next few decades to centuries could well be one of the principal controlling factors of the near-future climate.

In a metaphorical sense, carbon dioxide plays the role of a suspended cable of a great bridge of coevolution, starting at the earth's dawn, crossing the present, and leading into the future. CO_2 not only spans periods of geologic time but joins together both climate and life on earth in all eras.[7]

Carbon dioxide amounts are not the only trace greenhouse gas concentrations that humans have been rapidly changing. Methane has increased in the atmosphere by almost 100% since 1800. This increase is due to a variety of reasons, and scientists can't be precise as to the quantitative contribution of the many culprits. However, we do know that methane is produced by biological processes in which bacteria in relatively oxygen-free places have access to organic matter. These anoxic locations can be the guts of termites or cows, as well as bogs, marshes, garbage dumps, landfills, or rice paddies. Some methane is also liberated in the process of extracting coal or transporting natural gas. Methane is some thirty times more effective at trapping infrared radiation than CO_2 but is only about 40% as important as CO_2 in the augmented greenhouse effect since 1850 because CO_2 is more prevalent in the atmosphere. Chlorofluorocarbons (CFCs), which are man-made chemicals used as spray can propellants, foam blowers, refrigerants, and in other industrial applications because of their relative inertness, are even more effective greenhouse gases, but are about 20% of the CO_2 greenhouse effect added since 1850. They are implicated in the destruction of stratospheric ozone.

Oxygen exists on the earth today primarily because of life. For several billion years, oxygen-producing algae or bacteria lived in a world with an atmosphere containing extremely little oxygen relative to today. Eventually, these one-celled chemical factories injected enough oxygen to allow its gradual buildup in the atmosphere, beginning perhaps 2 billion years ago.[8] In the process, these bacteria and algae ruined their own environment since oxygen is a poison for many of them, and its buildup in the air relegated these organisms to anaerobic niches, such as swamps, soils, and, later on, the guts of cows, termites, or coral polyps.[9]

Oxygen molecules consist of two oxygen atoms, O_2. This molecule is relatively easily broken up by very short wavelength sunlight known as ultraviolet radiation. When oxygen molecules break up into oxygen atoms (O), they tend to recombine into a form of oxygen known as ozone (O_3), in which three oxygen atoms are combined in one molecule.[10] Ozone has the peculiar and important property of absorbing most of the sun's ultraviolet radiation. It does this in the upper part of the atmosphere known as the stratosphere (between about 10 and 50 kilometres – 6 to 30 miles – above the earth's surface). In the process of absorbing the ultraviolet energy, which is about 3% of the total radiant heat of the sun, the stratosphere is heated up greatly.

As life evolved into higher forms and crawled out of the oceans onto land during the last billion years, oxygen was present in sufficient quantities for the ozone in the stratosphere to protect the surface of the earth from much of the sun's ultraviolet radiation. Since ultraviolet light damages the DNA molecules that are the building blocks of life, and since most of life near the surface evolved during the past half billion years, our evolution has been dependent upon the ozone layer shielding us from harmful solar ultraviolet radiation. At the same time, ozone is part of the earth's greenhouse effect, although not as important a greenhouse gas as CO_2 or methane. Ozone in the lower atmosphere, however, is not helpful to life because it is very reactive and can damage plant or lung tissues. It is created as a pollutant in photochemical smog in cities like Los Angeles. In the lower atmosphere ozone heats the climate both as a greenhouse gas and by directly absorbing some sunlight.

Finally, several other trace greenhouse gases, such as nitrous oxide (laughing gas), carbon tetrachloride, and a number of other very minor gases can also have significant greenhouse properties. When all the human-produced trace greenhouse gases other than CO_2 are added up and their increases projected into the future, their collective greenhouse effect may add between 50% and 150% to the increase in greenhouse effect expected from CO_2 alone. Thus, whereas each of these non-CO_2 trace greenhouse gases may not account for more than 20% of the added greenhouse effect in the greenhouse century, together they are likely to double the global warming from CO_2 alone. Put metaphorically, when all these nickels and dimes are added together, it appears they will net about half a dollar.[11]

Little controversy exists among those knowledgeable in atmospheric science over the greenhouse effect as a scientific proposition. Nor is there much disagreement that CO_2 has increased by 25% and that other lesser trace greenhouse gases, such as methane, CFCs, nitrogen oxides, and so on have increased by even greater percentages because of human activities over the past century. There is also little doubt that if human trends in population, industrial, and economic growth continue into the next century, there will be substantial increases in the greenhouse properties of the earth's atmosphere, which are virtually certain to create environmental change. The controversies begin when we try to put numbers on the amount of warming, its timing, and its implications for ecosystems and society. Most controversial, of course, is what to do about this prospect.

In August 1988, Colorado Senator Timothy Wirth introduced a comprehensive, controversial, and far-reaching piece of legislation. (A similar bill was introduced in 1989 in the House by Rhode Island representative Claudine Schneider.) The Wirth bill, entitled the National Energy Policy Act of 1988, called for controls on industrial and agricultural emissions producing greenhouse gases; strong regulations to ensure energy efficiency; controls on deforestation; curbs on population growth excesses; and increased funding for the development of less-polluting energy alternatives, including solar, hydro, and modern, passively safe nuclear power (i.e., reactors that can't melt down and spill out radioactivity). Such a comprehensive bill was bound to step on many interests, and it did.

Wirth called for a committee hearing a week later, inviting a Department of Energy (DOE) official, an Environmental Protection Agency (EPA) official, a National Oceanic and Atmospheric Administration (NOAA) scientist, a NASA scientist, and me to testify. Donna Fitzpatrick of the Department of Energy argued that it was premature to take any action in response to the prospect of the greenhouse effect, for after all, there was too much uncertainty to risk damaging the economic viability of American industries. The EPA official took a cautiously opposite position, but one that stressed scientific uncertainty as well. My testimony was in sharp contrast to Donna Fitzpatrick's. "These scientific uncertainties must be reduced," she said, "before we commit the nation's economic future to drastic and potentially misplaced policy responses." I countered that we should not "use platitudes about scientific uncertainty to evade the need to act now" and argued that global warming is not an esoteric issue to be debated only by Washington bureaucrats and academic scientists, but is now a public policy issue that belongs high on national and international agendas. It deserves a place on the U.S. agenda along with foreign policy, taxes, and the competitiveness of American products. In fact, it is part of all those high-profile issues. This sharp exchange was well covered in the press.[12]

In the middle of the proceedings, sometime after most of the formal written statements had been submitted for the *Congressional Record,* Senator Bill Bradley of New Jersey took his seat. He thumbed through the written testimonies of the various witnesses and soon was recognized by the hearings chairman, Senator Wirth. Bradley noted that in my formal testimony I said that the public shouldn't be put off by scientific uncertainty, but rather should ask scientists fundamental questions: what can happen, what is the likelihood it could happen, how did you determine the odds, and what does it mean? He then asked the three scientists still on the panel to kindly answer those questions.

My colleagues from the National Oceanic and Atmospheric Administration and NASA tried their best to answer his questions. There could be unprecedented climate change into the twenty-first century, one said, and no prudent society would want to risk that outcome without at least working hard to understand better what might happen (scientists always manage, somehow, to recommend more research). The climate has already warmed up about 0.5° C (1° F) this century, we think, said the other scientist. In fifty years, it could be 5° C (9° F) warmer. "Could you help me understand why one, two, or three degrees can have such a startling impact on our environment?" Bradley asked. "The commonsense notion would be if the temperature every day up over a year is 1° warmer, what difference does that make?" A few excerpts from the transcript of the exchange that followed might be helpful in answering his politically important question.

DR. SCHNEIDER: Let's start with the simplest thing that everybody can relate to. Supposing you take the temperature in Washington, D.C., as it has been over the last 100 years and you simply, as we mentioned earlier, assume the climate is like dice. And you roll them and there are various odds of events. One of the events that we worry about is a heat wave, the kind we are in right now. What is the probability in Washington, D.C., of having five days in a row or more in July with the afternoon temperature greater than 95 degrees? Well, at present those numbers . . . are about 1 in 6 from the data that I have seen.

If you raise the temperature by 3 degrees Fahrenheit, . . . those odds go from 1 in 6 to something like 1 in 2. Now, what that means for air stagnation, for the health of especially elderly or other vulnerable citizens, is one of the things that EPA has been looking into in their studies. And what that might mean for the loads for air conditioning which would affect the need to either build new power plants or have stronger energy efficiencies are also important. So, that we can relate to quite easily.

I have done the same sort of calculation for Dallas, but we used 100 degrees. What is the odds today of having five days in a row above 100 degrees in Dallas? And it is about 1 in 3. If you increase by 3 degrees Fahrenheit, which is not nearly as far as the 3 to 5 degrees Celsius that Senator Wirth was talking about, then the odds go from 1 in 3 to 2 in 3. . . .

What else does it mean? Well, it also means that perhaps on the good side that the growing seasons would increase because you would get less commensurate runs of cold days. And presumably that means growing seasons would increase.

You can also calculate what happens to the evaporation from soils and from watersheds if the temperature increases. And, indeed, a number of people are doing that—hydrologists are using these kinds of numbers. And the typical kind of thing that you see is that the Colorado River basin might have 10 to 40 percent less runoff under this kind of scenario.

Other people have calculated what it might do to sea level. If you warm the oceans, you are simply going to do what you do with any fluid in a thermometer. If you heat the fluid in a thermometer, it will rise up the tube simply because it expands when it gets hotter. Well, in this case the fluid is the oceans and the tubes are the continental margins. So, if you heat the oceans, they are going to rise, and that is without melting any ice anywhere, just simply heating the oceans. And most suggestions are that sea level could rise—and you wanted probabilities—oh, maybe there is a 10 percent chance that they won't rise and 10 percent they will rise 5 feet in the next 100 years, and sort of everything else in between works out based on those calculations. So, a few degrees can be significant in that way. . . .

So, a few degrees temperature really can be very significant not just for the globe, but if that temperature change is distributed nonuniformly as we anticipate, then we can get all kinds of changes in regional climates that could affect crops and water supplies and health and so forth.

Robert Watson from NASA and Daniel Albritton from NOAA also tried to answer Bradley's questions. They stressed the need to narrow scientific uncertainty and were at the time pressing to obtain funds to conduct yet another giant assessment of global warming. However, they did not use uncertainty as an excuse to avoid political action. Again, from the hearing transcript:

DR. WATSON: Many of these gases that are predicted to cause a global warming have very long atmospheric lifetimes—nitrous oxide, the fluorocarbons, CO_2. Once up there, their consequences will be here for decades or centuries. Therefore, in many cases you will have to trust theoretical models, and we have to get as much circumstantial evidence as we can and direct evidence to support those theories.

So, I think the real message [is] you cannot wait to see a large temperature change with consequences for sea level, agricultural productivity, natural ecosystems changing. You are going to have to have a combination of evidence such as global temperature trends, which are not perfect but certainly suggestive, and the theoretical models.

SEN. BRADLEY: . . . That answer would not be understood by any of my constituents [laughter] who would be faced with higher electricity rates in order to deal with the situation or higher gasoline prices in order to deal with the situation. So, what I need to know from you, since you are the experts—or tell me two or three things that are happening now that are evident to more people than theoreticians that would lend some basis for why there has to be changes in behavior.

DR. WATSON: I'm not sure we should give you evidence of changes if we haven't got evidence for changes. The fact that we predict change, as we did in ozone, I believe is why people should be concerned. Scientific uncertainty is a thing that scares me more than absolute knowledge. The fact that we have said that climate could possibly change dramatically and quickly has me concerned. The Antarctic ozone thing came as a surprise.

What if the oceans have more than one state of ocean circulation? It could be they would no longer take up CO_2 at the rate they currently do. They might liberate CO_2 into the atmosphere. And then you would get a feedback mechanism by which climate would change at a much faster rate than we currently predict.

So, the thing is scientific uncertainty that a situation might arise that would be completely unhealthy is a thing that I think one has to get across to constituents. If I personally have not seen change, that does not mean to say I am not concerned.

I interrupted to dispute that we have not seen climate change this century. I went on that we can't be certain that the very warm year of 1988 is directly a result of the greenhouse effect nor can we directly attribute to the greenhouse effect the unusual summer heat in the Southeast in 1986; the devastation from hot weather in the corn belt in August 1983; nor the killing temperatures in the late spring and early summer of 1980 in Texas, Oklahoma, Arkansas, and Missouri. It is quite possible that these events are simply the random occurrences of perverse nature. However, it is no easier to determine whether these kinds of events are connected to the observed 0.5° C (1° F) or so global warming of the past 100 years than it is to determine if two snake eyes in a row prove a pair of dice to be loaded. The more we roll, the surer we are of the true odds. The longer we wait to establish scientific certainty that the greenhouse effect has descended, the greater the dose of climate change we and the rest of living things on this planet will have to adapt to over the next several generations.

How important is a degree of temperature change? The ice age, in which mile-high ice sheets covered most of Canada, parts of the Midwest Northeastern United States, some of Northern Europe, and most of Scandanavia and Britain, ended some 10,000 years ago. The global average temperature at the height of the ice age some 18,000 years ago was only 3–5° C (5.4–9° F) colder than the present temperature. It took 10,000 years or so for the planet to recover from that ice age, an event that revamped the ecological face of the world. Humans are putting pollutants into the atmosphere at such a rate that we could be changing the climate on a sustained basis some ten to one hundred times faster on average than nature has since the height of the last ice age. A degree or two temperature change is not a trivial number in global terms and it usually takes nature hundreds to thousands of years to bring it about on her own. We may be doing that in decades.

I'm sure the constituents of New Jersey understand the damages from storms coming along the coast pushing water up against the shore in storm surges. Sea level appears to rise perhaps as much as 3 to 5 metres (10 to 15 feet), inundating coastal areas and causing millions (and sometimes billions) of dollars of damage before the floodwaters recede hours later. How far each storm surge penetrates inland, and therefore how much damage it does, depends on the height of the sea level when the storm hits. If the storm hits at high tide, it is likely to do much more damage than at low tide. Similarly, if the world average sea level increases half a metre or so (a few feet), then the likelihood of storm surges damaging some part of a coastline every, say, thirty years might change to that level of damage occurring, say, every fifteen years. That could radically alter permissible land use.

It is very hard to determine what happens at any one coastal spot because shorelines at some parts of the world are rising and others sinking. Nevertheless, the typical estimate is that the sea level of the world is apparently 10 to 15 cm (4 to 6 inches) higher now than it was a century ago. (It has gone up nearly twice that much locally at New York City, for example, because the East Coast is sinking.) This global sea level rise is roughly consistent with the observations that the world has warmed up some 0.5° C (1° F) over the past century, although cause and effect is not certain. And, at the same time, there have been 25% or greater increases in the buildup of some of the greenhouse gases. Whether this piece of circumstantial evidence proves that the 0.5° C (1° F) global warming was caused by the increases in greenhouse gases is still intensely debated.[13]

Not everybody understands or accepts the twentieth-century record of global climatic change. For example, about two dozen scientists from the American Association for the Advancement of Science (AAAS) conducted a two-year long study of the effect of climate change on U.S. water resources.[14] About half a dozen of them, myself included, travelled to Sacramento, California, in January 1989 to present to California water officials the results and implications of this study. We pointed out the finding suggesting that global temperatures are about 0.5° C (1° F) warmer now *on average* than 100 years ago. We also said that specific locations can vary substantially from global trends, with some areas warming much more than average and others cooling substantially.

Officials responsible for water supplies in the emotionally charged and politically volatile atmosphere of California water politics were not particularly receptive to the notion that their beliefs and practices, which tend to ignore the prospect of substantial greenhouse warming or alteration to future water supplies, might not be adequate any more for planning for California's needs. One of the key officials of the California group had prepared a chart showing that seventeen rural stations in California had experienced a slight cooling trend over the past few decades, but a comparable number of thermometers in California cities gave very different results. The city stations showed a warming trend somewhat comparable to the global one I had shown them earlier. The California results prove, one of the state planners said, that the warming trend upon which the AAAS scientists were basing their greenhouse effect arguments was an artifact of the unnatural heating trends in cities. I responded that the so-called urban heat island issue had long been known to the people who put together the global records. Moreover, they don't use seventeen stations; hundreds of times more observations are included in the global records. Indeed, a major international debate was taking place at that time over how to correct for urban heating effects that creep into the global temperature records. These corrections had already been applied to existing global records, changing them somewhat, but still leaving about 0.5° C (1° F) warming as a best estimate for the global trend rather than a 0.8° C (1.5° F) warming that would be implied if you simply added up all the thermometer readings without the latest correction for the increased heating caused by cities that had grown up around the thermometers.[15]

One of the California water officials then dropped his claim about no global trend and argued that regardless of what the globe is doing, California isn't warming and therefore it is an exception, and they could ignore the global record. "What fraction of the earth's surface do you think California represents?" I said, hoping to make them realize that what happens in much less than 1% of the planet is virtually irrelevant to determining a global temperature trend that is still only a degree or so in magnitude. "Well, maybe California is not very large," said one of the participants, "but until it happens here, why should we worry about what the globe is doing?" "If you look at a map of the world," I went on, "and you plotted the average temperature changes over the last decade or two, you would find continental-sized areas that cooled a bit and comparably sized areas that warmed. The net effect of this warming and cooling mosaic has been a global average warming of about 1° F." Short-term, local cooling trends do not disprove global trends. Local cooling should be expected over the globe by chance—as long as the global temperature changes are still restricted to a degree or less. However, if the global warming increases to several degrees or more over the next few decades into the greenhouse century, then regional cooling trends of up to a degree or so that randomly occur around the planet will eventually be overwhelmed by the global warming. At the same time, random warming fluctuations will be substantially boosted and could by chance be soon felt in places like California, regardless of what has happened there in the recent past.

Therefore, local officials who insist on denying that global change is significant for them until it occurs unambiguously in their backyard are simply postponing the day when they will need to respond to those changes. And that postponement could lead to irreversible losses, particularly if new systems were to be built now based upon the current assumption of climate as usual when in fact there is a much-better-than-even chance that climate statistics will be substantially different in the future. If everyone has to have global long-term trends proved beyond doubt in their district before taking global warming seriously, then I'm afraid many places on the planet will be forced to adapt to rapid change without preparation. And these changes could well grow to unprecedented levels long before the greenhouse century has run its course.

The thermometers of the world are widely scattered and not entirely reliable, but when a team of NASA scientists in the United States and (independently) British researchers in England added thousands of them together, they showed the decade of the 1980s to be the warmest one on record. When average temperatures are higher, it seems logical to expect that heat waves will intensify. The constituents of New Jersey can thus be told that warmer global temperature probably means increased probability of extreme heat waves: times when crops can be damaged, water supplies reduced, air pollution crises created by increased power production to meet the demand for electricity for air conditioners, and even heat-related deaths, particularly among the poor, elderly, and weak.

When it is hotter and also drier, there is a much greater chance of forest fires, particularly in the western United States. Drought and heat in 1987 caused severe and prolonged fire damage to northern California, Oregon, and other parts of the West. 1988 saw major fire damage to Yellowstone National Park. Increased likelihood of forest fires is another consequence that could well occur.

The United States is the principal grain exporter in the world. But if global warming lengthens the growing season in Siberia and Canada, moving the corn belt northward, for example, then the United States may lose the comparative advantage it has enjoyed over most of the rest of the world as the prime agricultural exporter. Given the massive deficit in the balance of trade between the United States and its trading partners, to reduce the one area of trade surplus that is so important to the American economy could be a serious national loss. That too, I suspect, would be clearly tangible to most constituents, as is the fact that 1988 was the first year in modern memory in which the U.S. produced less grain than it consumed.

In Chicago in hot weather people flock to the beaches on Lake Michigan. Prolonged heat could well lead to increases in evaporation from the Great Lakes, dropping the levels by perhaps a metre (several feet) or so. Aside from the potential threat to inland shipping from such a possibility, no nearby lake along Lake Shore Drive would be quite tangible to the residents of the Windy City.

In summary, while it is true that much uncertainty attends detailed predictions of specific climatic changes, a strong and building case of circumstantial evidence shows that global warming is already afoot, and if current economic, energy, and population growth trends remain unaltered (or accelerate), we will have unprecedented, centuries-long climatic changes upon us much more rapidly than we can avert them. Although one can't prove beyond doubt that the greenhouse century has already begun, I strongly suspect that by the year 2000 increasing numbers of people will point to the 1980s as the time the global warming signal emerged from the natural background of climatic noise.

Although as far as the global climate is concerned, it is only a matter of degrees, a few degrees can be a very big deal—not so much from the temperature rise itself, but from all the other possible effects that can accompany a global warming, such as sea level rise, greater hurricane strength, and increased intensity of heat waves that were spelled out at the Senate Committee on Energy and Natural Resources hearing. Senators Bradley and Wirth listened carefully and interacted frequently with the scientist witnesses. The greenhouse effect was rising, it seemed, on the political agenda.

Earlier in the hearing that day Arkansas Senator Dale Bumpers asked the witnesses to take off their scientific hats and tell the committee what they would do to deal with the situation if they had the power. The transcript reads:

SEN. BUMPERS: Now, if we could elect you king for ten years, and you could deal with this problem in a very autocratic way, what would be the first thing you would do?

MR. WAYLAND: [EPA witness.] I'm not sure who is going to get coronated here first.

SEN. BUMPERS: We will make it God, if you want to. [laughter] Do you read Far Side? [laughter] Do you remember the cartoon of the dinosaurs' convention?

DR. SCHNEIDER: Yes.

SEN. BUMPERS: The ice age is coming. We are surrounded by predators and we have a brain the size of a pea. [laughter] See, we have a brain the size of a cantaloupe and the political will the size of a pea. [laughter] So, I am going to put them on all fours with you and you just tell us how you would deal with it.

DR. SCHNEIDER: Since I am not a federal employee and not subject to the Hatch Act [which prohibits federal employees from uttering political opinions], I don't mind playing king first. [laughter] A lot of the things that I would do are exactly what is in the 1988 Energy Policy Act. The first thing I would do is I would look very, very hard at how we can improve the energy efficiency of both this country and the rest of the world, the single most important item.

SEN. BUMPERS: Well, you certainly wouldn't favour reducing the CAFE standards [federally required petrol mileage standards for new cars] then, would you?

DR. SCHNEIDER: No, I would not.

SEN. BUMPERS: And you wouldn't either, would you, Mr. Wayland?

MR. WAYLAND: Senator, I have to say that that is not an area of EPA's responsibility . . . [laughter] . . . responsibility to know the answer to that question.

SEN. BUMPERS: Go ahead, Dr. Schneider. I want you to continue with this. I want you to talk about reforestation, energy conservation, energy efficiency. I want you to sort of go through those on a priority basis for me.

I very much enjoyed this exchange, particularly the typical refusal of an agency official to comment on any policy issue outside his agency's responsibility, no matter how obvious the question. It was also amusing to have Senator Bumpers try to telegraph my answers to him in advance, but then he is one of the senators who has been concerned about environmental damages from atmospheric change for over a decade and was entitled, I believe, to prompt us with his list.

Finally, because the hearing was directed as much at policy responses as scientific assessments, I felt a special obligation to make clear that when those of us on the panel who are scientists moved beyond statements of scientific evidence into issues of policy choice, we had passed beyond our official expertise into the realm of political values. For that reason I felt obligated to make that issue explicit at the hearing:

> DR. SCHNEIDER: But one postscript on what Senator Bradley said. Earlier he was suggesting that the science may be sufficient to justify action. I just wanted to make very clear—as a scientist, I have to do this and for my colleagues as well—that whether we should take action, of course, is a value judgment about whether we fear more having the future descend on us with rapid change or whether we fear more investing present resources to hedge against that. In my value system, I believe in insurance. I am conservative in that sense. I think a little insurance that, as Senator Wirth said, also has high leverage, makes good sense. But that, of course, is not a judgment that is scientific.

On May 8, 1989 the issue of global warming took on a new political character in front of a Senate Commerce Committee hearing chaired by Tennessee Senator Albert Gore, Jr. This hearing was to examine scientific research issues. The final panel, consisting of Jerry Mahlman from NOAA, James Hansen from NASA and me, became the focus of a national controversy when Gore discovered that the Office of Management and Budget (OMB) of the 100-day-old Bush administration had altered Hansen's testimony so as to water down the strength of his conclusions,[16] over Hansen's objections. Jim was particularly angry that they altered his scientific opinions, rather than his policy opinions—the latter, Hansen conceded, was the government's right.[17] Gore—joined by Senator Wirth— excoriated the Bush administration for censorship and demanded that they keep their 1988 election campaign promises to solve the greenhouse effect with "the White House effect"—what Wirth called the "whitewash effect" at the hearing. As of the spring of 1989 it was unclear whether Congress-Administration cooperation or confrontation was to mark the political future of the global warming issue. In November of 1989 it still remained unclear, as the U.S., joined by Japan and the United Kingdom, earned unwelcomed headlines in the world press by opposing specific measures to control CO_2 emissions at an international meeting held that month in Noordwijk, Holland. And by July 1990, with White House Chief of Staff John Sununu opposing U.S. support of European initiatives to control CO_2 emissions, political confrontation seems increasingly likely.

What, then, might the coming greenhouse century look like to the residents of the world? In particular, will the sea level rise in New York, the lake level fall in Chicago, the river traffic dry up on the Mississippi, the bayous continue to disappear in Louisiana, the temperatures repeatedly sear the inhabitants of Dallas and Phoenix, smoke clouds fill the skies near Denver and San Francisco, and water shortages in the Sierra Nevada and the Rockies turn the lawns and swimming pools of Los Angeles brown and dry? Will northern Canada and Siberia become the new breadbaskets with shiploads of their grain plying ice-free waters of the now-frozen North? Will there be more frequent devastating floods in Indonesia, India, or (as there were in 1988) Bangladesh and threats to the security of developing nations? Will trade in coal reduce to a trickle and the deserts of the world become energy parks covered with solar collectors?

Next we'll address briefly the history of the earth's climate, since it is the backdrop against which to calibrate the importance of human-induced changes. Then, I will sketch a number of plausible scenarios for the greenhouse century, and later address the strengths and weaknesses of the scientific evidence behind them. After that we'll examine how this topic is communicated—or miscommunicated—to the public-at-large by the media, environmental groups, industrial advocates, and scientists. Ultimately, I will return to the most important and urgent question that any responsible citizen concerned with the future of the planet must confront: What can or should we do about the greenhouse effect?

3

Fire or Ice?

In order to assess the seriousness of human interventions in the natural system, it is increasingly necessary to discover how nature has changed the climate; for we can only understand the importance of the environmental impact of human activity by understanding the history of natural changes in the earth's atmosphere, oceans, ice, and life. The largest changes in recent geologic history are the cycles of ice ages and warmer interglacial periods. As dramatic as an ice age is, it results from a planetary cooling of only about 5° C (9° F)—which is comparable to the magnitude of change that humans may bring about in the greenhouse century. Going further back, some 100 million years to the age of the dinosaurs, temperature changes were as large or larger than we project for the greenhouse century. These vast changes are correlated with major species extinctions or migrations, and generally took place much more slowly than the changes humans may cause over the next several decades.

Certainly, as we'll see, there are many still-unanswered questions about why the ice ages came and went. However, we do understand a great deal about these changes. The problem is whether we believe we know enough to act for the future. As I noted thirteen years ago in *The Genesis Strategy*, "The dilemma rests, metaphorically, in our need to gaze into a very dirty crystal ball; but the tough judgment to be made here is precisely how long we should clean the glass before acting on what we believe we see inside."[1] Before detailing the methods scientists use to estimate human-induced climate change, an obvious question emerges first: what caused the climatic changes of the past?

The climates of the past billion years, when most biological evolution has taken place, have been some 10°–15° C (18–27° F) warmer than our present climate and probably not much more than 5° C (9° F) colder. A rapid change from the present average surface temperature of the earth of about 14° C (57° F) to, say, 20° C (68° F) would have an enormous impact on species extinction and evolution rates. Gradual temperature changes of that magnitude have already been associated with radical changes in the ecology of the earth. Such changes are large in their effects on individual species, but from the perspective of the overall existence of life on earth, even a 15° C (27° F) temperature change is not threatening. For example, 100 million years ago dinosaurs roamed a planet some 15° C warmer than today, and tropical plant and animal fossils have been found in high-latitude locations such as Alaska.[2] The great cooling of the past 100 million years has seen sea levels fall some 300 metres (1,000 feet) or more, permanent ice accumulate at the poles, life zones of warm- and cold-loving species evolve, and many species go extinct as they lost their ecological niches. Nevertheless, life seems to have survived these and even greater catastrophic environmental events, such as the hypothesized collision of a large asteroid with the Earth at the end of the dinosaur era 65 million years ago.[3] The dust and smoke generated in the aftermath of such a collision could well have obscured sunlight for months, curtailing photosynthesis, depleting all ozone, and dropping temperatures to near or below freezing on a planet that at that time had no permanent ice.[4] Many species suffered extinction in the few-million-year interval at the end of the Mesozoic era,[5] but neither this nor other great extinction events have threatened the vast populations of bacteria, most forms of plant life, and many of the insects and smaller animals.

In today's concern over a general global warming trend, there does not appear to be any serious concern about a radical breakdown in climate stability, such as would result in an ice-covered earth or a runaway greenhouse effect (in which most of the carbon in the carbonate rocks would end up in the atmosphere and the planet's surface would become a Venus-like oven). No one I know has demonstrated even the remotest possibility that any disturbance from human events over the next hundreds of years (or even known extraterrestrial events over millions of years) could create that level of instability. But before we take too much comfort in that very optimistic view of our planet's habitability for some forms of life, we must remind ourselves that an ice age is only 5° C (9° F) or so colder than present, and change of that magnitude would be an immense stress for us and many other living things. Since an increase of 5° C (9° F) is the magnitude of change that humans may bring on themselves and other inhabitants of the earth over the next century, it seems important to examine the causes, and implications, of the many cycles over the millennia of global cooling and warming that we know as the ice ages, or glacials, and the warmer interglacials. Our understanding of these phenomena would provide much credibility for the prediction of comparably sized climate changes in the near future. That is because many of the same laws of climatic behavior should apply as well to cooling events as to warming from greenhouse gas increases.[6]

Human beings not radically different from us evolved some 2 to 4 million years ago, probably in Africa and the Middle East. About the same time, the earth's climate completed its gradual 65-million-year transition from the end of the era of the dinosaurs, when there was little or no permanent ice anywhere on the planet, to a planet permanently ice-covered in polar regions. About 2 to 3 million years ago, an amazing phenomenon began to develop: periodic growth and decay of additional major ice sheets outside the polar regions. These ice sheets, mostly in North America and Europe, started to wax and wane with time scales of roughly 40,000 years. Then about 800 thousand years ago they took on a cycle time of roughly one hundred thousand years, and a few shorter periods too.[7] By 300 thousand years ago, the glacial/interglacial cycles grew in intensity and began a fairly rhythmic pattern. Interglacials are not all alike, and not much is known about them. Nevertheless, it is thought that a typical interglacial exists for roughly 10,000 years (for example, the one preceding our present interglacial lasted from about 130,000 to about 120,000 years ago and may have been about 1° or 2° C [1.8° to 3.6° F] warmer than the present one). Interglacials are followed by a gradual (but not uniform) increase in polar ice accumulation for some 80,000–90,000 years, then by a rapid termination in which the 90,000 or so years of ice accumulation disappears in about 10,000 years. The relatively abrupt disappearance of the nonpermanent ice sheets gives way to the warm interglacials, which last about as long as the terminations. The temperature of these interglacials (such as the one we're living in now) is about 5° C (9° F) warmer on a global average basis than the temperature at the glacial extreme some 15,000 years earlier. The fact that the present interglacial is already about 10,000 years old has led some scientists to predict the onset of the next ice age. Indeed, if human activities do not interfere with the natural climate cycle, that transition is probably already underway—but would take many tens of thousands of years to accomplish fully.

Twenty thousand years ago, during the height of the most recent glacial, ice sheets stretched from New York state across the Great Lakes to Wisconsin northward through most of Canada. There, mile-high domes of ice tied up so much fresh water that the oceans were some 100 metres (more than 300 feet) lower than today. Vast tracts of continental shelves, now underwater, were exposed to the air. One could walk on dry land from Siberia to Alaska and New Guinea to Australia (as many did). (Today one can sometimes walk from Siberia to Alaska in winter, but only on the sea ice.)

Twenty thousand years is merely a blink in geologic time, and not many more blinks in biological evolution. If we compressed the 4.6 billion years of earth history into six metaphorical days of planetary creation, then the last 20,000 years would be but a few seconds. The time since American independence or the founding of Australia marks only 1% of that 20,000-year period. Since the last glacial, the great ice sheets have melted; oceans have risen hundreds of feet, flooding vast amounts of land; wide expanses of prairies have moved; forests have changed their floral mix; deserts have grown; animals have evolved or become extinct; and—not least—as the ice retreated, human civilization has flourished.

How is it that scientists know how long an ice age or interglacial period lasts, what the relative temperature differences are, or what

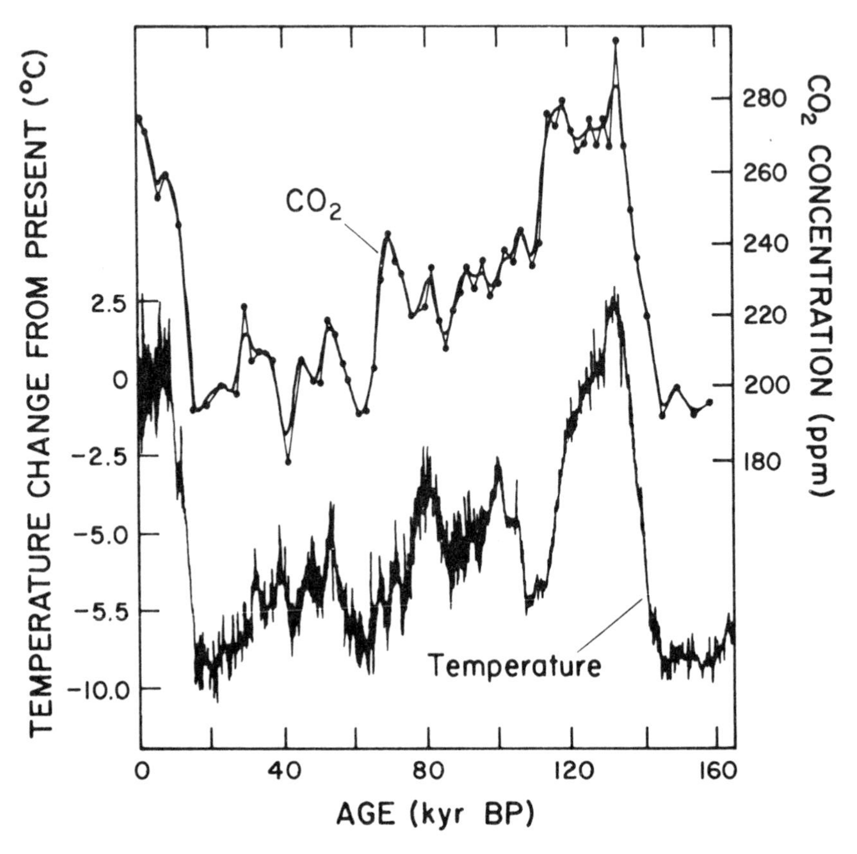

Figure 2. Air bubbles trapped in ancient polar ice sheets can be analysed to determine the changing composition of the atmosphere over hundreds of thousands of years. Such analyses at Vostok in Antarctica show that carbon dioxide concentrations were about 25–30% lower in glacial times than at interglacial periods over the past 160,000 years. Local temperatures (in Antarctica) at the extreme glacial times (about 20,000 and 150,000 years ago) were about 10° C (18° F) colder than at interglacial times. [Source: Barnola, J. M.; Raynaud, D.; Korotkevich, Y. S.; and Lorius, C. 1987. Vostok Ice Core Provides 160,000-year Record of Atmospheric CO_2, Nature *329:408–414.]*

the composition of the atmosphere may have been during the ice ages or previous interglacials? Figure 2 shows a complete glacial cycle inferred from measuring the chemical composition of ice deposited in Antarctica.[8] Because the massive ice dome on this frozen south polar continent is made up of compressed snow that has accumulated over the past several hundred thousand years, Antarctica is an ideal place to study the history of our planet's atmosphere. As the snow is squeezed by the weight of additional precipitation on top, it slowly turns to ice. As the ice forms, the air that surrounds the snowflakes is trapped into bubbles and held prisoner in the ice until it eventually is discharged into the oceans—typically several hundred thousand years after deposition. By drilling cores down thousands of metres into the ice, scientists can bring back samples of the kind of air that was breathed by our Neanderthal ancestors. Furthermore, the nature of the ice tells us a great deal about the temperature at which the snowfall took place.[9]

One of the most fascinating discoveries in earth science in the 1980s was the remarkably close correlation between the temperature of the air inferred from the chemical composition of the snow,[10] and the direct measurement of the amount of carbon dioxide (CO_2) present in the earth's atmosphere at the time the snow was compressed into ice and the air surrounding it trapped in bubbles. Before the Industrial Revolution and for much of the present interglacial, the CO_2 content of the atmosphere was about 280 parts per million (ppm). (In 1989, the CO_2 content of the atmosphere will average a little over 350 ppm, or about 0.035%.) This 280 ppm figure is comparable to the CO_2 composition seen in the previous interglacial period (120,000–130,000 years ago).[11] Scientists have also found that over the 80,000–90,000 years in which an ice age builds, CO_2 concentration tracks temperature variation fairly closely, reaching extremes at the last glacial maximums 18,000 and 140,000 years ago. At the time of glacial maximums, atmospheric CO_2 concentration appears to have been about 190 ppm, some 30% less than the preindustrial atmospheric CO_2 concentration of 280 ppm. Methane, another important trace greenhouse gas, was also found to track the temperature quite closely over the past two glacial events.

This remarkable story of earth history is also a paradigm for the kind of global consciousness and international interaction needed as we face the global scope of climate problems in the near future. Even in the dark days of East-West confrontation in the early 1980s, Russian, French, and American scientists and technicians in Antarctica were working together to produce this scientifically important picture of the earth's history. Russians had drilled down some 2,000 metres (6,500 feet) to remove core samples from the centre of the Antarctic ice sheet at a station known as Vostok. At nearly 3,500 metres above sea level, Vostok has the world record for cold temperature, about −89° C (−128° F). The Russians, however, lacked the ready capacity to make the sophisticated chemical analyses needed to translate this glacial library of atmospheric history into quantifications of temperature or CO_2 content, so the French glaciological laboratory in Grenoble, ably run by its knowledgeable, personable, and multilingual scientific director, Claude Lorius, agreed to perform the necessary laboratory work. Yet another problem, however, for both the French and the Russians, was the logistics of flying hundreds of sections of ice with a total length of more than a mile in frozen state to France. Therefore, while NATO and Warsaw Pact politicians and diplomats traded angry charges in the world's media, Russian scientists handed precious chunks of ancient Antarctic ice to American pilots, who flew them intact and frozen to a French team for later laboratory analysis.

Physical evidence such as the chemical composition of ice sheets is only one technique that scientists use to reconstruct the earth's past. Many other techniques produce so-called proxy records from which past climatic or ecological conditions can be deduced. For example, proxies show how the character of the forests of North America has changed with the climate warming of the present interglacial, beginning some 10,000–11,000 years ago. Maps of the ranges of various tree and other plant species have been constructed from analyses of fossil pollen over this period. Because some of the pollen fell in lakes, sank to the bottom, and was trapped in sedimentary muds, scientists are able to unearth the changing evolution of vegetation patterns—and thus climate history—by taking core samples of these muds and painstakingly counting the numbers of grains of the different pollens present.

What researchers found from analysing hundreds of cores was that 11,000 years ago, spruce trees, which now hug the Arctic rim as part of the vast boreal forests of conifers, were girding the great ice sheet that covered southern Canada. These trees survive best where summer temperatures are not too warm and moisture in the soil is adequate. On the other hand, broad-leafed hardwood forests, such as oak—which prefer warmer environments and are very prevalent today across the Midwest, the northern Atlantic states, and southern Canada—were greatly reduced. During the ice age, hardwoods survived in favourable southern locations, such as North Carolina, but not until the retreat of the ice and the warmth of the interglacial did they advance northward to establish their dominance across most of the eastern United States and southern Canada. Climates favoured by hardwoods were also appropriate for the kinds of crops later grown by settlers in Ontario, Quebec, New England, and the Midwest—many of which are still grown today. It is obvious from these findings that the transition from glacial to interglacial time revamped the ecological face of North America—indeed of the world—and provided the landscape that first allowed the North American Indians, and later the European colonists, to multiply and prosper.

At the same time, however, this time of ecological transition resulted in the extinction of many animal species. Brown University palaeoclimatologist Thompson Webb, who makes this kind of proxy study of forest history, has told me that the southern spruce forests of the ice age were relatively open and did not form a closed, dense canopy.[12] Grasses and shrubs filled in the spaces between widely scattered trees, allowing annual animal migrations and a variety of subhabitats within this open parkland. Then, as the interglacial set in, not only did the spruce move northward but they were also transformed into a closed, dense forest. The land they vacated became either a hardwood zone or grassland, and the ecosystem supported by the open forest was radically altered as a result. Animal extinctions following the retreat of the ice age were numerous and may have been caused by the reorganization of the vegetation into ecosystems that could not support them.[13]

On a global basis, the world warmed up some 5° C (9° F) from the beginning of the ice age's rapid retreat some 15,000 years ago to the middle of our present interglacial some 5,000 years ago. However, locally, near the ice caps in the northeastern United States, temperatures increased more—by some 10°–20° C (18°–36° F) over that 10,000-year period. From the migration of the hardwood forests northward as the climate warmed, we can calculate the natural average rate of temperature change in the northeastern United States as having been approximately 1°–2° C per 1,000 years[14]—a change that radically altered the natural resources and species balance of this area. Most tree species were able to survive the slowly changing climate, but faster temperature changes would put forests out of equilibrium with the changing climate, causing difficult-to-predict sequences of forest succession.[15]

The most commonly accepted theory of the cause of the glacial/interglacial cycles is an astronomical one relating climate change to changes in the tilt of the earth's axis and its orbit around the sun. James Croll, a Scottish geologist, recognized late in the nineteenth century that the earth's orbit was not fixed for all time, and he showed mathematically how changing orbits could change climate. This idea was first speculated on by Joseph Alphonse Adhémar in 1842.[16] Despite the fact that today's maps indelibly place the tropics of Cancer and Capricorn at plus or minus 23.5° latitude, the axis of the earth is not immutably tilted at 23.5° away from the plane of our orbit. It is now well known that the tugs on the earth from the gravitational forces of neighbouring planets cause the tilt of the axis to cycle between about 24° and 22.5° every 41,000 years or so. Also, the now somewhat elliptical orbit of the earth around the sun varies between more eccentric orbits and almost circular orbits roughly every 100,000 to 400,000 years. In addition, the North Pole points away from the sun when it is nearest the sun (i.e., the perihelion of the ellipse—in January at present), but this, too, cycles roughly every 20,000 years.

These perturbations to the earth's orbit about the sun do not cause a radical change in the total amount of solar heat received by the earth over a year—in fact, no more than a couple of tenths of a percent. But they do create a very substantial difference in the equator-to-pole and seasonal distribution of incoming sunlight.[17] For example, 9,000 years ago the amount of sunlight coming into the Northern Hemisphere in the summer was about 5%–10% more than at present, with a comparable decrease of sunlight in winter. Thus, the primary effect of the variation in the earth's orbital elements is to change the nature of the latitudinal and seasonal cycle of the solar heating of the earth's climate.

In the 1920s Milutin Milankovich, a Serbian mathematician, substantially improved the mathematical and physical components of the Croll-Adhémar theory by suggesting that ice ages build up when there is a relative reduction in summer radiation received in the high latitudes of the Northern Hemisphere. This, he postulated, would permit snow that fell in the winter and spring to remain throughout the summer. Even though the winter may have been warmer at these times, warmer winters can actually mean more snowfall in high latitudes, since the total amount of moisture the atmosphere retains is greater at warmer (but still below-freezing) temperatures. Snowfall simply requires temperatures below freezing. Temperatures much below freezing actually produce very little snowfall, as in Antarctica, which is a desert. The only reason massive ice sheets exist permanently in Antarctica is that virtually no ice melts in summer. Of course, warmer temperatures in summer, if above freezing, induce substantially greater amounts of melting. Milankovich also postulated that after glaciers built up, they would disappear tens of thousands of years later as more sunlight would come in during the summer, causing the glaciers finally to melt.

Other theories as to the cause of glacial/interglacial cycles abound, some based on such factors as massive volcanic periods.[18] Explosive volcanic eruptions throw veils of dust as well as sulphur gases high in the atmosphere. This stratospheric haze (mainly due to sulphuric acid droplets formed from sulphur gases) blocks some sunlight from reaching the earth's surface. Some scientists have theorized that sufficiently enhanced volcanism over long-enough periods could block out enough sunlight to allow glacier buildup. Both the astronomical theories of Croll and Milankovich and the volcanic theory fall into a causal category called external forcings—that is, these factors are external to the climatic system and would presumably occur regardless of the state of the weather.

Other causal hypotheses are known as internal oscillations.[19] This class of causal factors is based on the idea that the major subcomponents of the climatic system can exchange energy, momentum, and materials. These subsystems of the climate system include the atmosphere, the oceans (both upper and deep oceans), the sea ice, the ice sheets, the land surface, and the biota. The atmosphere by itself—that is, if somehow oceanic influence were removed—responds very quickly to disturbances, usually adjusting within a matter of months. The upper oceans have characteristic response times on the order of years, whereas the deep oceans take centuries to millennia to complete their changes. Ice sheets take millennia to tens of thousands of years to wax and wane, and biological systems can change substantially over the time frame of months to millennia.

The land is not static either. Continental drift—actually the movement of the earth's so-called tectonic plates, which cover the globe fitted together like pieces of a puzzle and include ocean floor as well as landmasses—occurs on a time scale of tens of millions of years; for example, India migrated from about 20° S latitude to collide with Asia (thus building the Himalayas) between about 60 million and 20 million years ago. However, important changes in elevation of the land can occur much faster. The weight of an ice sheet is sufficient to depress bedrock by as much as a kilometre in 10,000 years or so. After the glacier retreats, the land rises steadily back to where it was before the weight of the glacier depressed it. This process, called isostatic rebound, is occurring today in northeastern Canada and the United States. During Henry Hudson's exploration of Hudson Bay in 1610, his crew painted a mark on a shoreline rock at about the height of the ship's deck. That mark is now tens of feet higher, visible proof of isostatic rebound. (When Canada's rebound is complete, Hudson Bay will be almost completely drained—if the next ice age hasn't started by then.) Such rebounds make sea levels at particular locations appear to rise or fall, regardless of whether the absolute volume of the oceans has changed. For example, sea level near New York City has gone up over a foot since 1880, whereas sea level in Sitka, Alaska, has decreased a comparable amount.

The internal oscillation theory of glacial/interglacial cycling involves the scientific idea of feedback—evident in many smaller-scale living systems—in which any process basically creates conditions that cause change in one system to influence other systems, which, in turn, feed back change into the first system. Such a cycle might begin with a relatively warm earth in which abundant evaporation from the oceans takes place. This evaporation in turn creates storm systems that dump heavy snows in the high latitudes of Canada and Scandinavia in winter and spring—presumably at higher elevations. Sooner or later, enough snow accumulates so that it does not melt off in summer. Snow reflects much more sunlight than exposed land or water, and the increased reflection of sunlight from permanent snowfields (termed ice-albedo-feedback) causes further local cooling of the land surfaces. Such cooling does not yet overwhelm the warm oceans, which continue to pump massive amounts of snow during the winters of this relatively early glacial buildup time. Over a period of 10,000 years or so, the ice sheets grow big enough to begin to flow south. They eventually grow so large that cold air pours off the continents, cooling the oceans, eventually freezing the northern reaches, and creating a large cap of sea ice throughout much of the high-latitude seas. The circulation of the ocean, which normally involves the sinking of cold water in high latitudes and its replacement by warm water from below or from lower latitudes, is stopped. This sinking allows the inflow of warm surface currents that in turn warm the air above and help maintain relatively ice-free conditions. The critical role of an ice-free, open ocean in the climate balance is obvious when we consider Europe, which is at the latitude of Siberia and Hudson Bay but has a vastly milder climate because the Gulf Stream keeps the North Atlantic relatively ice free.

Continuing the cycle, the change in ocean circulation and the expansion of sea ice causes the northern half of the Northern Hemisphere to become a frozen wasteland. Snowfall decreases but still exceeds summer melt over the frigid ice sheets. Cold winds blow vast quantities of dust from the arid and relatively unvegetated landscape south of the ice sheets, blocking out some of the sunlight. The ice sheets thrive in this cold environment, continuing to creep inexorably to the south, covering most of Britain and Scandinavia, some of northern Europe, most of Canada, and parts of the Midwest and northeastern United States.

Sea level and rainfall decrease as ocean evaporation is reduced. The decreased sea level exposes shelf areas to erosion, which increases nutrient runoff into the ocean. This nutrient runoff, and perhaps also the enhanced fall of mineral dust, in turn increases the productivity of calcite-shelled phytoplankton. Since phytoplankton productivity works like photosynthesis and takes up CO_2, the oceans' uptake of atmospheric CO_2 increases. Atmospheric CO_2 concentration drops by approximately one-third, with most of the carbon from the air now trapped in the oceans. In addition, some atmospheric CO_2 is removed and the carbon is trapped in dead plant matter buried under glaciers or frozen ground or stored in bogs. Tropical forests decline, presumably from decreased rainfall and decreased atmospheric CO_2.

Recall that Antarctic ice core sampling has shown that changes in the methane content of the atmosphere are closely associated with changes in CO_2 and climate. Warmer climates have had one-third more CO_2 and twice as much methane as cooler climates. Gordon MacDonald at the MITRE Corporation has speculated that changes in temperature or sea level could mediate the atmospheric level of this biologically produced gas, much of which is trapped below permanently frozen ground in the Arctic tundra. Bog ecologist Lee Klinger from the University of Colorado and now associated with the National Center for Atmospheric Research (NCAR) has argued that bogs contain much dead organic matter and could very well have been a major storehouse for carbon as the climate cooled, thus helping explain the decrease in methane and carbon dioxide found with colder climates.[20]

Eventually, according to the internal oscillation theory of climate cycles, the ice sheets grow so big and flow so far south that their leading edges can no longer be sustained. First of all, they are starved of fresh snow because the oceans have cooled down too much. Furthermore, the sun is simply too hot at those midlatitudes, and the southern edges of the glaciers begin to melt. At the same time, the massive weight of the ice depresses the bedrock, thereby lowering the altitude of the ice sheets. Since lower altitudes are warmer than higher altitudes, the upper parts of the depressed ice sheet also begin to melt. At the same time, a lower-altitude ice sheet is less effective at blocking the flow of warm air over itself than is a higher ice sheet. The combined negative feedback effects of these processes begin the second part of the glacial/interglacial cycle, in which ice sheets wane. In North America, great lakes formed at the southern edges of the glaciers from rapidly broken-off chunks of ice, and vast quantities of meltwater rushed down the Mississippi basin into the Gulf of Mexico.

Cores of ocean mud taken by research vessels show that a meltwater spike occurred some 11,000 years ago in the Gulf. In one imaginative interpretation, the pioneering palaeoclimatologist Cesare Emiliani from the University of Miami went so far as to suggest that this meltwater rush might explain the legend of the rapid disappearance of the lost city of Atlantis, since sea levels rose about 100 metres (330 feet) some time between 15,000 and 8,000 years ago.[21]

Continuing with our internal oscillation theory, with the continental shelves now once again covered with water, and with dead plant material in bogs available for decomposition on dry land (and methane previously trapped under frozen land able to seep out), atmospheric CO_2 increases by 30% and methane by 100%, enhancing the greenhouse effect and accelerating climate warming. Meanwhile, during the last phases of the ice age, the cold northern oceans evaporate relatively little water, thereby continuing to starve the ice sheets from their northern end. The combination of all these factors results in rapid deglaciation.

This parable is a condensation of many scientists' proposed mechanisms to explain various components of the ice ages.[22] But it is not necessary to argue that the actual observed glacial/interglacial signal (such as seen in Figure 2) must be either an internally or externally caused climate change. It is also possible it is a combination of both. For example, when my NCAR colleague David Pollard was at Cal Tech, he developed a simple climate model that was able to reproduce the sawtooth pattern of glacial/interglacial cycles by including both internal processes and the external forcing effects of the earth's orbital variations.[23] Indeed, some mathematically inclined climatologists have suggested that the Milankovich astronomical theory can account for a number of features of the glacial/interglacial cycle (especially the 20,000- and 40,000-year components), but that in order to get the 100,000-year beat that appears to characterize the last million years of glacial/interglacial cycles (and in order to explain the very rapid termination and the slow buildup of ice ages), it is also necessary to invoke internal mechanisms such as those given in the parable.[24] Personally, I am very impressed with the strong correlation between fluctuations in the volume of ice on earth and certain configurations of the earth's orbit—particularly cycles with 41,000- and about 20,000-year periods—which themselves depend to some extent on the 100,000-year eccentricity variations.

The strongest evidence showing that the earth's climate has varied with roughly 23,000-, 41,000-, and 100,000-year cycles over the past million years was obtained by James Hays from Columbia University, John Imbrie from Brown University, and Nicholas Shackleton from Cambridge University, England, who in 1976 published an important analysis of the fossil contents of deep sea cores taken from ocean sediments in several parts of the world.[25] Different kinds of species can be shown to exist in varying abundance in different climatic conditions. Hays and his colleagues painstakingly counted the kinds of planktonic species that fell to the sediments at each depth in the core—increasing depths representing increasing age—and were able to establish that the changes in climate, as represented by the changes in warm- and cold-loving species, correlated remarkably with the cycles of the earth's orbit. For the 23,000- and 41,000-year cycles it can be shown that major changes in seasonal climates should occur.[26] On the other hand, so little change in the annual average incoming amount of solar radiation every 100,000 years is associated with the change in the eccentricity of the earth's orbit that I am inclined to think that this part of the glacial/interglacial cycles in the recent geologic past must have a substantial component related to the complex internal dynamics of the interactions of atmosphere, oceans, ice, land, and biological systems. Even more perplexing is the finding that the 100,000-year cycle is strong only in the past million years and was not dominant in the ice age cycles of the previous million years. Much progress has been made in the last ten years both on the reconstruction of climatic patterns and the identification of plausible mechanisms of change. But it is my opinion that the glacial/interglacial-cycles enigma is not yet completely solved and that more exciting discoveries remain.

North America was not the only place to experience the kinds of changes in climate and ecosystems described above. Europe saw similar tree-species migrations of thousands of kilometres and similar changes in forest composition. Africa experienced periods of intense aridity, and the American Southwest and the Great Basin were marked by periods of greater rainfall after the extreme of the last glacial period[27]. These conditions changed, just as the forests of North America changed, during the transition from the ice age to the interglacial.

Not all climatic changes proceed at the globally averaged rate, though, especially in specific regions. By 11,000 years ago the harsh glacial conditions that kept Britain a treeless tundra and western Europe a relatively uninhabited cold frontier had largely given way to warmer species. For example, warm-loving beetles moved into Britain and appeared to replace the arctic-loving species that had inhabited the area for many tens of thousands of years. Then suddenly, perhaps within 100 years some 10,800 years ago, something dramatic happened in which near-glacial conditions returned to the North Atlantic and northwestern Europe until about the year 10,200, when the ice age finally lost its last toe-hold.[28] This period of rapid climatic change, known as the Younger Dryas after a small, beautiful flower found in cold terrain, also seems to show up in the fossil records of other parts of the world, but not nearly as strongly or dramatically as in western Europe. Although the Younger Dryas signal was strongest in Europe, it was also significant in the maritime provinces of Canada, but has not been detected in the fossil record over most of North America. Its occurrence reminds us that although the bulk of geologic time suggests relatively slow change, nature has thrown a few surprises into the record.

The Younger Dryas is now a topic of intense research.[29] Since evidence suggests that near-ice-age conditions returned to Europe in less than a century, clearly such a rapid change could not be caused by the slow alterations in external factors such as astronomical cycles or internal factors such as the depression or rebound of bedrock below continental glaciers. Furthermore, as just noted, the largest signal of the Younger Dryas mini-ice age was in northern Europe and the maritime provinces of Canada. Although some evidence of climatic fluctuation has been unearthed elsewhere in the world, the magnitude is not nearly as large as in Europe, suggesting that conditions in the North Atlantic ocean might help explain the Younger Dryas.

One decade-old explanation, which has returned to popularity, is that the rapid melting of the Canadian and European ice sheets caused the flooding of meltwater into the North Atlantic after the ocean had essentially rewarmed. William Ruddiman and Andrew McIntyre, of Lamont-Doherty Geological Observatory, have used fossil plankton in the ocean sediments to map out the history of warm and cold water masses. These maps show the margin of icy water at the height of the last ice age to be on a line stretching roughly from New York to northern Spain. A line near the east coast of Greenland shows the icy water margin today. Another line, further out but still fairly close to Greenland, shows the icy water edge near the end of the ice age 11,000 years ago. A fourth line, nearly as far south as that of the maximum ice age extent, shows the readvance of cold ocean water in the Younger Dryas.[30] Since sea ice forms when the air above the water is very cold and when the surface waters are relatively less salty, it has been suggested that the rapid influx of fresh water from melting ice on land was responsible for a meltwater cap that promoted the rapid formation of sea ice in the winter. This ice then blocked the escape of heat from the oceans that, in turn, could no longer be blown downstream by the winds to keep Europe warm.

The reason salt is so crucial to the story of the Younger Dryas is that salty water is heavier and also has a lower freezing point than fresh water. The relatively salty surface water present in the North Atlantic today helps promote a vigorous circulation in winter, when the heavy, cold, salty water is forced to sink. This sinking invites northward penetration of warmer Gulf Stream water from the south and keeps the North Atlantic relatively ice free and Europe habitable. The reason the North Atlantic is so salty is that the average amount of precipitation that falls on the ocean or runs off the land into the Atlantic basin is less than the amount of water that evaporates from the Atlantic. On a global level, this excess evaporation over precipitation is compensated for by the reverse situation in the Pacific Ocean basin. Small changes in the salt balance of either ocean could trigger large changes in ocean water circulation, sea ice cover, and climate. Recently, sophisticated analyses of the chemical composition of the fossil shells of bottom-dwelling plankton buried 10,000 to 11,000 years ago suggest that a radical change in the circulation pattern in the North Atlantic ocean did occur at that time.[31]

In today's pattern of ocean water circulation, warm water from the Gulf Stream flows out of the Gulf of Mexico and up along the northeastern United States into the North Atlantic, keeping its waters warm past Greenland up to northern Scandinavia. There, cold air off Greenland or out of the Arctic chills the water in the region of the Norwegian Sea in winter. Because salty water increases its density as it cools, it sinks to the bottom, where it spreads out and flows at abyssal depths into the South Atlantic, under the Indian Ocean, and into the Pacific, rising near the Aleutians. From there, the current flows near the surface through Indonesia north of Australia, south of the Cape of Good Hope, back into the South Atlantic, across the equator, and into the North Atlantic. This pattern of circulation, which takes more than a thousand years to complete, depends critically on the salinity of the waters in various parts of the ocean. Salinity determines the density of the water and whether it will sink, causing warm water to replace it and heat to be released to the atmosphere, or freeze, blocking the flow of heat from the ocean to the air. Chemical analyses performed on the fossil shells of ancient phytoplankton suggest that both at the height of the last ice age and during the Younger Dryas, the present pattern of ocean circulation was radically altered: The Gulf Stream flowed far south of its present reach, and sea ice covered much of the North Atlantic that is now ice free.

Chemical analyses of the remains of fossil plankton at various depths in the oceans suggest that water reaching the bottom of the North Atlantic in today's interglacial mode had its origin in the Norwegian Sea area. However, in glacial times and during the Younger Dryas, the chemical fingerprints in the bottom-dwelling plankton suggest that the origin of much of this bottom water was the Weddell Sea in Antarctica, implying a radical shift in the vertical circulation pattern of the ocean during these colder times. Such a shift is consistent with the idea that the North Atlantic had a vast expansion of sea ice, which is also consistent with the evidence for extreme cold across Europe and the maritime provinces. Since delicate changes of only a few percent in the relative saltiness of the Atlantic and the Pacific oceans could cause such a flip-flop in ocean circulation, scientists now have an intense interest in understanding the Younger Dryas.

Wallace Broecker, an idea-generating geochemist from the Lamont-Doherty Geological Observatory of Columbia University, has reinterpreted data on the Younger Dryas to hypothesize the source of the fresh meltwater that set in motion the rapid climate cooling. To many scientists, fossil shorelines in the Great Lakes area suggested that a vast lake, much larger than today's Great Lakes, existed 11,000 years ago from water melting at the southern end of the great North American ice sheet. The existence of such a lake was consistent with evidence from ocean cores of the runoff near the Gulf of Mexico. Broecker theorized that about 10,800 years ago (the beginning of the Younger Dryas), an eastern lobe of the melting ice sheet broke that had been serving as a wall channelling meltwater down the Mississippi. The loss of this eastern restraint allowed a massive diversion of fresh glacial meltwater down the St. Lawrence River into the North Atlantic Ocean. Such an influx of fresh water would have radically reduced the surface salinity of the North Atlantic, permitted the rapid formation of sea ice, and disrupted the circulation pattern of the deep Atlantic.

But Broecker wasn't content to simply theorize on the cause of the Younger Dryas and its relation to ocean circulation. He went on to connect this phenomenon with the greenhouse effect presently being created by human pollution. Citing the Younger Dryas as a nasty surprise foisted by nature upon Europe just after the ice age seemed to have ended, Broecker wondered aloud at scientific meetings, in journal articles, as a witness before congressional committees, and in a number of media stories whether incredibly fast (geologically speaking) climate changes from the greenhouse effect could alter precipitation and evaporation patterns around the world enough to affect the salinity of the North Atlantic and thereby oceanic circulation?[32] Could the enhancement of the greenhouse gas envelope around the earth cause rapid cooling of the North Atlantic and a subsequent cooling catastrophe in Europe and the maritime provinces in Canada while the rest of the planet actually got warmer?

I do not wish to leave the impression, nor would Wally Broecker, that this scenario is highly probable or that it will take place in the next decade or two. The Younger Dryas was almost certainly not caused by a rapid pulse of greenhouse gases, and today there is no remnant of the North American ice sheet or great palaeolake poised to flood the North Atlantic with fresh water. But the Younger Dryas does dramatically remind us that the climatic system is a very complex set of interacting subcomponents whose workings are far from being fully understood. Quite simply, the bottom line is that if we disturb the climatic environment by as much as nature has from the last ice age to the present interglacial, and we do it some ten to fifty times faster, then nasty surprises, such as a radical and potentially catastrophic shift in ocean currents, are plausible. And the faster we alter the climate, the greater the likelihood of surprises. Unfortunately, I can't even venture a guess at what probability such an event could occur—but it is by no means zero. It may take another decade or two before scientists are able to more reliably pin down cause and effect in the Younger Dryas. Could it be that by that time we will have irreversibly committed the world to climatic changes large enough that a number of unforeseen surprises are virtually inevitable? That is the nature of the climate lottery we are now playing.

* * *

One postulated surprise is the arrival of another ice age before the year 2000 as a result of greenhouse gases.[33] This doomsday prediction has been pressed by two very passionate individuals, John Hamaker, a mechanical engineer who formulated the idea, and Don Weaver, an organic farmer who became his disciple, and their followers who formed groups with names such as the Earth Regeneration Society and the Institute for a Future. These groups call for the construction of billions of dollars worth of rock-grinding equipment to produce mineral dust to remineralize the soils. The essence of their theory is that during an interglacial, the enhanced productivity of the forests and agriculture depletes the soil of minerals, thereby allowing atmospheric CO_2 to build up as the productivity of the world's ecosystems declines with declining mineral content. The CO_2 buildup—and here is where most atmospheric scientists that I have talked to think this theory is way off line—leads to an ice age in as little as 100 years!

Earlier, recall, I wrote that increasing temperatures result in increased winter snowfall, provided winter temperatures are still below freezing. Hamaker and Weaver push this point to the extreme. They also resurrect a fifty-year-old theory of British climatologist Sir George Simpson that suggests that increasing temperature will increase evaporation from the oceans, transport the excess to the poles, and increase cloud cover in the high latitudes, thereby increasing the albedo (reflective capacity) of those regions, causing the rapid buildup of snow and ice.[34] One of the problems with this theory is that water vapour might not be transported to the poles with greenhouse warming since atmospheric circulation could weaken. Another problem is simply that the theory neglects the fact that increasing temperatures are by no means certain to increase average cloud cover locally, since cloud cover depends on relative humidity and not on the absolute amount of water vapour in the air. There is much more water vapour in the air on a cloudless day in New York City in July than on a snowy day in Antarctica, simply because higher temperatures permit more water vapour to be in the air. More importantly, I believe, increasing temperatures from an enhanced greenhouse effect will very rapidly melt snow in high latitudes where temperatures are above freezing—as they are today in summer over virtually all the unglaciated landmass in the Northern Hemisphere. Furthermore, nature performs a counterexperiment every year as the seasonal cycle generally leads to a decrease, not an increase, in snow cover in late spring and summer in high latitudes. Finally, on longer time scales, nature has performed another experiment for us in which the slow changes of the earth's orbit around the sun have changed the amount of sunlight entering the atmosphere between winter and summer. Ice ages seem to disappear when summer radiation is enhanced, as occurred from 15,000 to about 5,000 years ago. This summertime heating certainly does not seem to have built up the glaciers, but rather to have knocked them down.

In any case, in dealing with a system as vast and complex as the earth's climate, speculations of all sorts are difficult to rule out absolutely.[35] Most important, both for a general theory of climate and for the practical issue of forecasting climatic events in the greenhouse century, is the question of how we can validate the tools that we use to make forecasts.

By now, readers may be seriously concerned that our ignorance of the detailed cause-and-effect relationships in the climate system is so appalling that we should have no confidence in any of our forecasting methods. Therefore, it is with some pleasure that I turn to one major success story in climatology.

One should not infer from the discussion so far that the present interglacial has been a uniform climate all over the world with conditions precisely as they are in the twentieth century. Temperatures in the warmest part of this interglacial some 5,000–9,000 years ago appear to have been about 2° C (3.6° F) warmer in summer than they are today. Nevertheless, this temperature difference is still much less than temperature changes during the glacial-to-interglacial transition and much less than most projections of greenhouse-gas-induced climate change late in the greenhouse century. Temperature variations within the present interglacial are one of nature's experiments that climatologists need to decipher and then apply as validation tests of their climatic models.

One of the more scenic drives in North America can be found by heading east from Alliance, Nebraska, for about 200 miles along State Route 2 toward the sleepy rural town of Broken Bow.

This easy half-day trip through the Nebraska Sandhills reveals many vistas of rolling, grassy hills, grazing cows, growing wheat, wind-powered irrigation pumps, railroad hopper cars filled with coal from Wyoming moving east to the Midwest, and empty hopper cars returning west to be refilled. Seeing this productive agricultural region raised some questions in my mind as I drove through: Why are they called sandhills since they are largely covered with grass? When, in fact, were the Sandhills last sandy? I put that question to friends at the University of Nebraska later that day. Eventually I was introduced to a geologist who explained that some 3,000 to 8,000 years ago the sandhills were drifting dunes, and that paleontological and archaeological explorations showed vastly reduced vegetation and animal presence during that very dry period.[36] That's funny, I thought to myself, for 3,000–8,000 years ago is the period climatologists call the climatic optimum—a phase when summer temperatures were about 2° C (3.6° F) warmer than at present.

Fossil pollen studies suggest that during this warmer time the grasses of the high-plains prairie, now largely found west of the Missouri River, had migrated deep into the corn belt, with a bulge of prairie extending all the way to central Indiana. This bulge, when plotted on a map, justifies its nickname, the prairie peninsula. Clearly, a return to such dry conditions in the heartland of America would be a serious blow to the agricultural potential of the region and to the food-exporting capacity of the nation. Moreover, such a change has been associated with a seemingly small change of only a few degrees in summer temperatures. Could such a prairie peninsula recur if the planet warmed up several degrees over the next few decades—this time from the human pollutants enhancing the greenhouse effect?

This issue was given prominent consideration a decade ago by my now-retired colleague William Kellogg. His map of the climatic optimum, derived from proxy data, stimulated much research on the reconstruction of the climate of the past 10,000 years (see Figure 3).[37] As Kellogg's map shows, the climatic optimum was a time of substantially increased moisture in North Africa and the Middle East. Today the very mention of the Middle East conjures up images of camels, conflict, and controversy over scarce resources in a largely arid setting. But the Sahara dunes and the small, closed lakes of the region today are not at

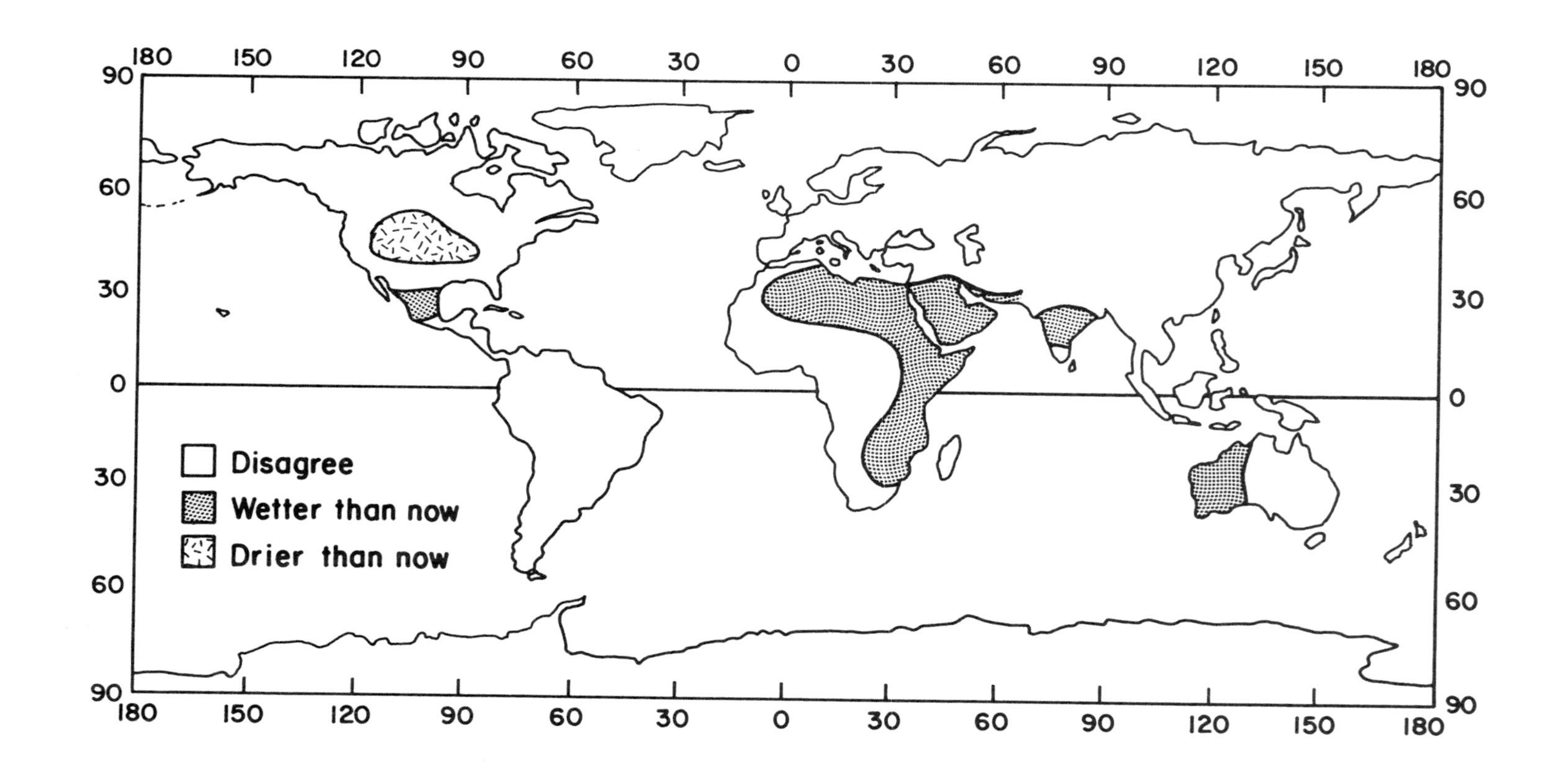
Disagree
Wetter than now
Drier than now
180
150
120
90
60
30
0
30
60
90
120
150
180
90
60
30
0
30
60
90

all representative of the steppelike vegetation and vast inland lakes that thrived 5,000 to 10,000 years ago. Expeditions to the Sahara by geologists and geographers have revealed ancient lake shorelines of much greater dimensions than today. While dunes covered Nebraska, civilization was springing up from a wetter and more fertile crescent in the Levant.

Kellogg surveyed the palaeoclimatic literature to make this map in the hopes of finding a geographic scenario of regional climate change associated with warming. His premise was that if the climatic optimum were a few degrees warmer than the present with different distribution of wet and dry regions, then perhaps we could use it to forecast climate change in the next couple of decades, which most climate modellers project to be about 2° C warmer than heretofore in the twentieth century. However, the climatic optimum was almost certainly not caused by a substantial increase in the greenhouse effect, and thus the detailed nature of its regional climatic features cannot automatically be taken as a metaphor for a greenhouse century caused by industrial pollutants (especially in Canada, where remnants of the great ice sheet persisted until 6,000 years ago). Nevertheless, Kellogg's work certainly made it clear that substantial changes in the character of regional climates so vital to agriculture, forestry, water supply, and natural ecosystems can be expected with planetary temperature changes of only a degree or so. Several years after Kellogg investigated the possibility of using palaeoclimatic maps as meta-

Figure 3. In the absence of direct evidence of how a CO_2-induced climatic warming might affect regional distribution of aridity or increased wetness, William Kellogg looked to the middle of the present interglacial, some 5,000 to 9,000 years ago, when (according to inferences made from a number of climatic proxies) summer temperatures were apparently a few degrees warmer than at present. While probably not caused by CO_2, the results he mapped could provide warm-earth analogies. Anthropologist Karl Butzer also produced such a map, and Kellogg combined on to another map (seen here) the areas of agreement and disagreement between these two paleoclimate interpretations. Note that although the authors' reconstructions disagree over most of the land area, they do agree that there was increased dryness in central North America and increased wetness in parts of tropical and subtropical Africa, Asia, and Australia. Agreement between Kellogg and Butzer does not, of course, necessarily imply correctness, as there is a possibility that these maps could agree or disagree by chance. [Source: Kellogg, William W., and Schware, R. 1981. Climate Change and Society: Consequences of Increasing Atmospheric Carbon Dioxide *(Boulder: Westview Press), Figure C.3.]*

phors for a warm earth, other scientists were able to explain fairly convincingly how nature created those changes.

As can be seen on Kellogg's map, the period of the climatic optimum included wetter and drier regions. Note that Africa and parts of Asia were wetter during the climatic optimum than today. John Kutzbach, an interdisciplinary meteorologist from the University of Wisconsin, had the idea that this period might well be explained by the theory of glacial/interglacial cycles being related to changes in the earth's orbit. Moreover, he felt he had a tool that could test his idea and had access to data to verify his test method. Kutzbach proposed that increased solar radiation in the northern summer season 9,000 years ago was sufficient to heat up the interior of continents, particularly in the tropics and subtropics, by several degrees Celsius. Because land has less heat-storing ability than the oceans, it warms up faster than oceans from the redistribution of sunlight caused by the orbital changes. If the land warmed up more than the ocean in the summers, Kutzbach suggested, the temperature differential would enhance the driving mechanism of summer monsoon rainfalls.

Monsoon rains are caused by the rapid rise in land temperature relative to ocean temperature in summers, when the sun is overhead, which causes a pressure drop over the land. That pressure difference causes winds to flow from the Arabian Sea to the Indian subcontinent, and from the central Atlantic to the Sahel (west and north central Africa). Since winds originating over water have a high moisture content, when they flow over heated or elevated land the air begins to rise and its moisture is squeezed out in crop-sustaining monsoon rains.

Kutzbach felt that the monsoonal rains would have been enhanced during the climatic optimum because of the extra solar radiation in the summer season and proposed to test this idea by changing the distribution of solar energy in a mathematical climate model available at the National Center for Atmospheric Research. This three-dimensional community climate model (CCM) of the earth's atmosphere was modified by Kutzbach and colleagues to account for the increased summer heating 9,000 years ago. As predicted, the model generated warmer temperatures by a few degrees Celsius over the continental interiors of Africa and Asia, and summer monsoon rains were enhanced.[38] From this model result, Kutzbach predicted that larger lakes should have existed in the interior of Africa and that greater runoff should have occurred from Asian and African rivers.

To validate his findings, Kutzbach joined with a number of colleagues in an international programme to document how the earth's climate had changed over the past 18,000 years. Palaeogeographer Alayne Street-Perrott from Cambridge University, England, spent many years digging in the palaeosoils of central Africa identifying the boundaries of ancient lakes through plant and animal fossils. By using chemical dating techniques, Street-Perrott was also able to define the relative aridity of central Africa over the last 18,000 years. Her findings were in startling agreement with Kutzbach's CCM results, which showed that the relative wetness in Africa was controlled to a large degree by the earth's orbital variations.[39] Further evidence supporting the notion of orbitally induced monsoons was provided by oceanographers who had taken core samples from the ocean floor near the mouths of certain Indian rivers. These samples suggested substantially enhanced river runoff 9,000 years ago.[40] Palaeoecologists, such as Brown University's Thompson Webb and the University of Minnesota's Margaret Davis, have been attempting to reconstruct the vegetation and climatic history of large portions of the earth over the past 20,000 years. The findings of this interdisciplinary scientific team effort are proving invaluable to those of us who are trying to explain how both the climate and the ecology of the planet have changed from the last ice age to the present interglacial period.[41]

This research is not simply an academic exercise. Kutzbach and his colleagues have done much more than to simply explain what caused the climate change that helped civilization to flourish in Africa 5,000–10,000 years ago. Their findings have major policy implications in regard to the question of greenhouse effect. The ability of computer models to reproduce regional climatic responses to the changes in solar radiation brought about by variations in the earth's orbit lends a major element of confidence to the reliability of these models as forecasting tools for the future climate resulting from increasing greenhouse gases.

The warmest time of our millennium — the so-called medieval optimum, when Norse colonies were set up in Greenland and Vikings patrolled the Atlantic region—was probably about a degree or so warmer than the present. Within a few hundred years, however, these colonies were frozen to death, and Europe and North America entered a period known as the Little Ice Age, in which Northern Hemispheric temperatures appear to have been perhaps 1°–2° C colder than in the twentieth century.[42] This period of a few hundred years with temperatures a degree or so above or below the average for the millennium has had impact on historical events.[43] But climate variability can have much greater swings, and much greater human impact, over shorter time periods, such as devastating droughts lasting a decade or more, or a very cold summer, or a few frost days that wreak havoc with crops.

The Anasazi cliff dwellings of southwestern Colorado's Mesa Verde National Park stand in tranquil beauty like 100-foot-high bas-relief art etched into the vertical yellow walls of the quiet canyon. The dwellings are lonely today, unoccupied by their builders for more than 700 years. The Anasazi Indians practised agriculture on the land above the canyon and felled trees, using some of the logs in the construction of their houses. Archaeologists have long been interested in why the Anasazi abandoned this magnificent setting so abruptly. Was it plague, invasion, or famine that drove them out?

The age of the Anasazi cliff dwellings has been determined by tree ring analysis.[44] Because every year's weather is somewhat different and because a combination of weather elements influences the growth rate of trees, a unique sequence of tree rings is created by the differing weather from year to year. By examining the intricate pattern of ring widths of living trees, overlapping these patterns with those of dead trees that go back hundreds of years further, and overlapping those patterns with tree ring patterns of even older dead trees, tree ring scientists—dendrochronologists—have been able to literally count back ring by ring, year by year, to the trees used by the Anasazi in their buildings. By this method the exodus from Mesa Verde has been dated at about the year 1280. By going back further, it was possible not only to date various stages of construction, but also to correlate the widths of the tree rings with climatic conditions. It appeared that the collapse of the Anasazi civilization coincided with a substantial drought period of a decade or so. This seemed an obvious explanation for the flight of the Indians to other locations, such as northern New Mexico and Arizona.

But dendrochronologists looked back even further, discovering other periods in which tree ring sequences suggested limited growth probably caused by drought. Such earlier drought periods did not cause an abandonment of the cliff dwellings. So the obvious question is, Why could some droughts be endured but the final one not? No certain answers are available, but I have heard interesting theories to explain the evidence. For example, archaeologists can count pottery fragments or fossil remains of seeds to infer something of the population size and agricultural enterprise over the time of occupation. Such evidence has led to the suggestion that the Anasazi population was lower during earlier droughts than the final one. It is possible that the many good years of weather that preceded the ultimate Anasazi catastrophe increased carrying capacity of the limited farmland and allowed a concurrent expansion in population. Although bad weather had struck before, even the reduced carrying capacity of the land had been adequate to sustain the lesser population. When disaster finally struck around 1280, the cliff-dwelling civilization collapsed from its lack of resilience to climatic extremes.

It is not at all certain that the story just given is fact, but the message is nonetheless quite general: human vulnerability to changes in climate depends as much on how our societies are organized as on the nature of the climate change itself. We know that sea level rises have reshaped the coastlines and that receding ice has been chased north by advancing boreal forests while the ecological vacuum has been filled by encroaching hardwoods. Temperature changes of a few degrees Celsius or less on a global basis have been associated with substantial changes in the regional patterns of rainfall, and species of animals and vegetation either migrated, adapted, or became extinct in response to such changes. When humans were living beyond the carrying capacity of the land in bad years, they suffered and moved. But no longer is mass migration such an easy option for the more than 5 billion human inhabitants of planet Earth. We are locked tightly in national boundaries, and some billion of our kind live very close to the nutritional margins of survival. Civilization is still embedded in the environment.[45]

When environmental conditions change slowly, adaptation is

much easier. The more rapid the change and the less flexible the society, the more difficult it is to deal with such change. A climate change of a degree or two per thousand years, typical of that experienced since the waning of the great glaciers some 10,000–20,000 years ago, would create new habitats and regional climates. A change at those natural rates, particularly if we had some advance warning of their character, would probably prove tractable for the adaptation of most societies. (Of course, some activities would have to be abandoned in some places, although new opportunities could be created in others.) However, our capacity to adapt to changes much faster than natural rates is certainly less assured. And the resilience of natural ecosystems that have taken 10,000 years to become established is even more doubtful.

We have had about 10,000 years in our present interglacial, a period the geologists call the Holocene. A number of scientists have suggested that the demise of this interglacial is due and that we will be heading in a thousand years or so back toward the next ice age, a century or two after CO_2 pollution has finally dissolved in the ocean. But the record of the glacial cycles over the past million years suggests that glacial/interglacial cycles follow a sawtooth pattern: ice sheets develop slowly and end rapidly. This sharp recovery then leads to a short interglacial, and the cycle repeats. While some events (for example, the Younger Dryas or the Little Ice Age) in climate history do suggest that regional changes more rapid than 1°–2° C per thousand years have occurred occasionally, it still appears more likely that the slide into the next ice age will take 10,000 years or more. Massive glaciers, such as those that built up over Canada and Scandinavia in the last ice age, simply cannot be constructed in a decade, a century, or even 1,000 years. At typical Canadian snowfalls of some 1 metre (3 feet) per year, it would take 3,000 years to build back the great ice sheet to its 3-kilometre height—and that only if we make the totally unreasonable assumption that no snow melted in the summer. The cause of the rapid (10,000-year) destruction of the North American and European ice sheets every 100,000 years or so for the past million years is still a matter of debate, and scientists still aren't sure of the precise physical mechanisms of this 100,000-year glacial/interglacial cycle. Whatever the exact cause or causes, it is known that nature can change the climate substantially, that CO_2 and methane concentrations in the air seem to move in concert with temperature trends, and that humans and other living things will be dramatically affected along the way.

Humans are not simply passengers holding a temporary ticket on planet Earth's ride through the galaxy. We are actively altering the surface of the land and the composition of the atmosphere. These factors affect the natural flows of energy and materials around the planet and in turn are altering the climate. And while it usually takes nature thousands of years to create several degrees of temperature change on a globally sustained basis, human beings can do so in a century or less.

4

The Case for Climate Stability

Objections to predictions of major global climate changes from increasing greenhouse gases often come from those who argue that somehow nature has a buffering mechanism or a negative feedback process that ameliorates any kind of climate change. A buildup of clouds that screen out sunlight and cool a planet in response to the warming of that planet is the most commonly cited example of such a negative feedback process. This example is largely a physical feedback process since increasing the temperature of the oceans is virtually certain to increase evaporation rates. However, some important feedback processes can be attributed to life itself. Is it possible that life is not only a spectator of the earth's physical and chemical environment but is an active control mechanism as well? Could it be that all life taken together could serve the role of negative feedback, and thus nurturer of the planet? In short, is there a Goddess of the Earth? This is not a fanciful question, but one that has spurred a major debate over what has been called the Gaia hypothesis.

In 1976 NASA sent a Viking spacecraft to land on Mars and dig a soil sample to test for evidence of life. James Lovelock, a brilliant, interdisciplinary, and iconoclastic British scientist, had been invited by NASA in the late 1960s to help design life-search experiments for the Mars lander.[1] He and U.S. microbiologist

Lynn Margulis (and a few years earlier, philosopher Dian Hitchcock) had concluded from available earthbound observations of the composition of the Martian atmosphere that no life would be found because gases that are highly chemically reactive—such as oxygen, methane, and ammonia—are present on Earth and essentially absent from Mars. The abundance of these gases on Earth is well understood to be a direct consequence of a single factor: life. Without life, such gases could not exist in anything but the minutest amounts because they would react with the planet's surface materials and with each other and therefore be removed from the atmosphere very quickly. Photosynthesizing green plants produce oxygen (and before plants evolved on land some half billion years ago, oxygen was—and still is—produced by algae in the oceans). Methane, ammonia, and other such reactive gases are produced on Earth by microbial activity in such places as soils, bogs, the oceans, and the guts of animals.[2] The continuous production of these chemically unstable products on Earth keeps them at an atmospheric level way above that of a dead planet like Mars.

Telescopic observations had already showed that Mars lacked these reactive gases, and thus Lovelock and his colleagues concluded that it would be futile to search for life. Lovelock somewhat sarcastically recalls his impression of work at NASA on the biologic detection experiment:

> Here I discovered an extraordinary dichotomy. Seen at first hand, the quality of engineering and physical sciences of the NASA institution was often so competent as to achieve an exquisite beauty of its own. By contrast (with some very noticeable exceptions), the quality of the life sciences was primitive and steeped in ignorance. It was almost as if a group of the finest engineers were asked to design an automatic roving vehicle that could cross the Sahara Desert. When they had done this they were then required to design an automatic fishing rod and line to mount on the vehicle to catch the fish that swam among the sand dunes.[3]

Lovelock and Margulis were not content merely to predict the absence of life on Mars. They argued that life is an active control agent able to change the colour of the earth's surface, the composition of its atmosphere, and the salinity of the oceans so as to make the earth more fit for life.[4] As Lovelock put it: "In other words, the air we breathe can be thought of like the fur of a cat and the shell of a snail, not living but made by living cells so as to protect them against an unfavourable environment."[5] Lovelock and Margulis called this theory the Gaia hypothesis based on the suggestion of Lovelock's neighbour, Nobel prize-winning novelist William Golding. Gaia or Ge was the Greek mother earth goddess, and thus it seemed appropriate to name this planetary-scale nurturing entity after her.

Lovelock and Margulis's challenge to conventional wisdom was not only that life influenced the physical and chemical environment—something that had been well known and well accepted for years[6]—but that life served as an active feedback mechanism to damp out planetary changes that otherwise would have taken place. The title of Lovelock and Margulis's paper on the Gaia hypothesis used the term *homeostasis,*[7] meaning the ability to maintain a stable state. Comparing life on earth to a warm-blooded animal with homeostatic body temperature control, Lovelock and Margulis argued that life has mechanisms to cool the planet before it overheats or to warm it up if it is cooling down too much. The principal balancing mechanism they cite is the ability of certain kinds of phytoplankton—tiny photosynthesizing aquatic creatures—to take carbon dioxide out of the waters and incorporate it into their bodies in the form of calcium carbonate shells. When these plankton die they sink to the bottom of the ocean and a significant fraction becomes incorporated into the sediments as carbonate rocks (for example, limestone), effectively removing CO_2 from the upper oceans and the atmosphere (remember, the lower the CO_2 content of the atmosphere, the weaker the greenhouse effect). In addition, when land plants decay, bacterial decomposition of their remains increases the amount of CO_2 in the soil, thereby accelerating the rate at which minerals in the soil can be bonded into carbonate chemicals that eventually wash into the sea, where the carbon is removed as sediments.[8] Lovelock and Margulis have used such an argument to help explain how life helps solve what has come to be known as the faint early sun paradox.

According to mainstream astrophysical theory, stars like our sun follow a sequence in which they grow in size and intensity as they age. Some 4 billion years ago, when the first living cells appeared on earth, the sun is believed to have emitted about

25%–30% less radiant heat than it does today. If the sun emitted that much less energy, then why was the young earth not a frozen ball? It appears from fossil rocks dating back 3.8–3.9 billion years that liquid water was present at least in a few places at that time. That, plus the fossil evidence of life at that time, strongly suggests that at least some parts of the earth had surface temperatures not very different from today—certainly not much below freezing or much above 35° C (95° F). With 25%–30% less solar energy arriving, the planet should have been much colder and perhaps ice covered. Two decades ago Carl Sagan and George Mullen proposed a solution to the faint early sun paradox: the presence of a supergreenhouse effect.[9] They thought that ammonia or methane, which are very effective absorbers of infrared radiation, could have warmed the climate sufficiently even if they were present at concentrations of only 1 molecule per 10,000. This clever suggestion was later attacked on the grounds that ammonia and methane are chemically too reactive and could not have been continually resupplied to the atmosphere in sufficient quantities.

Despite the debate over the lifetime of these gases in the air, the supergreenhouse idea was a good one and stuck. A few years later other researchers suggested that a supergreenhouse effect was indeed the solution to the faint early sun paradox, but that the more likely greenhouse gas was CO_2, not ammonia or methane.[10] The idea was that in the earlier parts of earth history volcanic activity was high. Such activity emits considerable amounts of CO_2 from inside the earth. This extra CO_2 enhanced the greenhouse effect and helped keep the planetary climate warm despite the faint sun. But, if this were so, it poses a second problem: how did the earth get rid of all that extra CO_2 over the next 4 billion years as the sun warmed up? If the CO_2 were not somehow removed, the earth's temperature would be too hot today. In other words, is there a stabilizing or negative feedback mechanism that maintains planetary homeostasis despite the large increase in solar energy over earth's history?

Lovelock and Margulis suggested that as the planet warmed up, life was the agent for removing the primordial CO_2 through the well-accepted biologically controlled mechanism of carbonate mineral formation in the soils and the seas. But some critics have asked how life could have been doing that for the first 3 billion years of earth's history because plankton-secreting carbonate shells did not evolve in the oceans until only about 1 billion years ago. Furthermore, it took 500 million years more before land plants flourished on the continents and could have pumped CO_2 out of the air and into the soils.

An alternative hypothesis for planetary CO_2-climate control over the whole of earth's history was proposed by three geochemists then at the University of Michigan, James Walker, P. Hays, and James Kasting—a mechanism now known by the acronym WHAK after the authors' initials.[11] WHAK recognized that a warm planet with a water surface evaporates more water than a cool one, based on the well-known principle that warm water evaporates faster than cold water. Since what goes up must come down—as long as it doesn't leak into space—the water vapour evaporated from a warm planet must fall as rain or snow. Therefore, as the faint sun grew hotter over time, so did the earth get warmer. A warmer earth evaporated (and therefore precipitated) more water, which, in turn, should have meant greater weathering of minerals on land. This inorganic weathering of exposed rocks produced carbonate sediments that removed CO_2 from the air, thereby helping to stabilize the climate in the face of the warming sun. In other words, the WHAK mechanism was an inorganic alternative to climatic homeostasis by life. Kasting (now at Penn State), and two California NASA colleagues, Owen Toon and James Pollack, recently suggested that although life is an important component of climate regulation, as Lovelock and Margulis had suggested, it is not the dominant component and certainly had not been dominant throughout most of early earth history.[12] The NASA scientists argued that even if life did not exist, inorganic chemical reactions in the oceans would occur to make carbonate sediments and remove CO_2. Then by the time the planet warmed up some 10° C (18° F) more, the rock cycles would have done the job in the absence of life. Of course, given this scenario, the planet might be 10° C hotter than it is today—and 10° C (18° F) is about as hot as it was in the age of the dinosaurs about 100 million years ago.[13]

A raging debate broke out in the mid-1980s between Gaians arguing for the dominance of life control systems and geochemists pushing inorganic mechanisms as primary; another debate ensued between Gaians and Darwinian biologists, the latter arguing that evolution takes place by natural selection at the species level for the good of individual species, not at a planetary scale for the overall good of all of life as the Gaians implied.[14] The first major scientific society meeting on this subject convened in March 1988 at an American Geophysical Union Chapman Conference. All the factions came together in San Diego for a fascinating and dramatic week. Gaia came in for a rough time from Stanford University population biologist Paul Ehrlich, who pointed out that not all effects of life on the environment are benign. Others complained that the Gaians had never clearly defined what they meant by control that is "optimum for life."[15] Nevertheless, the Gaians described a number of fascinating pathways whereby trace chemical emissions by living things could change the composition of the atmosphere, the brightness of clouds, or other leverage points that affect climate.

In the mid-1980s Lovelock visited the University of Washington, where he met atmospheric scientist Robert Charlson, who has long been interested in how aerosols (that is, small particles suspended in the air) interact with the radiant energy coming both from the sun and from terrestrial sources. Perhaps the most important role played by these particles is in forming the nuclei of droplets that make up clouds. Also involved was Meinrat Andreae, who was at Florida State University at the time. Andreae is a specialist in the sulphur cycle and had been focusing on how an obscure sulphur chemical produced by certain kinds of phytoplankton in the ocean might be affecting the amount of sulphur gases in the remote marine atmosphere. It was later discovered that this gas, dimethyl sulphide (DMS), could account for a significant fraction of nuclei that allow cloud water droplets to form in the very pure air in remote parts of the oceans. The team of Lovelock, Charlson, and Andreae assessed a very interesting possible biological role in the climate system. Their argument went something like this: Dimethyl sulphide is produced by certain kinds of phytoplankton in the upper layers of the oceans. Some of the gas leaks from the oceans into the atmosphere, where it is rapidly converted to sulphur dioxide. Later it is transformed into sulphuric acid particles that can act as cloud condensation nuclei upon which drops form.

Many years earlier, Australian scientist Sean Twomey had postulated that an increase in the number of particles in the atmosphere would take the same amount of liquid water in a cloud and divide it up into a greater number of droplets.[16] He argued that the more droplets in a cloud of a given size and given amount of water, the less sunlight would be transmitted through to the earth's surface and the more would be reflected back to space. (When a large drop is split up into several smaller droplets, more surface area is created, and it is these droplet surfaces that reflect the sunlight.) The higher the reflectivity, or albedo, the less solar radiation is absorbed by the earth. Therefore, Twomey argued, aerosol particles not only can interfere with the radiant energy penetrating through the clear sky but also could affect the climate by changing the relative brightness of clouds. This would mean that our present climate might be dependent on phytoplankton, which are responsible for keeping cloud droplets small over the unpolluted oceans. If the phytoplankton were killed off, the earth would warm.

Andreae and Charlson built on this idea, bringing in climate theorist Stephen Warren from the University of Washington. Warren had worked with climate models to evaluate how changes in the amount of radiant energy absorbed by the planet could affect the climate. Steve helped the team by calculating how much a given amount of change in the number of cloud condensation nuclei over the oceans would affect climate. He concluded that if a change in phytoplankton productivity of dimethyl sulphide (DMS) caused an increase in the number of nuclei by 30%, then this could increase the cloud albedo by 0.02 (about 4%).

Charlson and Lovelock then considered whether the phytoplankton might be involved in a climatic feedback loop. Increasing cloud albedo would mean less light available for photosynthesis and colder ocean temperatures, which might slow down the production of DMS. They argued that this chain of events would be a negative feedback mechanism serving to stabilize the climate. However, the DMS scientists conceded at the San Diego Gaia meeting that it wasn't at all clear whether a change in climate would cause an increase or decrease in DMS production by phytoplankton since some species produce DMS, whereas others consume it. The complications of this biogeochemical feedback mechanism are immense; nevertheless, the possibility that biota could play an important role in climate stability is still well demonstrated by this example.[17]

A few years later, Charlson worked with a group of French scientists who were analysing the chemicals trapped in ancient ice from Antarctica. They made the startling discovery that sulphate chemicals in an ice core at Vostok occurred with a regular pattern over 150,000 years of earth history. In particular, it seemed that the sulphate concentration built up every 23,000 years, a cycle coincident with a change in the earth's orbit that causes the North Pole to point toward the sun when the Earth is closest to it. Charlson tentatively interpreted this 23,000-year sulphate cycle as support for the possibility that some of the ice age changes could be influenced by the fact that every 23,000 years more sunlight came into the tropical regions because of the earth's orbital cycle. Changes in incoming solar radiation to the tropics could modify some phytoplankton activity, thereby altering the DMS production.

Steve Warren recently updated the debate in a letter to me:

> The latest evidence is that the feedback loop is positive, so that this mechanism might exacerbate the greenhouse warming rather than mitigate it. The evidence comes from ice cores. DMS does not last long enough in the atmosphere to reach the interior of Antarctica, but its two oxidation products (methane sulphonic acid and sulphate) do, and both are found in much higher concentrations in ice from the last ice age than in modern snow.

The ice core also showed that another chemical deposit, calcium carbonate, also had a roughly cyclic pattern, this time with variations every 40,000 years. This cycle is coincident with the period of the change in the tilt angle of the earth's axis relative to the plane of its orbit. The greater the tilt angle the more pronounced the seasonal heating extremes are, especially at high latitudes. One possible interpretation of this calcium carbonate signal in Antarctica is that the sea level would fluctuate to some extent (tens of metres) with the amount of the earth's tilt, thereby uncovering more of the continental shelves whenever ice built up and sea levels dropped. This newly exposed land, rich in calcium carbonate sediments deposited from decaying phytoplankton when the shelves were covered with oceans, would now be free to erode. Blowing calcium carbonate dust from exposed continental shelves could reach Antarctica and produce the

40,000-year periodicity evidenced in the ice cores. It is far from certain that either of these somewhat speculative explanations of the 23,000-year sulphate and 40,000-year calcium carbonate cycles in the polar cap arise from these hypothesized causes. Nevertheless, these phenomena are partly biological in origin; although they may not prove homeostasis or negative feedback is at work, they clearly demonstrate the importance of connecting the biosphere to the geophysical history of the planet.[18]

Another important feature that also demands consideration of a biological connection to climate stability can be found in ice cores. Changes in the amount of carbon dioxide closely track the temperature changes inferred over the past 150,000 years. However, so does methane, which is twenty to thirty times more potent as a greenhouse gas than CO_2 (on a molecule-for-molecule basis). Since methane is produced largely by microbial decomposition of dead organic matter in bogs, garbage dumps, swamps, and in the guts of termites and animals, the close correlation between change in methane, CO_2, and temperature implies that biological processes have been involved. A large amount of methane is buried in the earth, trapped under permafrost in a form of sediment known as clathrate, a complex of methane with water molecules found mostly in northern continental shelves and tundra regions. Clathrates exist under high pressures and cold temperatures, so methane can be trapped within sedimentary material in geologic formations.

Gordon MacDonald, a geophysicist from the MITRE Corporation, has suggested that it takes from 500 to 1,000 years for the wave of warming temperatures that occur as the ice ages wane to penetrate the tundra and permafrost into the clathrates.[19] This warming could release methane to the air, providing a feedback mechanism that would further accelerate the end of the ice age. This idea could also explain the increased concentration of methane found in the ice cores when the climate is warmer. MacDonald also argues that the change in sea level associated with climate change alters the weight of the ocean column on top of continental shelf sediments, which in turn can change the rate at which these submarine clathrates let methane into the oceans and eventually the air.

Another biological feedback mechanism may involve bog formation, in which vast quantities of organic matter are stored during glacial periods.[20] All these are speculative ideas and require considerably more testing. Nevertheless, they are fascinating examples of how biological processes could, through admittedly complex pathways, be an important part of the explanation of large climatic changes that have occurred in the past.

Let's return to the issue of the Gaia hypothesis and its early reception in scientific circles. It is ironic that it took nearly two decades from the introduction of the hypothesis to have the meeting in San Diego of its supporters and detractors from the scientific establishment.[21] When Gaia was first introduced, it was embraced primarily by two radical opposites: on the one hand by what one might term ecofreaks (people looking for oneness in nature), and on the other hand, by polluters (those hoping that the Gaia hypothesis would allow them to continue to pollute with impunity since the great earth goddess would mitigate their impacts).[22] The San Diego meeting did not resolve the question of the direction or the amount of control that life exerts over the planetary climate, but it did lift the debate to the level at which it belongs: in the hands of those in the scientific community with knowledge and desire to test ideas and let the evidence lead where it may.

However, no one, especially Lovelock, argued that Gaia (should she exist) could completely undo the ravages of human alterations to the planet that are now taking place in a matter of decades. The time frame for whatever planetary homeostasis may exist is certainly not as rapid as that in which we are disrupting the environment. Nor are all biological feedback processes certain to be stabilizing. In fact, life could accelerate changes in the direction they were already headed. For example, there is about as much carbon in the atmosphere in the form of CO_2 as there is in living matter (mostly in trees). But there is several times more carbon in the soils stored as dead organic matter (known as necromass). Bacteria eventually help decompose some of this carbon-laden necromass into greenhouse gases such as methane, nitrous oxide (N_2O), and CO_2. Biologist George Woodwell has argued that this decomposition process will speed up if the soils get warmer, thereby emitting more greenhouse gases and enhancing the original warming. Such a positive feedback would, if it proves true, be a counterexample to climatic homeostasis from living processes.[23]

The seriousness of our alterations to the planet are in better perspective when viewed against the backdrop of the mutual evolution of living and inorganic components over 4 billion years of earth history. Mutual change, in which two separate species evolve differently because of each other's presence is known as coevolution, a term coined by population biologists Paul Ehrlich and Peter Raven while studying the mutual influence of butterflies and their host plants that led to genetic changes in both.[24] It took some 2 billion years for the bacteria and algae of the earth to build up the oxygen that eventually made our lives possible. It took 2 billion years more for multicelled creatures to evolve and branch into plant and animal kingdoms. One hundred million years ago, dinosaurs reigned over a planet whose average temperatures were some 10° to 15° C (18° to 27° F) warmer than present. No permanent ice existed at the poles that we can detect from the historical geologic record from that period. Then the continents drifted and the map was rearranged: the Antarctic continent isolated itself at the South Pole, India drifted northward across the equator and eventually collided with Asia, the Atlantic Ocean grew, the Isthmus of Panama rose, the Tibetan plateau uplifted, sea level fell some 300 metres (1,000 feet or so), the planet cooled, permanent ice formed, and the diverse habitats we enjoy today evolved. The extent to which life, inexorable geological forces, or some combination of both caused these changes is still debated. But the coevolution of climate and life on a planetary scale usually is measured in tens of millions of years, a snail's pace compared to the breakneck speed of human impacts on earth. Therefore, even if part of that physical and biological coevolution is homeostatic (that is, if Gaia exists), this mechanism is unlikely to negate rapid human impacts instantaneously.

The Gaia hypothesis provokes us to ask a broader fundamental question: how stable is the climate? Most evidence suggests that the climate has fluctuated between rather wide limits (plus or minus 15° C–27° F) for hundreds of millions of years. While these limits are large enough to have had major influence on species' extinction and evolution and large enough to encompass glacial and interglacial ages, there does not seem to be much chance that Earth is vulnerable to a runaway greenhouse effect as on Venus in which the oceans would boil away or to a cold catastrophe as on Mars, no matter what we do in the next century. Still, climate changes as great as an ice age would almost certainly be disastrous for humanity if they occurred rapidly. And, as I have repeatedly argued, humans may cause damage on this scale in as short a time frame as the next century.

5

Climate Prediction

Can we predict how much the earth's surface temperature will rise given a certain increase in a trace greenhouse gas such as CO_2? Two approaches to this question are open to scientists: mathematical modelling using supercomputers and studying the past for clues to the future. Let's begin with a glance at the past.

LOOKING BACKWARDS. We suspect that in the ancient past—some 100 million years ago, when dinosaurs roamed a planet approximately 10°–15° C warmer than today's—the CO_2 concentration in the atmosphere was many times larger than now.[1] But continents were in different places then, and the ocean bottom had a different shape. Unfortunately, there is no time in the earth's history when the geography matched that of today and for which we know precisely the earth's temperature and the amounts of CO_2 in the atmosphere. Yet we do know that climates of the ancient past were quite different from that of the present, and we strongly suspect that the amount of carbon dioxide may also have been different in those times. Many climatologists argue that attempting to estimate, however crudely, both global temperatures and CO_2 concentrations will yield information about the relationship between these two variables, information we can apply to predictions about our climatic future. The most prominent proponent of this approach to estimating the climate's sensitivity to changes in the atmospheric composition of CO_2 is the Russian climatologist Mikhail Budyko.

In April 1971, I attended a lecture to a packed house of geoscientists in Washington, D.C., by William Kellogg, then a division director at NCAR. His topic was "Man's Impact on Climate." I had anticipated this talk with great excitement because this was my newly chosen field; I had begun a postdoctoral fellowship in this area only a few months earlier, despite the fact that my doctorate was in plasma physics.

Following the talk, I went up to speak with Kellogg, to tell him of some calculations on the effects of pollution on climate that I was performing with Ichtiaque Rasool at the Goddard Institute for Space Studies. After a few minutes, he asked me a question that was to change my life: Would I will be willing to attend a three-week meeting he was organizing in Stockholm that summer that would help define the field of inadvertent climate modification? He needed a young postdoctoral fellow to help him write the report of the conference.[2] He hoped I would consider doing the job. Would I consider it! This meeting could well create a new field—how could I resist?

Mikhail Budyko, then the director of the Main Geophysical Observatory in Leningrad, was one of the most interesting and impressive scientists at that meeting. He was the first person to point out to me the importance of social assumptions in studying the greenhouse effect, referring to the need to forecast future fossil-energy consumption as "people problems." Budyko had developed an ingenious but simple climatic model that he used not only to predict the future but also to understand great climatic changes in the past. He insisted that for climatologists to validate their mathematical climatic models—the second approach to trying to predict future climate—we would have to test them on the very different palaeoclimatic situations of the ancient earth. His breadth and wisdom had a major impact on my thinking, and thus it was a great disappointment to me that he later became, temporarily at least, under apparent restriction in the Soviet Union. His international travel ceased, and in a few years his situation became so difficult that, for example, he was reduced to communicating with me by personal courier.[3]

Fortunately, in the 1980s Professor Budyko once again rose in prestige and importance in the Soviet Union—and certainly in visibility to Westerners. Then, in September 1987, the ultimate good news came: in this era of *glasnost,* Budyko was to travel to Washington with a Russian delegation and wanted to meet me, if possible, in October. We arranged our second meeting together in sixteen years. This time there was no need of special go-betweens; we had frank and open conversations at dinner, in halls, in the lobby—and without the usual nonscientist chaperone, who most of us had always presumed was a KGB tail. How amazing did the positive influence of *glasnost* seem to me!

In the lobby of the hotel, Budyko provided an autographed copy of his most recent book. In it, he had reviewed and extended substantially his analyses of ancient paleoclimates, begun several decades earlier. The book included a graph that was stunning: on one axis was a series of dots representing temperatures of the earth at different ancient times, and on the other axis was the corresponding amount of carbon dioxide present in the atmosphere at each of those times. He drew a straight line through these dots, and the fit was so spectacular that virtually every dot touched the line.[4] The conclusion? Increases in atmospheric CO_2 were correlated with global temperature increases—our computer models of the future were suggesting the same conclusion. "This is empirical proof of the greenhouse effect," he told me. "The actual earth proves that carbon dioxide will change surface temperature much as the models suggest. Look," he added excitedly, pointing at the graph in his book. "In the past when carbon dioxide was twice what it is now, the earth was 3° C warmer. This is precisely the middle of the range suggested by present climate models." He asserted that he could also reconstruct what the regional pattern of climatic changes would be for a globe 1-degree, 3-degrees, 5-degrees warmer, and so on. He looked at me expecting an excited and enthusiastic response—but my face revealed scepticism.

It was hard for me to accept his basic premise—that we can know the globally averaged surface temperature of the earth to better than several degrees Celsius for any time period up to tens of millions of years ago. And I knew it was even more difficult to determine precisely how much carbon dioxide was in the atmosphere at those times; we have no direct measurements but only inferences based on the chemical composition of sediments or assumptions about the rate of the ocean floor's spread. It seemed inconceivable to me that we could pin down simultaneous temperature and carbon dioxide amounts precisely enough to display them as dots on a graph, as Budyko had done, let alone expect that they would all neatly fall out along a straight line. Budyko explained that the Soviets must expend vastly greater efforts than Americans in reconstructing the earth's past because the Russians have no supercomputers, and thus lack the luxury of relying on models. I agreed that some of us in the United States did rely too heavily on our computer models while underemphasizing our study of the earth's rich climatic history. But I was doubtful—and remain so—that, regardless of the methods we use, we could rely on the kind of precise estimates Budyko had showed me.

Furthermore, I wasn't convinced that knowledge of regional climatic differences over the past several million years, at times when temperatures were probably a few degrees warmer, could translate into the changing regional climates that will evolve over the next hundred years. Of course, I agreed that his was an important approach that needed further testing and debate, but I judged it too nonquantitative to guide, for instance, agricultural responses to future climate change—something Budyko told me the Soviets were already seriously contemplating.

"You are too cautious," Budyko advised, remarking that a scientist should be willing to stick his neck out and stand up for what he considers probable. I do believe the greenhouse effect will substantially change climate, and I agree that it is likely that there was more CO_2 during warmer times in the ancient past (and I also stick my neck out often!). But we simply do not know the ancient past well enough to use it in making quantitative predictions about how sensitive the earth's temperature will be to the unprecedented rapid buildup of greenhouse gases now under way.

Professor Budyko seemed disappointed in my response, and I felt quite badly that this long-anticipated reacquaintance had proved somewhat tense at times. But I certainly came away with the two things I expected from any meeting with Budyko: excitement and stimulation. When I left that meeting, I was convinced that what I had learned from Budyko sixteen years earlier was still true: it is crucial that we check our estimates of the sensitivity of climate changes in the future by looking backward. But what I had learned in between had taught me that drawing precise conclusions would not be an easy task or one soon performed.

REGARDING THE RECORD. Despite my reservations about precise inferences, there are historic events for which we have excellent quantitative knowledge. These are the climate variations that have taken place since major meteorological instruments were invented and long-time series of measurements began to be recorded. The thermometer is many hundreds of years old, and some existing records of temperature measurements start with the Renaissance. But the earliest thermometers were rare and largely limited to the advanced cities of Europe and America. Not until the late nineteenth century was anything approaching a crude worldwide network of temperature measurement in place. Now, literally thousands of stations can be used to construct the earth's temperature and study such important climatic phenomena as twentieth-century temperature trends or the large temperature differences associated with the seasonal cycle. The ability of present-day climatic models to reproduce these large and reasonably well-measured changes will be important components of validating those models' performance. But before we judge the predictive power of the models, let's assess the reliability of our available records of the past.

Climatic change occurs over many time scales—from seasonally to periods spanning tens to hundreds of millions of years, from locally to globally. And there have been large variations—such as the glacial/interglacial cycles of the past 3 million years or so—that have cycle times of a few tens to a few hundreds of thousands of years. The temperature difference between a warm palaeoclimatic period in the Mesozoic era some hundred million years ago and the present is estimated to be 10° to 15° C (18° to 27° F) globally. However, the largest natural change since the Mesozoic era actually recurs annually between winter and summer. The difference in surface air temperature averaged over the Northern Hemisphere between winter and summer is now some 15° C (27° F). In the Southern Hemisphere, this difference is only about 5° C (9° F) because the large amount of ocean in the Southern Hemisphere, acting as a great heat-storage tank, reduces the seasonal response to the annual cycle of the sun. The seasons are, of course, caused by the annual cycle of solar radiation driven by the earth-sun orbital geometry. This process is known as seasonal forcing. Other factors considered external to the climate system can similarly force climatic changes.

The difference in monthly averaged temperature from one year to the next, the so-called interannual variability, is much smaller than the seasonal cycle, averaging from only a few tenths of a degree on a hemispheric basis to perhaps as much as a degree or two on a continental scale. Thus, on a hemispheric scale, the seasonal signal (the clear data we can be sure of) is more than ten times the noise (background factors that could obscure the signal) of interannual variability. For precipitation in some regions, however, the noise of interannual variability such as droughts and floods can be much greater than the signal of the seasons in such places as northern Europe or the U.S. Northeast. For other regions—say, southern California or the monsoon belts—the seasonal cycle of precipitation is quite marked. However, in such places interannual variations in precipitation can still be quite large, and they affect agricultural productivity, water supplies, water quality, human health, and natural ecosystems.

Although the hemispheric average of interannual temperature variability may be no more than a half degree Celsius, local variability can be greater. For example, warm or cold winters can differ locally by many degrees from long-term averages, and interannual-precipitation variability can be many times different from average precipitation amounts. Now, if increasing greenhouse gases were to warm hemispheric or global temperature a few degrees, this large-scale change could be distributed nonuniformly in space and time—as in previously observed trends. For example, recent analyses by James Hansen and Sergei Lebedeff at NASA show high-latitude temperature trends for this century that are larger than at mid or low latitudes.[5] Thus, trends in annual, global records could be enhanced or even have opposite signs in some regions or in some seasons.

However, reconstructing the earth's temperature during the past century is not simply a matter of adding up the trends measured by all existing thermometers. Many records are not reliable because they are not continuous, thermometers having been moved in many cases from centres of cities to airports. Also, in some cases cities grew up around thermometers, perturbing the local climate by forming "urban heat islands." We must take considerable care, therefore, in ascertaining that each record is a valid measure of the average temperature of an area or that it has been corrected for such spurious effects as urban heating. Furthermore, we must recognize that thermometers are not uniformly distributed around the world but are concentrated in highly populated regions of industrial countries and that remote land and ocean places often lack them. This irregular distribution samples climate changes nonuniformly, potentially biasing deductions regarding actual temperature trends over specific latitude zones.

Inevitably, such spatial sampling errors creep into all records of hemispheric and global temperature trends, and it requires some care to minimize them. Such groups as those led by James Hansen at the Goddard Institute of Space Studies (GISS) and by T. M. L. Wigley at the Climatic Research Unit (CRU) in East Anglia, England, have assembled thousands of temperature records over the past century (see Figure 4), attempting to account for the nonuniform sampling by calculating for each thermometer an annual temperature anomaly (variation) from the long-term mean taken over the lifetime of the record.[6] They then apportion the anomalies for each year to nearby grid points in order to minimize the biases produced by unequal thermometer distribution. But their techniques cannot entirely eliminate spatial sampling errors and some other biases. Indeed, they (especially GISS) have recently been attacked for showing too great a trend over the United States by a few tenths of a degree. A debate over the credibility of those trends—or their relevance to the greenhouse effect—erupted in 1988, especially in the media. It is discussed at length in Chapter 7.

INTERPRETING THE RECORD. Despite the pitfalls and the fact that records in various countries contain individual fluctuations from year to year, both the GISS and CRU reconstructions, even after recent corrections for urban heating, agree on the two salient features: an approximately 0.5° C (1° F) warming trend in global average over the past century and the six warmest years bunched in the 1980s (1988, 1987, and 1981 being the warmest years

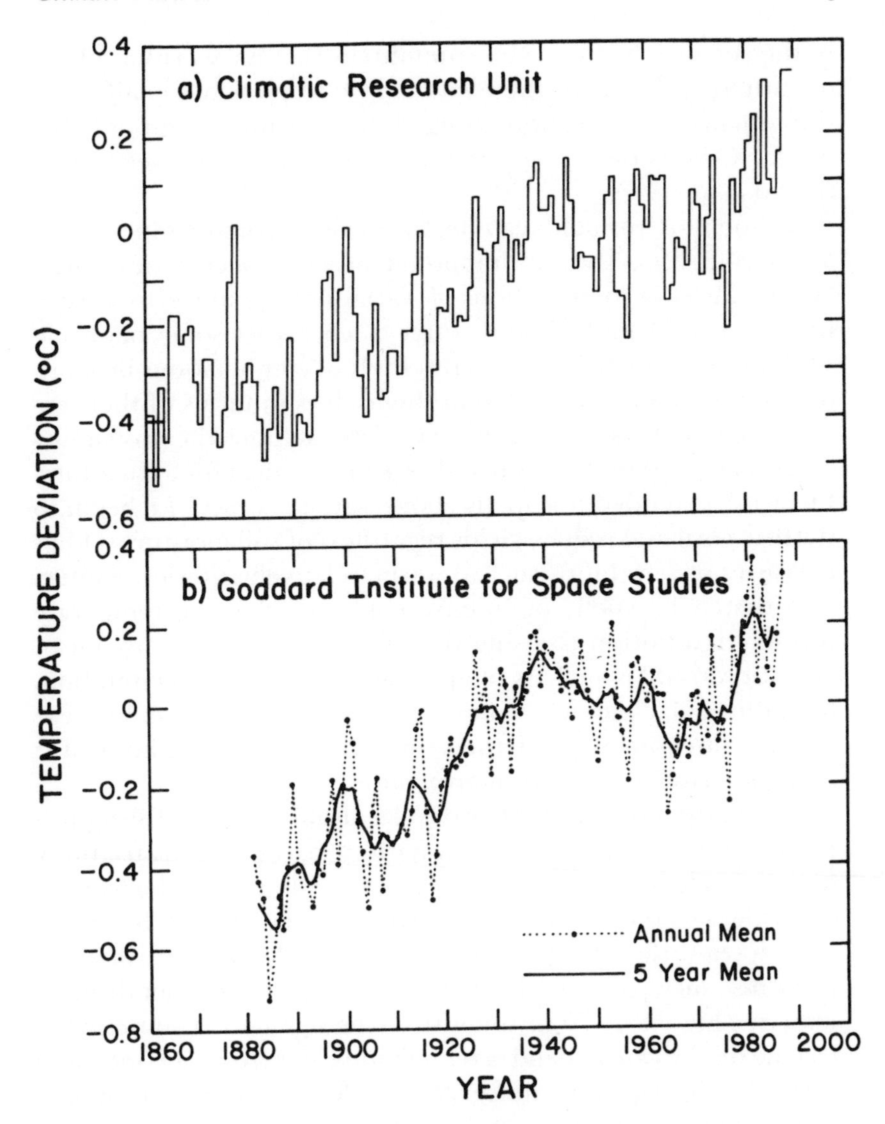

Figure 4. A comparison of the global surface temperature trends of the past 100 years constructed from land and island stations and ocean surface temperature data sets at (A) the Climatic Research Unit and from a similar set of stations (minus the ocean surface temperature data set) and at (B) the Goddard Institute for Space Studies. [Source: (A) Jones, P. D., and Wigley, T. M. L. Personal communication, 1988; (B) Hansen, J., and Lebedeff, S. 1988. Global Surface Air Temperatures: Update Through 1987, Geophys. Res. Lett. 15*:323–326.]*

in the record). None of the committees that have criticized the quantitative details of these climate-trend plots has challenged the conclusion of a century-long global warming trend of half a degree Celsius or so; only the detailed values have been questioned.[7]

Seasonal or annual mean hemispheric or zonal temperatures are, of course, not the only important measures of climatic change. We can also use combinations of "variability variables" to discern phenomena of interest and to try to construct some plausible scenarios of climatic change. Hundreds of combinations of short-term climatic variables present themselves in terms of their environmental or societal significance. For example, the probability of frost at a certain location and season is critical for agriculture. Likewise, extended hot spells above, say, 35° C (95° F), begin to sterilize corn and reduce yields regardless of soil moisture. Other measures of variability, such as warm, relatively snowless winters punctuated by spells of intense cold, can damage temperate forests. In addition to using the usual array of monthly mean climatic variables such as temperature, pressure, precipitation, humidity, solar radiation, and wind, we must also look to these climatic phenomena, with their time scales as short as hours, for their descriptive and predictive value.

The analysis of variability variables should help us distinguish predicted trends (signals) from natural background variability (noise). Such noise might be caused by vacillations of the troughs and ridges of the circumpolar vortex (the jet stream), by anomalies in sea-surface temperature patterns, and so on—vacillations that cause such patterns of anomalies as those in a Palmer drought map (see Figure 5). To distinguish a true signal of an anticipated change (in the case at hand, a temperature rise from CO_2 increase) from this background noise, we divide the suspected climate change by some measure of natural variability (for instance, the standard deviation of interannual temperatures). In this way, we can use standard statistical tests to determine the probability that

Figure 5. Drought severity as of July 9, 1988, showing the worst drought conditions in the United States since the dust bowl years of the 1930s. Even so, note that the severity of drought varies, with some parts of the country still having wetter than normal conditions. [Source: National Oceanic and Atmospheric Administration/U.S. Department of Agriculture Joint Agricultural-Weather Facility.]

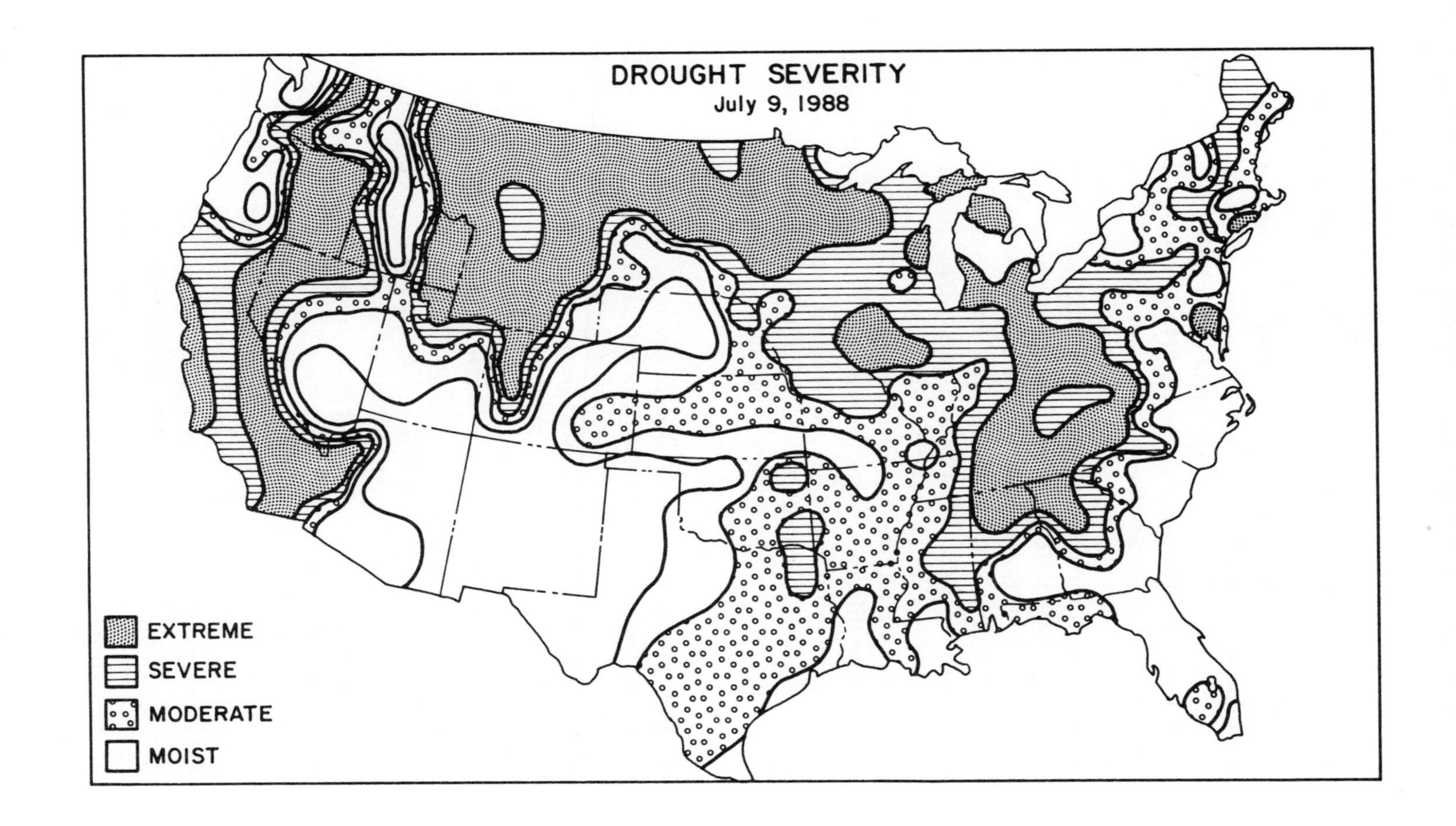
DROUGHT SEVERITY
July 9, 1988
EXTREME
SEVERE
MODERATE
MOIST

any hypothesized signal detected in both real data and climatic models is not noise.

Still, subjectivity enters the picture when we must evaluate the reliability of the signal-noise ratio. Some scientists might accept a low signal-to-noise ratio as evidence of a true climate change whereas others acknowledge the reality of a change only after a strong signal has persisted for a long time. Thus, two equally competent scientists viewing identical records and computing identical signal-to-noise ratios may arrive at different conclusions, one claiming to have detected a signal of a greenhouse effect already while the other argues that the signal is statistically insignificant. They often differ over the probability that the alleged signal has a known cause. Deeming any particular signal-to-noise ratio to be cause-and-effect or chance is really a judgment call, and different scientists will retain or shed their scepticism in different degrees and at different times. Anyone with a serious interest in climate prediction, especially with regard to the greenhouse gas signal, will do well to expect to encounter disagreement and even heated debate, a debate often magnified in the media.

MATHEMATICAL MODELS. There is no doubt that detecting the greenhouse signal in the twentieth-century temperature record is connected to empirical analysis. But concern for future climate change is actually more directly related to the validation of model forecasts. Some empirical techniques do attempt to forecast how a change in greenhouse gas composition of the atmosphere might affect the earth's climate.[8] However, although empirical methods can provide detailed regional descriptions and can tell us what the actual earth's climate has done, they suffer seriously from being based on climate *in the past,* and are not necessarily reliable for "what if" questions such as, "What if greenhouse gases were to double by 2050?"[9]

For making our predictions regarding greenhouse warming, we turn to climatic models. These are not laboratory models since no physical experiments could begin to approach the complexity of the real world. Rather, these models are *mathematical* simulations of the earth's current climate in which known physical laws are applied to the atmosphere, oceans, land, biota, and glaciers. The equations that represent these laws are solved in the most powerful computers available. Then, by means of computer programs, we alter the quantities of greenhouse gases, perform our computations again, and compare the results with the so-called control calculation—the original description of the current climate.

Many such models have been built over the past few decades, and all roughly agree that if CO_2 were to double (neglecting other trace gases), then the earth's surface temperature would eventually warm up somewhere between 1° and 5° C, with most of the recent estimates in the 3.0° to 5.5° C range.[10] Compare this with the difference between our present world average surface temperature and the ice age extreme of 18,000 years ago: the ice age was only about 5° C colder than our current average temperature. Also recall from Chapter 3 that a change of only 1° to 2° C in summer coincided with a considerably altered pattern of aridity (for example, drying of U.S. plains and wetter conditions in Africa some 6,000 years ago). Thus, a sustained global temperature change of more than 2° C would be a very substantial alteration and unprecedented in the era of human civilization.

Since mathematical models of the climate manipulated by fast supercomputers have become the primary tool in our attempt to forecast the climatic future, and since literally trillion-dollar decisions may hinge on their results, it seems essential to explain in some detail how these models work. What exactly lies inside these high-tech crystal balls, and what must we know to use and understand them?

THE CLIMATE SYSTEM. It is because the climate system is a physical entity that obeys known laws that it is, to a certain degree and with the correct tools, potentially predictable. Before we describe how the models do their predictive work, it is important to explain the system and its components and mechanisms as briefly as possible. Climate change results from complex interactions among the many components of the climate system—atmosphere, oceans, land, ice and snow, and terrestrial and marine biota.[11] The two fluid components of the system are the atmosphere and oceans. Each exhibits organized, large-scale circulation; less organized eddies; and random turbulence. However, the two react to disturbances on very different time scales; moreover, interactions between the two occur on many time and space scales. These sea-air interactions involve physical exchanges of energy as well as chemical exchanges—for example, evaporation from the oceans and runoff from the land to the sea.

Solar radiation is the main source of energy for the climate system. The components of the atmosphere—for example, water, dust particles, carbon dioxide, and ozone—directly affect the atmosphere's ability to absorb and transmit solar radiation. Further, aerosols (particles suspended in a gas) in the atmosphere help clouds to form and precipitation to fall. On the ocean side, the salinity of oceans affects water density and thus the oceans' circulation. And the air-surface exchange of such radiation absorbers as water vapour, carbon dioxide, and nitrogen and sulphur oxides is further influenced by such *physical* processes as winds, rainfall, and runoff, and such *biological* ones as forest metabolism, soil decomposition, and phytoplankton productivity.

The third component of the climate system, after the atmosphere and oceans, is the cryosphere—the extensive ice fields of Antarctica and Greenland plus other stretches of continental snow and ice and sea ice. The total area of continental snow and sea ice varies both seasonally and interannually, causing large annual variations in continental heating, upper-ocean mixing, and the energy exchange between the surface and atmosphere. The large continental ice sheets do not change rapidly enough to cause seasonal or yearly climatic anomalies, but they do play a major role in the climatic changes that occur over thousands of years such as those in the glacial/interglacial cycles.

The land and its biomass constitute the fourth component of the climate system. This component includes the slowly changing extent, position, and shape of the continents and the more rapidly varying characteristics of lakes, rivers, soil moisture, and vegetation. The land and its biomass are variable parts of the climate system on all time scales—seasonally, interannually, and across the millennia.

Interaction among the biota, air, sea, ice, and land with solar radiation, the energy that drives the climate system, accounts for its continual changes. For example, variations in the gaseous and particulate constituents of the atmosphere, along with changes in the earth's position relative to the sun, affect the amount and distribution of sunlight received. The oceans' temperature has a marked influence on the heating and moisture content of the atmosphere. The pattern of unreflected solar energy gives rise to temperature differences that drive the atmospheric circulation; in turn, wind friction and heat transfer drive the ocean circulation, which changes the oceanic surface temperatures that in turn alter air currents. The atmosphere and oceans are both influenced by the extent and thickness of the ice covering the land and sea, and by the shape and composition of the land surface itself. Since each of these changes has its own range of response times, the whole system evolves continuously, with some parts leading or lagging behind others.

Feedback loops between the system's interacting components are another basic feature. These loops amplify (positive feedback) or damp (negative feedback) perturbations. For example, any increase in the area of polar ice or snow resulting from a forced cooling means that more of the incoming solar radiation will be reflected, leaving less to be absorbed by the surface. An adequate snowfall further lowers surface temperature, increasing the ice and snow cover in a positive feedback loop. It is reasonable to expect, however, that the increasing snow cover and associated coldness of a continental interior would gradually limit the ability of the overlying atmosphere to import moisture into the region, eventually decreasing snowfall and limiting growth of the snow cover—in a negative feedback loop.

Not only can several processes interact to cause fluctuations in climate, but also those processes might act simultaneously or in a particular sequence over multiple time scales. On the seasonal and interannual scale, climatic anomalies result from interactions that have been noted but not explained. For example, approximately 25% of variability of droughts in the Great Plains have been correlated with the twenty-two-year "double sunspot cycle," which corresponds to one cycle in the reversing solar magnetic field,[12] but how sunspots might drive the earth's atmosphere is unknown. Recently, climate variations correlated to sunspots have been related to the direction of high-altitude winds, but again no clear physical mechanisms have been named. A relationship between small variations in the radiant energy output of the sun and variations in the earth's climate has been postulated but not confirmed; the same is true of recent theories linking climate and atmospheric opacity from volcanic eruptions, but this connection is physically better based.[13]

To this point, we have touched on only natural factors affecting climate, but scientists suspect that humans may be altering climate as well. For example, it has been suggested that overgrazing in the marginally arable Sahel has increased the reflectivity of the earth's surface and lowered the absorption of solar radiation. This would cause sinking air to replenish the loss of radiative heating, reducing rainfall and perhaps aggravating the persistent droughts of the 1970s. Tropical deforestation has also been implicated in at least regional climatic change, but there is considerable debate on this matter.[14]

Of concern for the immediate future, and at issue in this book, is the undisputed evidence that levels of the trace gases, carbon dioxide, methane, chlorofluorocarbons, and others are being increased in the atmosphere by human activities. Although only traces, these gases are efficient absorbers of radiation, and any substantial changes in their concentration are likely to affect the atmosphere's solar and infrared radiative balance and hence the climate. Current projections of fossil fuel consumption suggest that an equivalent doubling of atmospheric carbon dioxide may occur between the years 2020 and 2080. When climate modellers have tested the sensitivity of their climate model to an equivalent doubling of CO_2 (that is, when CO_2 and other trace greenhouse gases combined have a heat-trapping effect equal to a CO_2 doubling by itself), they've found that the effect of such an increase could warm the earth several degrees Celsius over the next fifty years.

WHAT THE MODELS CAN TELL US. Although the atmosphere-earth-ice-ocean system is complex, we can describe the known physical laws mathematically, at least in principle. In practice, however, solving these equations in full, explicit detail is impossible. First, the possible scales of motion in the atmospheric and oceanic components range from submolecular to global. Second, there are interactions among the different scales of motion. Finally, many motions are inherently unstable; small disturbances, for example, grow rapidly in size if conditions are favourable. Thus, the results of seemingly small differences between two similar atmospheric or oceanic states can cause eventual divergence of the two weather patterns. From theoretical considerations and practical experience, meteorologists have learned that predicting the details of weather beyond about ten days is impossible, even using the best possible observations of current weather conditions.[15]

Nevertheless, there are undaunted people who produce—and sell—detailed forecasts of the daily weather *years* in advance, marketing them for a few dollars to the public, typically at checkout counters of supermarkets. Others produce and sell long-range weather predictions exclusively to private clients such as agribusiness corporations, typically for tens of thousands of dollars. Depending on their bent, professional meteorologists are amused or angered by such entrepreneurs.[16]

Long-range *climate* predictions do not face a known theoretical barrier to potential skill. Meteorologists can have some success in predicting *average* conditions for an extended period. For example, let's take a baseball player's batting average as an analogy for predicting weather and climate. A hitter can have a single, a strikeout, a walk, a home run, and so on in some unique sequence during his individual times at bat; his batting average is the long-term average of his many appearances at bat over a period of time. While it's impossible to predict reliably what any specific hitter will do in four trips to the plate next July 23rd, it *is* possible to forecast his final batting average, even before the season starts, based on a knowledge of his history, the state of his health, his order in the lineup, and other similar factors. Prediction of weather and climate can be viewed in the same perspective. As we've seen, it's difficult to predict the weather in the short-term: one day it will rain, the next day will be dry, the next hot, and so on. However, a particular climate—say, moist and cool like Brussels in winter—is the result of all these individual weather events over a long period of time. As with forecasting a hitter's batting average, we can often forecast average climatic conditions based upon the anomaly patterns of ocean temperatures, the amount of incoming sunlight, the altitude of the site and its proximity to oceans or ice fields, and so on.

Another reason that climate predictions for longer periods may be possible is that the climate system is subject to predictable forcing processes that may be of overriding importance for some time scales. An example is the annual variation in the global distribution of solar radiation. The strength of this forcing causes the seasons to follow each other predictably, although relatively small differences—anomalies—in seasons are, of course, important from year to year. However, our inability to predict the seasonal *anomalies* with confidence should not blind us to the human significance of these quantitatively small fluctuations—small only in comparison to winter-to-summer differences. For example, the summer of 1988, with its heat and drought in the United States and flooding in Bangladesh, exhibited anomalies of considerable seriousness.[17]

While it is true that some forcing mechanisms are predictable, others, such as volcanic activity, are not. Also, the atmosphere is forced by such mechanisms as oceanic surface temperatures that themselves respond gradually to atmospheric forcing in a complicated feedback mechanism. Such feedbacks, noted earlier, are commonly called internal forcings of the climate system, in contrast to the straightforward external forcings, such as solar radiation. The internal forcings, such as the energy exchange between atmosphere and ocean, are the most commonly invoked explanation of short-term climatic anomalies. The episodic warming of the equatorial eastern Pacific Ocean off the coast of Peru every several years is known to create dramatic changes in climate. Torrential rainfall off the west coast of South America and uncommon drought and heat in Australia and Indonesia are two such events associated with these so-called *El Niño* conditions. It is also believed that such unusual patterns of Pacific Ocean temperatures help to steer the jet stream into irregular configurations that have created such painful anomalies as the recent coastal damage in southern California or the frigid early winter of 1983.[18] On the other hand, the out-of-position jet stream in the late spring and early summer of 1988 that steered storms away from the central United States and up into Canada was caused, some believe, by a reverse *El Niño*-type condition, where the equatorial Pacific Ocean was anomalously cold rather than hot as in the *El Niño* years. My NCAR colleagues Kevin Trenberth and Grant Branstator used a computer model to test this idea, with initially positive results.[19] In any case, observations over the past 100 years suggest that *El Niños* and their opposites tend to last for a year or two; then the ocean either goes back to normal or to the extreme opposite condition within a few years. Thus, while *El Niños* and reverse *El Niños* may substantially influence the temperature or weather patterns for a year or two, they are unlikely to cause climatic changes measured in many decades.

What is an internal fluctuation on long time scales may be a large external forcing on shorter ones, depending on what processes are included in the climatic system defined for an investigation.

Nevertheless, the presence of forcing implies that some aspects of climate may be predictable on those time scales where the forcing and its response are important. In fact, one would expect the degree of climate response and, hence, predictability, to be related to both the strength and the time period of forcing. Although this is true in general for external forcing, complicated feedback processes within the system can produce inscrutable results. Furthermore, internal interactions can cause internal oscillations on different time scales or even what is called chaotic behaviour.[20]

Where long time scales are concerned, many meteorologists believe that the climate system responds predictably to current boundary conditions (that is, external forcings) and has little memory of its history. In other words, the climate system will move to a unique, stable equilibrium state after a transient adjustment. Thus, such deterministic models provide an equilibrium snapshot appropriate for the time scale of the external forcing. The validity of such equilibrium snapshots, however, depends on the existence of a unique equilibrium state for any particular boundary condition, still a debated assumption for the earth's climate system. In other words, if we specified the amount of heat from the sun, the shape of the earth, the location of the continents, the amount of carbon dioxide in the atmosphere, and so on, a fully deterministic model with a unique equilibrium would have one and only one climate for that set of boundary conditions. However, as MIT theoretical meteorologist Edward Lorenz pointed out twenty years ago, it's possible that a very complicated system may exhibit what is known as chaotic behavior and have, perhaps, several temporary quasi-equilibrium states, even if all the external factors like sunlight, CO_2, and geography were held fixed. The two different modes of ocean circulation mentioned in the discussion of the Younger Dryas in Chapter 3 provide examples of the possibility that two separate, nonunique climatic states may exist. Until we are certain whether our climate system possesses a unique and stable equilibrium, there will always be debate over the extent to which even the long-term climatic future can be predicted.

In part, then, climate prediction, like forecasts involving other complex systems, is essentially extrapolation: we attempt to determine the future behavior of the climate system by observing its past and present behavior. The two tools open to forecasters are the empirical–statistical methods used to analyse existing (and, as we saw, imperfect) records and mathematical climate models. The pertinent question in this context, then, is, How fruitful are the models in providing data from which to extrapolate?

Climate models vary in their spatial resolution—that is, in how many dimensions they simulate and how great is the spatial detail they include. A simple model calculates only the average temperature of the earth, independent of time, as an energy balance between two factors: the earth's average reflectivity and the average greenhouse properties of the atmosphere. Such a model is zero dimensional; it collapses the real temperature distribution on the earth to a global average. In contrast, three-dimensional climate models reproduce the variation of temperature with latitude, longitude, and altitude. The most complex models, known as general-circulation models (GCMs), predict through three dimensions in space the time evolution of temperature plus humidity, wind, soil moisture, sea ice, and other variables. Although GCMs are usually more comprehensive than simpler models in their physical, chemical, and biological detail, they are also more expensive to design, run, and interpret.

The optimal level of complexity for a model depends on the problem and the resources available. More is not necessarily better. Often it makes sense to attack a problem first with a simple model and then to employ the results to guide research at higher resolution. In fact, deciding on how complex a model to use for a given task is more an intuitive judgment than a scientific choice subject to explicit, logical criteria. It often involves a tradeoff of completeness and accuracy for tractability and economy.[21]

And in fact even the most complex general-circulation model is limited in the amount of spatial detail it can resolve. No computer is fast enough to calculate climatic variables everywhere on the earth and in the atmosphere in a reasonable time. Rather, calculations are limited to the interactions of factors at widely spaced points of a three-dimensional grid at and above the surface, and such calculations can easily fall into the several hundred-hour range of supercomputer time.[22]

The size of the grid is another limitation of these models, with wide grid spacing creating a problem. Many climatic phenomena occur in areas smaller than those covered by an individual box in a grid. For example, clouds reflect much incident sunlight back to space. They also block the escape of infrared radiation from below, adding to the greenhouse effect. Therefore, they help to determine the temperature on the earth. Clearly, predicting changes in cloudiness is an essential part of climate simulation, but no GCM now has a grid fine enough to resolve individual clouds, which tend to cover a few kilometres rather than a few hundred kilometres.

A technique known as parameterization (short for parametric representation) accounts for subgrid-scale phenomena such as clouds. Parameterization, for example, involves searches through climatological data for relations between variables resolved by the grid and ones that are not. For instance, the average temperature and humidity over, say, the large area beneath one grid box the size of Colorado, can be related to the average cloudiness over the same area; to make the equation work one introduces a parameter or proportionality factor derived empirically from the cloudiness, temperature, and humidity data. Since a model can calculate the average temperature and humidity in a box from physical principles, the semiempirical parameterization predicts the average cloudiness in the box even though it cannot predict individual clouds. Despite this technique, the very issue of cloud effect and its significance for predictions has been actively studied for more than twenty years with no resolution in sight, and the parameterization of clouds is one of the major points of contention when climate modellers discuss how the climate may change into the greenhouse century.[23] I personally am pessimistic about our ability to observe small cloudiness changes accurately enough from space over the next decade or two to resolve this major uncertainty. Still, parameterization can be reasonably accurate, perhaps through luck alone, as a number of empirical validation experiments have shown.[24] Modellers, of course, strive to keep their parameterizations as physical, nonempirical, and scale-independent as practical. But at some point, the validity of parameterization and overall model performance must meet the tests of the real world, not only of first principles. The fact that many of these tests remain to be performed should suggest the limitations on GCMs.

THE FUTURE OF THE MODELS. At this stage, it is appropriate to ask about the prospects for improving parameterizations and the performance of general-circulation models. If a big improvement can be expected over the next few years, it would seem logical for the policy community to ask climate modellers to accelerate their efforts and provide answers more rapidly. But if the answers to our questions are decades away from the modellers' grasp, then of course the real world may provide answers before today's scientists do. Again, I am somewhat pessimistic. If we are to accurately forecast the regional climatic responses to increases in greenhouse gases, it is essential that we move the state-of-the-art rapidly beyond its current stage, running atmosphere and ocean models separately rather than coupling the two. Moreover, since biological factors—such as forests, grasslands, and peat coverage of the surface—have a major impact on the rates at which water is evaporated and solar energy absorbed, we must also include these ecosystem characteristics in our forecasting models. And should the climate change rapidly enough to alter ecosystems, then our models will need to incorporate these alterations as well. On a broad scale, our success in making climatic predictions rests on a timely effort to unite the scientific community and bring to bear on the problem the most original minds in diverse fields.[25]

At present, what climate models can do well is analyse the sensitivity of the climate to uncertain or unpredictable variables. In the case of carbon dioxide, it would be possible to construct plausible scenarios of economic, technological, and population growth to project the trend of CO_2 emission and model the climatic consequences. And it would be possible to vary such uncertain climatic factors as cloud feedback or forest-prairie border changes over a plausible range. At least the calculations would indicate *which* uncertain factor was most important in making the climate sensitive to carbon dioxide increases. One could then concentrate both disciplinary and interdisciplinary research on that factor—or those factors. Model results of this kind would also suggest the *range* of climatic futures to which ecosystems and societies may be forced to adapt and the necessary rates of adaptation. How we would, or should, respond to such information, of course, would be a political issue, one examined in other chapters of this book.

CREATING THE SCENARIO. We now have enough background to see how climatologists use climate models. However, a crucial variable that we have yet to determine is the precise amount of greenhouse gases in the atmosphere to be figured into our prediction efforts. As Budyko said in 1971, this is a question of *social* science, not natural science. It depends upon projections of human population, the per capita consumption of fossil fuel, deforestation rates, aforestation activities, and perhaps even countermeasures to deal with the extra carbon dioxide in the air (for instance, planting more trees). These projections further depend on the likelihood that alternative energy systems and conservation measures will be available, affordable, and socially accepted. Furthermore, trade in fuel carbon (for example, large-scale transfers from coal-rich to coal-poor nations) will depend not only on energy requirements and available alternatives but also on the economic health of the importing nations. This in turn will depend upon whether those nations have adequate capital resources to spend on energy rather than other precious strategic commodities—such as food or fertilizer as well as some other strategic materials such as weaponry.

Nearly two decades ago, John Holdren and Paul Ehrlich gave us a formula for determining the relationship between pollution and population.[26] Applied to the CO_2–energy issue, it gives the total emissions from energy technologies.

$$\text{total } CO_2 \text{ emission} = \frac{CO_2 \text{ emission}}{\text{technology}} \times \frac{\text{technology}}{\text{capita}} \times \text{total population size}$$

The first term to the right of the = sign represents engineering, the second standard of living, and the third demography.

We can make scenarios (as in Figure 6) that show alternative CO_2 futures based on assumed rates of growth in the use of fossil fuels. Most typical projections show a 1% to 2% annual growth in use rate for fossil fuel, implying a doubling of preindustrial CO_2 (to 600 ppm) in the twenty-first century. Carbon dioxide concentration has *already* increased by some 25% over the past century.[27] We know that CO_2 has increased by this amount simply because gas bubbles trapped in glacial ice near the North and South Pole can be dated by counting the annual layers of snow backward in time. The amount of CO_2 in the air in preindustrial

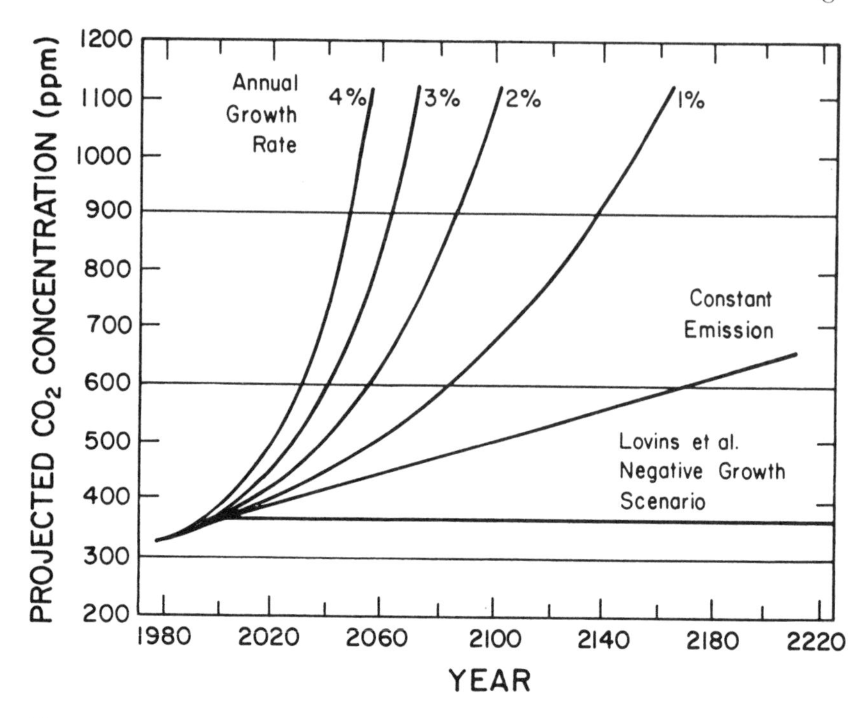

Figure 6. The extent to which CO_2-induced climatic change will prove significant in the future depends, of course, on the rate of injection of CO_2 into the atmosphere. This depends, in turn, on behavioural assumptions as to how much fossil fuel burning will take place. (This graph neglects biospheric effects.) Since the end of World War II, a world energy growth rate of about 5.3% per year occurred until the mid-1970s, the time of the OPEC price rises. Rates have come down substantially since then and hardly grew at all in the early 1980s. The figure shows projected CO_2 concentrations for different annual growth rates in fossil energy use, including one for the assumption that no increase in fossil energy use occurs (constant 1975 emission) and even a "negative growth scenario" in which energy growth after 1985 is assumed to be reduced by a fixed amount (0.2 terawatts [TW] per year, which is about 2% of present demand) each year. [Source: Lovins, A. B.; Lovins, L. H.; Krause, F.; and Bach, W. 1989. Least-Cost Energy: Solving the CO_2 Problem *(Snowmass, Colorado: Rocky Mountain Institute,* p. 10)

times is about 280 parts per million, which indicates a 25% increase to 350 parts per million in the late 1980s. That part of the greenhouse effect debate is virtually certain. Other greenhouse gases, such as the CFCs, methane, the nitrogen oxides, and so on, could, taken together, be as important as CO_2 in the future greenhouse effect.

Methane, for example, doubled between the glacial maximum

some 18,000 years ago and the preindustrial period and has nearly doubled again, presumably due to human activities. Methane is produced by bacteria that like to live in relatively oxygen-free places like the guts of animals or in rice paddies under irrigation or under permafrost in the Arctic or in landfills or waste tips. It also is natural gas and may be released in the process of coal mining or oil or gas exploration and transport. Methane has 20–30 times more greenhouse—i.e., infrared heat trapping—capacity per molecule than CO_2, and thus is an important greenhouse gas, accounting for perhaps as much as 20% of the human-induced greenhouse effect that we expect into the greenhouse century. However, since it only exists at about the 1 part per million level, less than 1% of CO_2, it is still not yet as important for climate as CO_2. However, it will not be easy to limit the expansion of rice paddy agriculture, animals like cows which produce great quantities of methane, or other activities as easily as it might be to control fossil fuel burning.[28]

CFCs, which are, of course, manmade compounds implicated in stratospheric ozone reduction, also have very powerful greenhouse properties. They exist in only 1 part per billion quantities, otherwise they would dominate the greenhouse effect. They may account for as much as 25% of carbon-dioxide induced greenhouse warming into the greenhouse century unless radical actions are taken to control their emissions.[29] These various greenhouse gases may have very complicated interactions. For example, carbon dioxide can cool the stratosphere which slows the air chemistry that destroys ozone. On the other hand, stratospheric cooling can create high-altitude clouds which interact with chlorofluorocarbons to create ozone holes. Methane can be produced or destroyed in the lower atmosphere at varying rates, depending upon other pollutants that might be present. It also can affect chemicals that control ozone formation near the surface. But because these other greenhouse gases have complicated biogeochemical interactions, for simplicity, I refer to all greenhouse gases taken together as *equivalent* CO_2. Note that this simplification makes the projection for CO_2 alone on Figure 6 an *underestimate* of the total greenhouse gas buildup by roughly a factor of 2.[30]

Once we have a plausible set of scenarios for how much CO_2 will be injected into the atmosphere, we must then identify and account for the interacting biogeochemical processes that control the global distribution and stocks of the carbon. One such process is the uptake by green plants (since CO_2 is the basis of photosynthesis, more CO_2 in the air means faster rates of photosynthesis where CO_2 is a limiting factor). Others are changes in the amount of forested area, changes in what is planted, and the impact of climate change on natural land and ocean ecosystems.

The slow removal of CO_2 from the atmosphere is largely accomplished through biological and chemical processes in the oceans, which take decades to centuries. Therefore, we must also take into account the rates at which climate change modifies mixing processes in the ocean. There is considerable uncertainty about just how much newly injected CO_2 will remain in the air over the next century, but most present estimates put this airborne fraction at about 50% of the total.[31]

The oceans are not the only areas of uncertainty in the carbon cycle, of course. Approximately the same amount of carbon (about 750 billion tons) is stored in the atmosphere as is stored in living plant matter on land, mostly in trees. Animals retain only a trivial amount of carbon by comparison, something like 1 to 2 billion tons, and the amount in humans is only a small percentage of that. Bacteria have about as much weight in carbon atoms as all the animals put together, and fungi have about half that amount. On the other hand, dead organic matter, mostly in soils, contains at least twice as much carbon as does the atmosphere. Thus, major potential exists for biological feedback processes to affect the amounts of carbon dioxide that might be injected into the air over the next century. For example, as suggested earlier, as CO_2 increases, green plants should incorporate more carbon dioxide into plant tissues through photosynthesis, perhaps reducing slightly the buildup of CO_2 that otherwise would have occurred in the atmosphere. Some have cited this negative feedback mechanism as a reason to mitigate our concern about the greenhouse effect. But remember that about twice as much carbon is stored in the soil as dead organic matter as is in either the living organic matter or the atmosphere. Since dead organic matter is slowly transformed into methane and CO_2 by soil microorganisms, the important question with respect to feedback is, How will the rate of this reversion of soil–organic carbon to atmospheric carbon dioxide be affected by global warming? George Woodwell, of the Woods Hole Institute, has argued that increasing the temperature in the soils by several degrees could substantially increase the activity rates of bacteria that convert dead organic matter into CO_2. This would be a positive feedback, since warming would increase the CO_2 produced in the soils, thereby further increasing the warming.[32] Dan Lashof, of the Environmental Protection Agency, referred to this large stock of soil-organic carbon as a "sleeping giant," with a potential for major positive feedback that could substantially accelerate the greenhouse effect once it really got going. Lashof went on to characterize more than a dozen biological feedback processes that might alter our present estimates of the sensitivity of the temperature to greenhouse gases added by human activities. He concluded that all these biological feedbacks operating together could, perhaps, *double* the sensitivity of the climatic system to initial injections of greenhouse gases.[33] However, at present it is impossible to verify this frightening possibility. That is because the time frame over which such processes would occur is decades to a century or so and these changes are outside of our experience.

Although all the living matter in the oceans contains only about 3 billion tons of carbon, ten thousand times that amount is dissolved in the oceans, mostly in nonliving form. And the rocks—in particular, the carbonate sediments in the continental crust and, to a lesser extent, in the ocean floor—contain something like 65 million billion tons of carbon, vast quantities compared to those in the atmosphere and in living and dead biota. While no one knowledgeable in geochemistry has suggested that these vast quantities of CO_2 in the oceans or sediments could be released soon into the earth's atmosphere, creating the kind of runaway greenhouse effect that exists on Venus, it is not currently clear just which feedback processes could work major surprises in the evolving carbon cycle. But it is safe to say that the more rapidly the climate warms up, the more likely it is that major feedback processes in the carbon cycle will significantly change current projections of greenhouse gas buildups late in the greenhouse century. It is critical that we accelerate interdisciplinary research on these biogeochemical cycling processes and their interactions with climate if we are to resolve these uncertainties any time soon. But the lack of clarity on these matters in no way alters the widespread consensus that CO_2 and other trace greenhouse gases are likely to double sometime in the next century.

TIME-RELATED UNCERTAINTIES. Most climate models produce a climate in stable equilibrium when they are forced by doubling of CO_2. In other words, if in the computations you change the amount of atmospheric CO_2 from the 1900 condition of about 300 parts per million to a future condition of 600 parts per million, most recent three-dimensional models show an equilibrium achieved eventually—if you wait long enough—at an average surface temperature warming of approximately 2.0° to 5° C (3.6° to 9° F).[34] The actual time this change takes to evolve will likely be characterized by a steadily increasing curve, probably like one of the middle ones in Figure 6. Remember that the oceans of the world have a very large capacity to store heat. As an example, in San Francisco the massive storage of heat in the oceans strongly moderates both the day-to-night and season-to-season temperature differences relative to those in the middle of the continent. Similarly, even if we were somehow able to magically double the carbon dioxide content of the atmosphere next month, the earth's temperature would not reach its new equilibrium value for a century or more. This delay in warming is a two-edged sword. On the one hand, it may seem like good news, for the more rapidly the climate changes the more ecological and societal damage is likely to occur. However, on the other edge of the blade are the facts that this "unrealized warming" (a term coined by James Hansen and his colleagues)[35] is absolutely inevitable and that whatever warming we have already experienced is by no means the last of it. Even if we were somehow to manage suddenly to control all CO_2 emissions, we could still expect another degree or so of warming while the climatic system catches up with the greenhouse forcing already liberated. Nor can we reliably predict at this time the emission or removal rates of CO_2, methane, and other greenhouse gases that are likely as time goes on.

All these uncertainties associated with the evolving reality of greenhouse forcing compel futurists to construct scenarios that try to bracket the most likely outcomes. Figure 7 shows three plausible scenarios based on high, medium, and low emission rates combined with the uncertainties associated with climate-sensitivity estimates and the delay associated with oceanic heat storage. It is difficult to assign a quantitative probability to any of these three scenarios, other than intuitively. In my estimation, the likelihood of climatic warming as great or greater than the

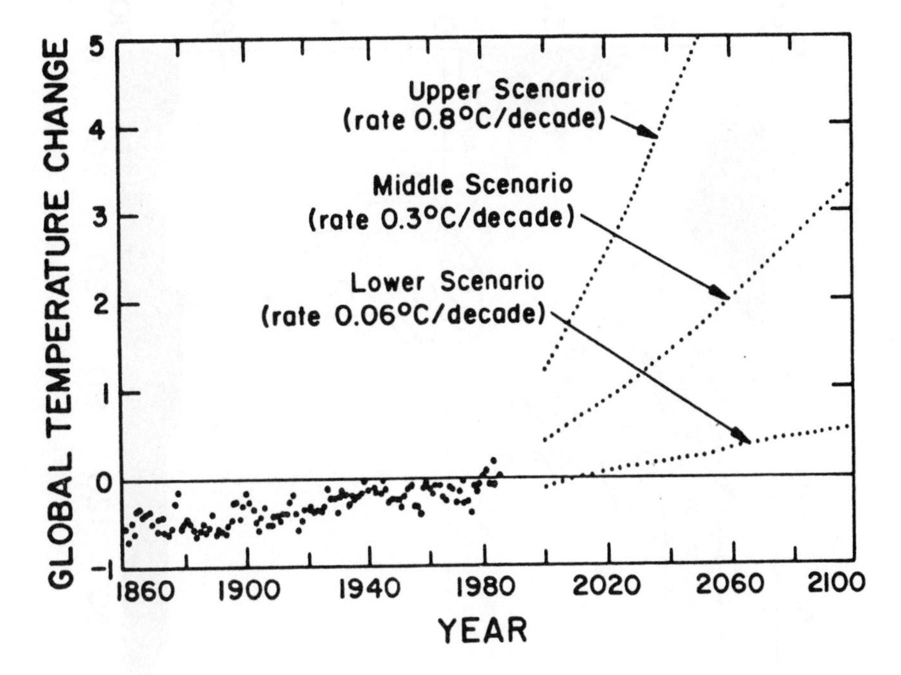

Figure 7. Three scenarios for global temperature change to the year 2100 derived from combining uncertainties in future trace greenhouse gas projections with uncertainties of modelling the climatic response to those projections. Sustained global temperature changes beyond 2° C (3.6° F) would be unprecedented during the era of human civilization. The middle to upper range represents climatic change at a pace ten to one hundred times faster than typical long-term natural average rates of change. [Source: Jäger, J. April, 1988. Developing Policies for Responding to Climatic Change, A Summary of the Discussions and Recommendations of the Workshops Held in Villach 28 September to 2 October 1987 *(WCIP-1, WMO/TD-No. 225).]*

high scenario or as low or lower than the low scenario are no more than, say, 10%. All but the slowest scenario shows a rapid, unprecedentedly large climatic change to which the environment and society will have to adapt.

REGIONAL PREDICTIONS. To estimate the societal importance of climatic changes, however, it is not so much global average temperature we need to study as the regional patterns of climatic change. Will it be drier in Iowa in 2010, hotter in India, wetter in Africa, more humid in New York, stormier in northern Australia, more forest-fire-prone in California, or flooded in Venice?

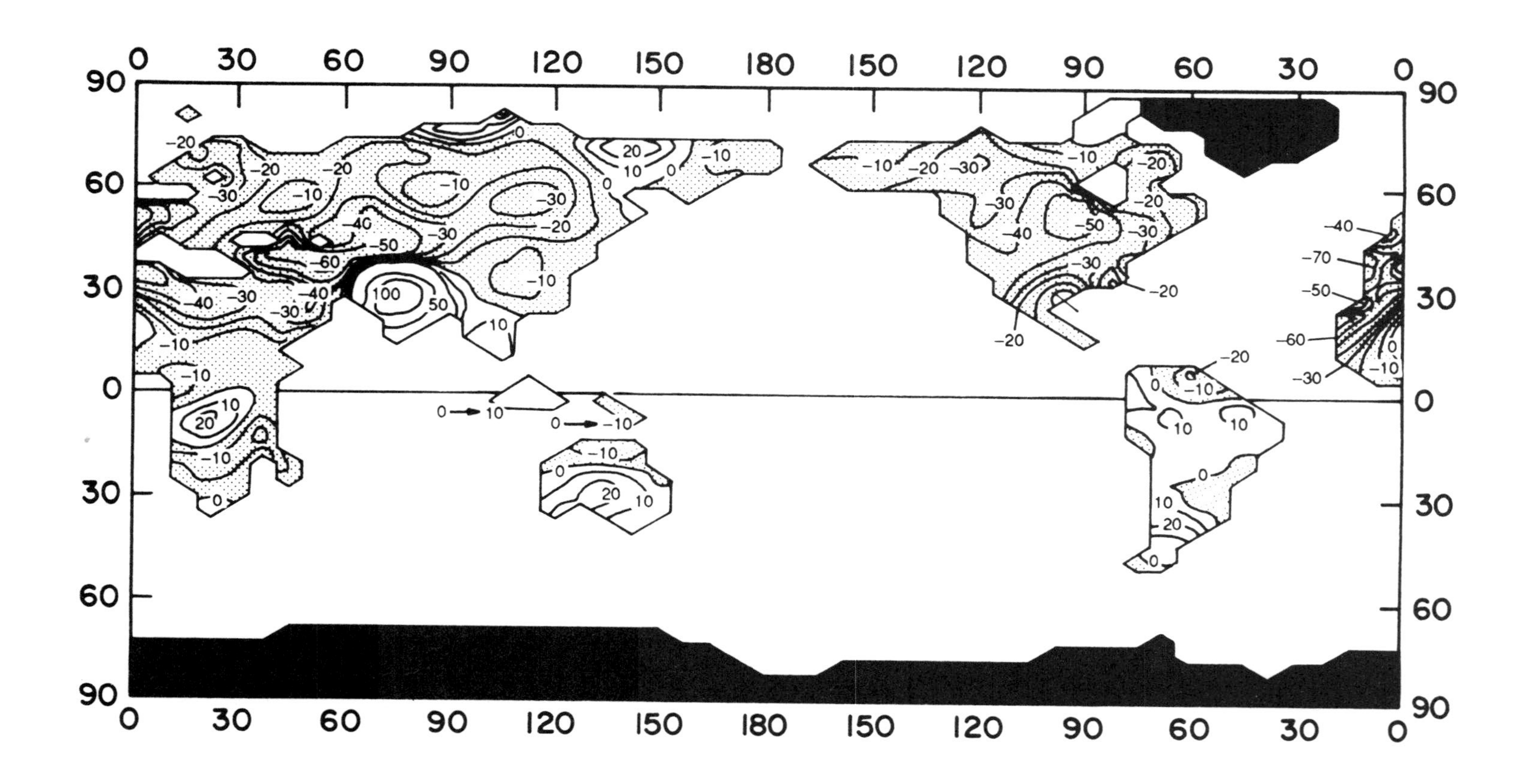

0
30
60
90
120
150
180
150
120
90
60
30
0
90
60
30
0
30
60
90

Unfortunately, making reliable predictions as to local to regional responses of variables such as temperature and rainfall requires climatic models of greater complexity and expense than are currently available. It's not that we haven't made preliminary calculations of these variables, but knowledgeable atmospheric scientists would agree that the regional predictions of state-of-the-art models are very unreliable. Nevertheless, as was inherent in Chapter 1, we do have a number of plausible regional scenarios of change, and we can cite some (but hardly conclusive) evidence that the following regional changes might well emerge over the next several decades:

- wetter subtropical monsoonal rain belts
- longer growing seasons in high latitudes
- wetter springtimes in high and middle latitudes
- drier midsummer conditions in some midlatitude areas (a potentially serious problem for the water supply and agriculture in major grain-producing nations such as the United States), but more intense tropical cyclones
- an increased probability of extreme heat waves (with possible health consequences for people and animals in already warm climates) with concomitant reduced probability of extreme cold snaps
- an increased likelihood of summertime vegetation fires in drier/hotter regions
- increased sea levels by as much as a metre or so (several feet) over the next hundred years

Figure 8 is an example of one computer model's forecast of the change in a very important variable, soil moisture, in the event of a CO_2 doubling.[36] This result, from Syukuro Manabe and

Figure 8. CO_2-induced change in soil moisture expressed as a percentage of soil moisture obtained from a computer model with doubled CO_2 compared to a control run with normal CO_2 amounts. Note the nonuniform response of this ecologically important variable to the uniform change in CO_2. [Source: Manabe, S., and Wetherald, R. 1986. Reduction in Summer Soil Wetness Induced by an Increase in Atmospheric Carbon Dioxide, Science *232: 626–628.]*

Richard Wetherald at NOAA's Geophysical Fluid Dynamics Laboratory in Princeton, New Jersey, suggests substantially drier conditions in the middle parts of midlatitudes, along with wetter summertime conditions in the subtropics—India, for example. Enhanced monsoons might seem a blessing to this highly populated region, whose very survival literally depends upon the summer monsoon rainfalls. However, increasing the reliability or intensity of monsoon rainfall also increases the probability of crop-, property-, and life-threatening floods, examples of which occurred in the summer of 1988, when one-quarter of Bangladesh was under water. What Figure 8 suggests is that, assuming this snapshot of climate change in a CO_2-doubled world is valid, regional climatic changes will not necessarily follow globally averaged alterations. Indeed, significant regional departures from the average occur. Given that, I view the issue of global climate change as a redistribution of climatic resources—and because of the central importance of climate to our well-being, such global change represents a case of redistributive justice. I believe that the main conclusion over the redistributive nature of climatic change will remain valid even if the details of what regions get wetter and which get drier change from model to model or as models are improved.

Let's compare the forecasts of five different climate models for change in the very important soil-moisture variable for the United States. On each of the maps in Figure 9 is plotted the difference between a model calculation with CO_2 doubled and present climate. (The five models represent the work of different scientists at different institutions.[37] William Kellogg and Zong-Ci Zhao compiled these data from published values in the scientific literature.) The shaded regions on the maps indicate increases in dryness for doubled CO_2.

Each model has produced quantitatively different results, but when all five models are composited, as in Figure 10, it can be seen that in a small circle in the central part of the United States all five models agree that soil moisture will drop. But we need to be sceptical in reading these maps, for the laws of probability tell us that even if the models had no predictive value whatever, the probability of all five being drier would be one-half multiplied by itself five times, or 1/32. In other words, if none of the computer models had predictive skill, then we would expect all five

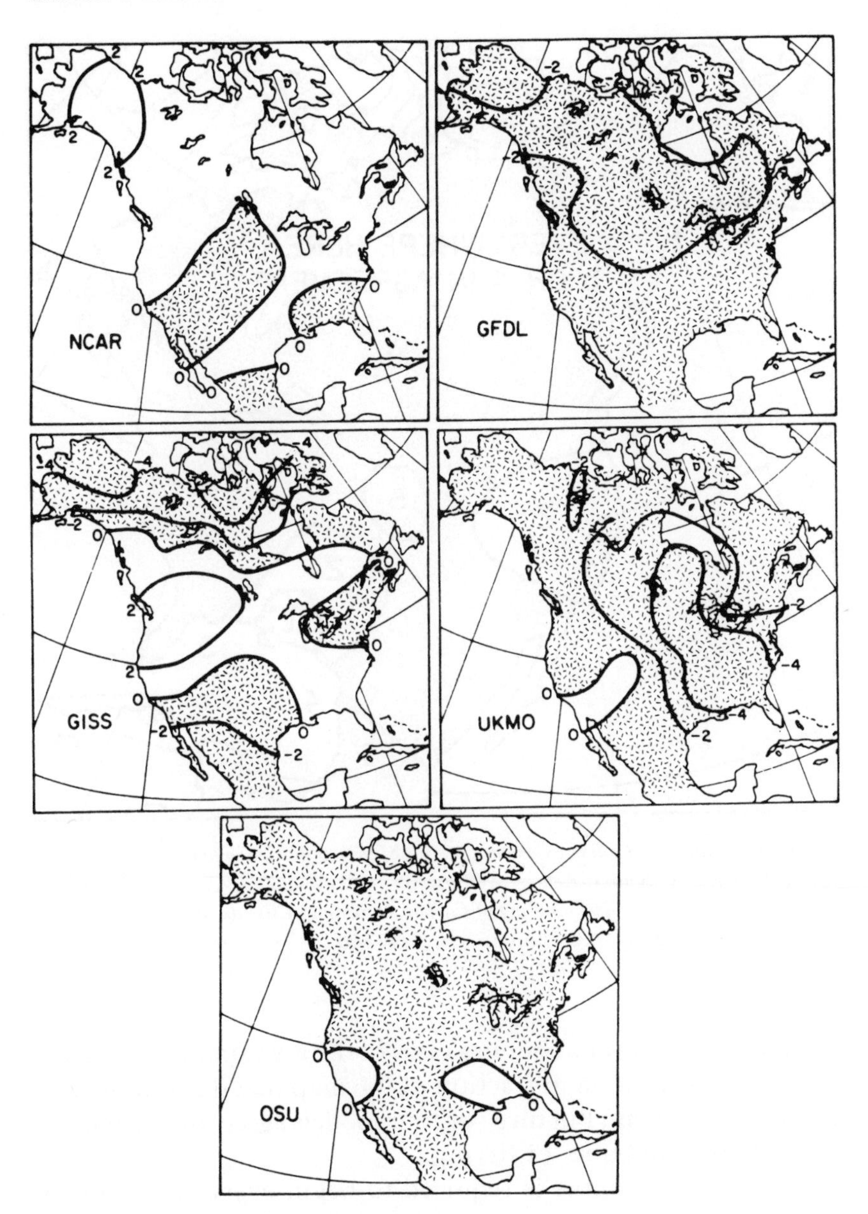

Figure 9. July soil moisture change for a CO_2 doubling minus control simulations for five different general-circulation models (in centimetres of soil moisture). Stippled regions show soil moisture decreases. [Source: Kellogg, William W., and Zhao, Zong-Ci. April, 1988. Sensitivity of Soil Moisture To Doubling of Carbon Dioxide in Climate Model Experiments. Part I: North America, J. of Climate *1:4, 348–366.]*

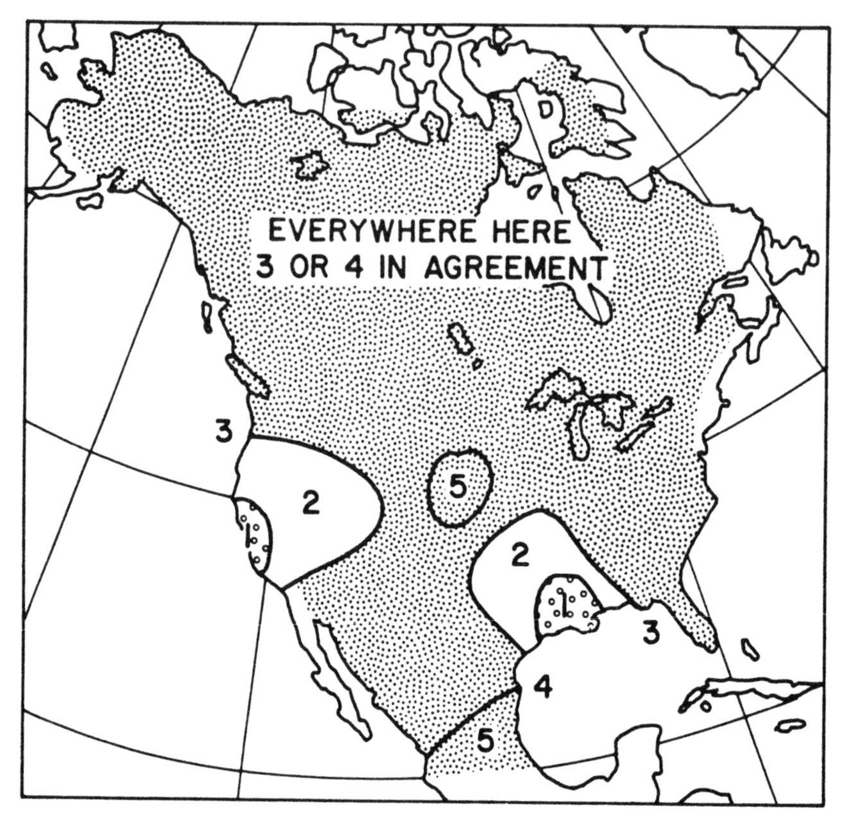

Figure 10. Sensitivity of soil moisture to doubling of carbon dioxide in five combined climate model experiments. [Source: Kellogg, William W., and Zhao, Zong-Ci. April, 1988. Sensitivity of Soil Moisture To Doubling of Carbon Dioxide in Climate Model Experiments. Part I: North America, J. of Climate *1:4, 348–366.]*

models to agree over about 1/32 (3%) of the area simply by chance. In fact, it is only a small fraction of the map that does show such five-way agreement. But three or four models agree on much more than chance amounts of drying.

Although a scientist is required to begin every analysis with scepticism, I believe that the models *do* have some skill and that the results of these maps are plausible. One reason that the models all agree on drying is that the central part of the United States will warm up substantially, by some 2° to 8° C (3.6° to 14.4° F). The models *disagree* on whether precipitation increases or decreases, but the combined results nevertheless strongly suggest a better-than-even probability for the drying scenario: an increase in evaporation associated with substantial warming in the summer months in midcontinents occurs in all models, regardless of what happens to rainfall. This suggests that the midcontinental drying scenario, while not quantitatively reliable in detail, is at least highly suggestive. Therefore, this scenario is useful in analysing how crop yields, forests, and lake levels might change. Although we cannot assign anything but intuitive numerical probabilities to any particular scenario, a fairly cautious assessment committee assembled by the U.S. National Research Council has referred to the midcontinental drying scenario as "likely in the long-term."[38] My own guess is that the likelihood of midcontinental drying is a better than even-odds gamble.

Another variable that climate models address is the change in the geographic distribution of temperatures. Once again there are substantial differences in the details of different models for a CO_2-doubled world as compared to the present, but all models agree that the largest temperature changes will eventually occur in the high latitudes. All these models produce snow and sea ice as part of their internal climate. The warming induced by CO_2 causes a melting of some of that snow and some sea ice. Melting sea ice exposes the air to the warm, dark ocean rather than the cold, bright ice surface. This causes dramatic temperature increases in the climate model in those regions.

The various GCMs predict not only that the mean temperature of the world will increase but also that temperature *differences* from place to place—from equator to pole and midcontinent to mid-ocean—will change dramatically. These temperature gradients are what create the pressure differences that are the driving forces behind the winds. The winds then blow on the oceans to mix the upper waters and to create the ocean currents, and the currents and oceanic mixing in turn influence the ocean temperatures that feed back on the winds. This system creates the regional climatic differences that make possible the ecological diversity of the world and our adaptation to it. Therefore, although each model may differ in its specific prediction of regional climatic anomalies, there is one general conclusion: regional climates—patterns of winds, rain, and so forth—are virtually certain to change as a result of global warming, since that global warming is not uniformly distributed around the world. This conclusion will only be reinforced during the time-evolving phase of climate changes.

A number of examples of this regional transient complication are evident from the few experiments that have been done in which atmospheric general circulation models are coupled to oceanic general circulation models. For example, Kirk Bryan and Syukuro Manabe and colleagues at NOAA in Princeton calculated a very surprising result in some recent model experiments. They found that increasing the amount of carbon dioxide instantaneously from the present value to a doubling, created immediate warming response for the global climate but actually caused a slight cooling in the high southern latitudes of the southern ocean.[39] Indeed, my colleague Starley Thompson and I were intrigued by this result, having earlier predicted that the southern oceans would warm only very slowly compared to the rest of the globe because of the large heat capacity of that oceanic part of the planet.[40] However, the NOAA scientists found an additional reason for their conclusion: that their ocean model had a dynamical response in which the upwelling of colder water from below retarded the warming in that region. To be sure, the globe would warm up on average, but some spots would warm up less or might even get cooler. This would distort temperature difference from place to place and could well imply that any time-evolving climate scenario could be quite different from the equilibrium doubling results. This could be especially true for Australia, since it sits between the rapidly warming tropical and slowly warming—or even cooling—southern oceans.

Another computer experiment was recently performed by my NCAR colleagues Warren Washington and Gerald Meehl.[41] They coupled their atmospheric general-circulation model to an ocean circulation model built by Albert Semtner, now at the naval postgraduate school at Monterey, California. They did not get a radical change in the deep southern oceans. However, they did find a shift in a semipermanent feature of the earth's climate: the so-called Icelandic low region in the North Atlantic in which many storm systems exist, especially in winter; this semipermanent feature directs northerly winds into Greenland and southwesterly winds into Britain. In the wintertime these southwesterly winds help to keep the British Isles bathed in the warmer air from the Atlantic rather than the colder air from the Norwegian Sea or the European continent. Washington and Meehl's calculation suggested that during the first twenty years of the transient run, the Icelandic low moved west, causing warming at its western side and cooling at the eastern, thereby depriving Britain of its accustomed frequency of southwesterly flow. Therefore, while the world warmed, for a decade or two at least the British Isles cooled.

I do not think that the details of either the NOAA or NCAR results are convincing. These are very early experiments with coupled atmosphere/ocean models. What they demonstrate is that models that connect the atmosphere, oceans, sea ice, and ecosystems will be necessary for credible forecasts of regional change. Furthermore, the more rapidly the greenhouse gases build up, the more difficult it will be to predict the transient response and the less likely it will be that the transient response will resemble the equilibrium calculations.[42] Thus, the faster we increase the greenhouse gas content of the atmosphere, the more likely it is that we will be unable to predict the regional details; therefore the more likely it is that we cannot identify regional impacts and the greater the chances of having major climatic surprises.

Changes in the average temperature or rainfall or soil moisture of some region at some season are not the only climatic variables of importance to people, plants, or animals. Biological systems are often much more sensitive to extreme events in the climate than to mean values. For example, freezing is an obvious threshold below which many plants are killed. Generally, increasing the average temperature a few degrees will probably not have much of an effect directly on, say, corn crops, unless it represents a change in the likelihood of extreme heat waves, which can be devastating to corn yields. Therefore, climate modellers have recently begun to ask their models not only how mean climatic conditions may change with increasing greenhouse gases, but also how climatic variability will change. David Rind and colleagues at the Goddard Institute for Space Studies (GISS) used the GISS model to investigate this question.[43] His preliminary results suggest that the temperature variability from one year to the next seems to go down, whereas the precipitation variability seems to go up across the United States as a result of CO_2 doubling.

With my NCAR colleagues Linda Mearns, Starley Thompson, and Larry McDaniel, I have performed such tests with a version of the NCAR model run by Warren Washington and Jerry Meehl.[44] We obtained a number of results that were similar but not identical to the GISS results. For the past fifteen years general-circulation models have been used for climate change experiments. Yet it is only in the past few years that changes in a few measures of climatic variability have been analysed. Therefore, it is too soon to draw even tentative conclusions as to how climatic variability might change with changes in long-term climatic averages. But nothing that we have found so far suggests that climate variability will change so much as to invalidate the earlier-stated concern that global average warming is very likely to increase the probability of extreme heat waves—in other words, to load the climatic dice with more hot faces.

It should be stressed again that, given the likely conclusion that regional climates will change, considerable uncertainty remains over the specific model-predicted regional features on land. The principal reasons for the uncertainty are twofold: the crude treatment in climatic models of land-surface (for example, biological and hydrological) processes and a neglect of the effects of the deep oceans. The latter would respond slowly—over many decades to centuries—to climatic warming at the surface, but would also act differentially (that is, nonuniformly in space and over time). That means that the oceans, like the forests, would be out of equilibrium with the atmosphere if greenhouse gases increase as rapidly as is typically projected and if climatic warming were to occur as quickly as 2° to 6° C (3.6° to 10.8° F) over the next century. Remember that this typical projection is ten to sixty times faster than the natural global average rate of temperature change, 1° to 2° C per millennium, seen from the end of the last ice age to the present warm period.

Furthermore, if the oceans are out of equilibrium with the atmosphere, then it will be hard to assign much credibility to specific regional forecasts such as those in Figures 8 through 10 until fully coupled atmosphere/ocean/biosphere models are tested and applied. Such coupling is a formidable scientific and computational task. Typically it takes one hundred hours of supercomputer time to run ten years of simulated climate, and we would need to do dozens and dozens of hundred-year runs with coupled atmosphere/ocean/ice/biota models. Such work in itself would take a decade or more to accomplish even if research support grew threefold over the current level. With lower levels of support it will take several decades longer to build a strong scientific consensus on the regional details of global warming.

VERIFYING THE MODELS. Given all the admitted problems with GCMs, it is appropriate to ask how we might verify the model predictions we have. Can we base trillion-dollar decisions about global economic development strategies on the projections of these admittedly cloudy crystal balls? The most perplexing question about climate models is to what extent they can be trusted to provide grounds for altering social policies, such as those governing carbon dioxide emissions.

The first verification method is checking the model's ability to simulate today's climate. The seasonal cycle is one critical test, because in the northern hemisphere these natural temperature changes are several times larger than those from an ice age to an interglacial period, or those related to any projected greenhouse warming. Global climate models do indeed map the seasonal cycle well, which suggests that they are on the right track.

As is well known, the annual migration of the sun causes the seasons. This creates, from one season to the next, a difference in the heating of the earth's surface many times greater than that we would expect even from a quadrupling of CO_2 and other trace gases. Therefore, the ability of a model to reproduce the tens of degrees difference between winter and summer is critical to its credibility for forecasting the much smaller climatic response to the radiative heating implied by increasing trace greenhouse gases. I recall describing with some pride to a congressional committee how well our climatic models were doing in reproducing the very large difference between winter and summer in the mid to high latitudes and the relatively small seasonal differences in the tropics. After the session, one congressman walked up to me and said, "Do you mean to tell me that you guys have spent millions of dollars of the taxpayers' money telling us that winters are cold and the summers are warm?" "Yes, sir," I replied enthusiastically, "and we're proud of it—for if we couldn't independently simulate the seasonal cycle well, I wouldn't have the nerve to stand before this committee and argue that rapidly changing climate from human pollution is a serious problem."

If the models had either missed or badly miscalculated a major feedback process occurring within a year's time, it is very unlikely that most atmospheric GCMs would be able to simulate the seasonal cycle of surface temperature nearly as well as they do.[45] This assures me that our estimates of our sensitivity to a doubling of CO_2 are likely to be accurate to a factor of 2 to 3.

On the other hand, I must admit, the seasonal cycle test is a validation only of what we could call "fast physics" (that is, involving processes occurring over a year or less). Clouds certainly fall into this category, since they form and dissipate within days. So, to some extent, does sea ice. However, a model's accuracy in simulating the seasonal cycle does not generalize to medium-scale or slow-scale processes. The decomposition of dead organic matter in the soil is a medium-scale process and deep ocean circulation is "slow physics." Unfortunately, it is the medium time scale—decades to centuries—over which CO_2 is expected to double, and that scale will be of most interest for the greenhouse century.

A second verification technique is to isolate individual physical components of the model, such as its parameterizations, and test them against reality (or at least a higher resolution model). For example, one can check whether the model's parameterized cloudiness matches the observed cloudiness of a particular box. But this technique cannot guarantee that the model treats the complex interactions of many individual model components properly. The model may be good at predicting average cloudiness but poor at representing cloud feedback. In that case, simulation of overall climatic response to, say, increased carbon dioxide is likely to be inaccurate. A model should reproduce to better than, say, 25% accuracy the flows of thermal energy among the atmosphere, surface, and space (as in Figure 1). Together, these energy flows comprise the well-established greenhouse effect on earth and constitute a formidable and necessary test for all models. A model's performance in simulating these energy flows is a test of its components.

A third method for assessing a model's overall, long-term simulation skill is to monitor the model's ability to reproduce the diverse climates of the ancient earth or even of other planets. Palaeoclimatic simulations of the Mesozoic era, glacial/interglacial cycles, or other extreme past climates help us to understand the coevolution of the earth's climate and living things. As verifications of climate models, however, they are also crucial to estimating both the climatic and biological future.[46] Assessing a model's ability to simulate major features of the geographic distribution of climatic differences between the last glacial maximum and the present or of the climatic optimum and the present can tell us much about the model's skill. However, we must temper our assessment with the recognition that such tests only help to validate fast physics. The redistribution of solar energy that is used to test the climatic optimum and the wind displacement owing to the imposed glaciers that we put in our ice age experiment both occur very quickly. Unfortunately, such validation experiments as the seasonal cycle or ice age "snapshots" are really *surrogate* validation tests. They can provide only circumstantial evidence. Remember: the models project climatic changes over the next century, a time scale for which there is not yet a clear surrogate validation test.

But the final validation method is not circumstantial. It simply involves checking the actual trend of the earth's temperature over the past hundred years, during which carbon dioxide and a number of other minor trace greenhouse gas species have increased, we know, by some 25% or more. Despite all the difficulties enumerated earlier, the approximately 0.5° C (1° F) increase in the global average temperature over the past century is roughly consistent with such trace gas increases, but somewhat too low. This merging of the climate predictors' two tools is our only direct test of greenhouse-effect changes on the medium time scale. However, as I explained earlier, the instrumental record is plagued by climatic noise, measurement problems, and the possibility that other causes of climate change could have accounted for some of the record, and these causes have not necessarily been properly accounted for.

The overall validation of climatic models thus depends on a constant appraisal and reappraisal of performance in all of the above categories plus an intense vigilance regarding other possible explanations for the models' results. Consider the significance of the 25% increase in carbon dioxide and greater increases of other trace greenhouse gases since the Industrial Revolution. Most climatic models predict that a warming of some 1° C should have occurred during the past century, but this is somewhat larger than the half degree warming actually observed. Possible explanations for the apparent discrepancy include the following:

- State-of-the-art models have overestimated trace gas greenhouse increases by about twice the actual amount.
- Competitive external forcings such as volcanic dust, changes in solar energy output, and regional tropospheric aerosols

from biological, agricultural, and industrial activity have not been properly accounted for.

- The large heat capacity of the oceans, which enables them to take up some of the heating of the greenhouse effect and delay, but not ultimately reduce, climate warming, has not been properly accounted for.
- Present models and inferences from observed climatic trends could both be correct, but models have typically been run for equivalent doubling of carbon dioxide, not for the 25% increase actually experienced.[47]
- The incomplete and nonuniform network of thermometers has underestimated actual global warming over this century.

Despite this litany of potential excuses, the twofold discrepancy between GCM predictions and observed global temperature trends is still fairly small. Most climatologists do not yet claim beyond a reasonable doubt that the observed temperature records have been caused by the greenhouse effect. It is possible that the circumstantial agreement to a factor of 2 between the observed trend and the predicted warming could still be a chance occurrence. Nevertheless, when taken together, validation examples provide strong circumstantial evidence that the current modeling of the sensitivity of global surface temperature change to increases in greenhouse gases at least over the last 100 years or so is probably valid within a factor of 2 to 3. The validation of regional predictions is more problematic, but not entirely uncertain. Another decade or two of observations should produce signal-to-noise ratios sufficiently obvious to enable scientists to agree as to the reliability of present estimates of global climatic sensitivity to increasing trace gases. This increase in data should transform the preponderance of evidence from circumstantial, as it now is, to direct. But waiting that long before acting entails serious risks. If we fail to take preventive actions until we are sure we have detected unequivocally a global warming signal in the temperature record, then we will have to adapt to a much larger dose of climatic change than if we began taking preventive actions now.

In 1989 the Marshall Institute, a Washington-based think tank that gained notoriety a few years earlier by advocacy for President Reagan's strategic defence initiative (SDI—popularly known as "Star Wars"), argued in the case of global warming that there were too many uncertainties in climate models, too many alternative explanations for the twentieth-century warming trend and such good prospects for quick (less than 5 years) research progress to justify any governmental policy response to the greenhouse gas buildups other than more research. Washington was rife with rumours that the Marshall Institute report had been influential behind-the-scenes at the White House, helping to bolster the Administration's reluctance to endorse specific CO_2 reduction targets being called for at international diplomatic and scientific meetings in 1989. Aside from the absurd overestimate of the likely rate of research progress (the Marshall Institute authors suggested only 3–5 years), the report did not mention the many validation measures already undertaken by climate modellers, including those listed above. Although the report often referred to the reservations and uncertainties that climate modellers normally note in their papers, it neglected to mention any validation measures, such as the model's simulation of the seasonal cycle, etc. Neither did the report say that whether to act politically, given any amount of uncertainty, is a value judgment. Instead it gave the impression that policy choice is a scientific question. One can only wonder whether the Marshall Institute believes that unprecedented global warming into the twenty-first century is more or less probable than a leak-proof strategic defence shield—and the Institute advocates spending billions of dollars annually on SDI. In the decade we are entering, we are likely to confirm whether or not our greenhouse effect calculations are approximately correct. If they are, the greenhouse century has already begun. Although I may one day have to eat these words, it is a good bet that by the end of the twentieth century most climatologists will look back to the 1980s as the decade when the greenhouse gas signal finally emerged from the noise of natural climate variability.

6

Assessing the Impact

In March of 1977, David Slade, a member of the Environmental Research Group at then, the Energy Research and Development Administration, a predecessor of the Department of Energy, held a landmark meeting in Miami, Florida. The purpose was to begin planning for a research programme to address the CO_2 issue. About fifty scientists representing all of the environmental disciplines attended. The meeting was to begin with a few-hour long general discussion among the participants. The discussion quickly became interdisciplinary as issues that involved marine, atmospheric, and biospheric information bounced back and forth between those representing the particular disciplines. The meeting chairman, Lester Machta, of NOAA, and Slade wisely adopted a *laissez faire* attitude and allowed this unusual free-wheeling discussion to continue. The few hours spilled over until afternoon and then well into the next day. By lunchtime of the next day, without recognizing it, the participants had designed the first comprehensive programme in the nation that dealt with the major environmental science aspects of what was later to be familiarly called the greenhouse effect research agenda. As an attendee in Miami, I helped draft a short recommendation that the planned DOE programme should become even more interdisciplinary by including a social science research component as well. With one exception, the general form of the new programme has been maintained to the present.

That exception began in 1979 when a quiet but important event took place in the evolution of thinking about global warming and how to cope with it. The U.S. Department of Energy sponsored a meeting that brought people of many disciplines together in Annapolis, Maryland.[1] Not only did climatologists, oceanographers, and other physical scientists get together with biologists, but also social scientists such as economists, political scientists, historians, anthropologists, and psychologists pondered an essential question: What if the climate changes?

Historians were invited to analyse major environmental disruptions and how society had dealt with them. For example, the famous potato blight that decimated Ireland in the late 1840s and spurred Irish emigration to the United States was compared to a simultaneous but much less disruptive blight in Holland and Belgium. The latter was less disruptive not so much because the blight was less serious but because those two countries had better developed economic and social protection mechanisms that enabled them to weather the difficulty without the massive starvation and immigration experienced in Ireland.

Psychologists at the meeting suggested that any reaction of people or politicians to the prospect or reality of climate change would depend not so much on the scientific facts but on the people's perception of the issues. We physical scientists had to find familiar metaphors and clearer ways to explain things, they told us. They also advised that, for maximum effectiveness, we couch the issue in terms relevant to the districts of individual politicians.

Political scientists thought about the alteration of climatic resources and the redistribution that it could imply for rich or poor countries. The economic capacity of a political unit to deal with change was deemed by Lester Lave, an economist at Carnegie-Mellon University, a major determinant in the amount of damage even an adverse climate change could wreak. This led to the general forecast that rich countries would be less harmed than poorer countries by change of any kind. Such economists as Lave or Jesse Ausubel, then at the National Academy of Sciences, said that many of the climatologists' projections of CO_2 increases were not rooted in the economic realities of massive coal use, food trade, and capital formation.

Eventually, this pioneering meeting was distilled into a small group of scientists from many disciplines under the chairmanship of the elder scientific statesman Roger Revelle, of Scripps Institute of Oceanography, and under the aegis of the American Association for the Advancement of Science (AAAS). I was a member of that group.

In 1980 AAAS put together a proposal for the U.S. Department of Energy (at its request) that would attempt to perform what we called integrated-scenario analysis. That jargon meant that climatologists would project climatic change, which would spark agronomists' projections of agricultural change, which would lead to hydrologists' projections of water supply change, health specialists' forecasts of health impacts, and so forth. Economists would then evaluate these costs, estimate the impact on the economy, and then feed back that information to ensure that the economies so perturbed by the climate could continue to produce fossil fuel pollution used in the first place by the climatologists in their climate scenario. We would each work independently at first, in what might be called a multidisciplinary way, then we would get together to try out our latest scenarios on each other and make sure they were internally consistent. In that process we would grow as a group to become truly interdisciplinary—that is, learn enough about each other's field so that we could actually synthesize solutions that were beyond the boundaries of our specific fields.[2]

David Slade, by then the head of the DOE office in charge of CO_2 research, which was sponsoring the effort, was very favourably inclined to the proposal. He verbally confirmed to David Burns of the AAAS that he was anxious to see it implemented. Then, in late January 1981, the Reagan administration took over. The DOE got a new secretary: a dentist, the former governor of South Carolina. When he called Slade in to explain why the DOE CO_2 programme was going to cost the taxpayers more than $10 million annually, Slade emphasized the uniqueness of the interdisciplinary effort. He told the secretary that the AAAS programme would permit the potential seriousness of the greenhouse effect to be evaluated even before we were certain about the physical and biological details. He went on to explain that the group would make economic and social impact assessments for various scenarios of changing climate and that these would be helpful for determining whether climate change would actually matter.

I was not personally at that meeting and cannot attest to the following story, but it is consistent with what happened shortly thereafter. My source told me that he was sitting near the secretary of energy and his deputy and overheard one of them quip, "What are we funding environmentalists and social scientists for?" These two groups had been targets for budget cutting in the early years of the Reagan administration.

Shortly after that meeting, Slade was relieved of his CO_2 project and replaced by a person formerly affiliated with the solar power satellite—a great solar collector in space that was thought to be a potential energy source for use on earth. It was to beam microwaves down to the earth's surface that had been produced from the primary energy of the sun intercepted out in space. As far as I could tell, this new project manager was a newcomer to the detailed debate over global warming. He and his people immediately rejected the AAAS proposal, arguing, among other things, that it was a waste of the taxpayer's money to ask "what if" questions before the physics and chemistry uncertainties had been resolved.

Word of the firing of the former CO_2 programme manager at DOE reached Congressman (now Senator) Albert Gore, Jr., from Tennessee. He, Congressman George Brown of California, and Congressman James Scheuer of New York held a hearing in March 1981 challenging the DOE on its firing of David Slade and the apparent rejection of impact-assessment research. The DOE came prepared for a fight—which they got. Four scientific witnesses were first called by the committee, to be followed by the DOE officials and then general discussion. One scientist was the usually sceptical economist Lester Lave; another the cautious head of the U.S. government's Geophysical Fluid Dynamics Laboratory, Joseph Smagorinsky; the third was the impeccably credentialed senior scientist Roger Revelle, who had led the scientific community in its very successful International Geophysical Year project in 1957–1958; and I was the fourth. Each of us explained, in turn, why he thought it was necessary to support all aspects of CO_2 research and why we were concerned that the DOE was about to veto the impact-assessment research represented by the AAAS proposal.

Roger Revelle led off the testimony. "Although the work inventory of weather and climate data is extensive, these have not often been reported in a form most useful for predicting biological and societal responses to climate change or climate variations. Improved historical and current climate data bases would be desirable. The form of these data bases should be determined by collaboration among biologists, agronomists, resource managers, social scientists, climatologists, statisticians, and data processing specialists. Different forms may be required for agriculture and for social sciences. In tropical agriculture, crop/climate data systems need to be built and tested that will include information on regional climate and agro-resources such as soil, topography, and socioeconomic conditions. . . . Biological monitoring of forests and other natural ecosystems should be undertaken on a continuing basis to identify and evaluate effects of increased carbon dioxide and climatic change. Such monitoring should be carried out both on a reconnaissance level and in greater detail at specially chosen sites to study species successions and other relationships."[3]

The next witness was Joseph Smagorinsky, then the chairman of a National Academy of Sciences committee that was preparing a report on CO_2 (one in a long line of reports that the academy had undertaken and continues to undertake). He described existing and planned national and international studies on the nature of the greenhouse effect and the need to develop and evaluate scenarios of future changes. Smagorinsky pointed out that the National Academy intended to "initiate and evaluate studies on the socioeconomic impacts of climatic changes such as on agricultural productivity, on health, on water resources, and on energy consumption—the sort of points covered by Professor Revelle." Smagorinsky concluded by noting that there were many uncertainties in the physical sciences and an important need to stress the validation of models. "Finally," he concluded, "despite the admitted existence of numerous uncertainties, the consensus on the nature and magnitude of the problem has remained remarkably constant throughout this long worldwide process of study and deliberation, in some instances going back to a century ago."

Next, it was my turn, and I addressed myself to the question, What if carbon dioxide increases and the climate changes? I asserted that this was necessarily an economic, social, and political question as well as a physical and biological one, and went on to argue that it would be irresponsible to sponsor applied research that focused on the physical and biological components of the problem without also asking those all-important "what if" questions. I concluded by pointing out that the AAAS/DOE effort initially sifted through a vast array of suggestions and that the steering group had come up with a series of tractable research proposals. We had boiled this down to five panel reports and ultimately distilled those into a single, final project suggestion: the highest priority item was integrated impact assessment based on formulating integrated scenarios. I then told the DOE representatives that I would welcome the explanations of why they were cancelling the programme.

But the big fireworks were yet to come. Lester Lave had had frequent disputes with environmentalists because he often argued that adaptation alone was the best response to the greenhouse effect,[4] but he was quite angry at what he now perceived to be the DOE's ideological cancellation of the AAAS proposal to do integrated-scenario analysis. "Carbon dioxide stands as a symbol now of our willingness to confront the future and especially of our willingness to consider problems that will not be manifest until after the next election," Lave began. "It will be a sad day when we decide that we just don't have the time or thoughtfulness to address those issues. I am disheartened by some of the changes I understand have taken place in DOE indicating they don't think they have the mandate, ability, or resources to address the CO_2 problem."

Lave called for a parallel approach, in which physical, biological, and social sciences would be pursued simultaneously while interacting with each other. The research "process can either be pursued serially, or it can be pursued in parallel. Pursuing it serially has some validity in that if we have no idea of what the effects of carbon dioxide are on the atmosphere and the oceans, then how in the world can you do anything else? However, if we wait until we get definitive answers as to the effect of carbon dioxide on the atmosphere and the ocean, we will not get to the second step in time to make adjustments. . . . That means that instead of trying to start at the beginning and through to the end slowly, that a better way of approaching this is to try and encircle the problem completely and then to close in from all sides. It is better to have a series of successively better approximations to the whole problem than to isolate one part and concentrate on it.

We learned this in the AAAS effort. . . . Given the attractiveness of fossil fuels, we are not going to stop using them in the next few decades. Thus, we had better worry about how adaptable are our institutions. Those kinds of social questions, the ones that allow us to take advantage of or at least not be terribly adversely affected by these changes, are the nature of the problem that the DOE seems to be turning its back on."

Lave had thrown down the gauntlet, and the DOE was ready to strike back. The lead DOE witness was a physicist, N. Douglas Pewitt, who had been present, I was told, at the meeting in which the secretary of energy had sacked Slade. Pewitt had hard lines and unusually harsh words for the AAAS proposal in general and the four previous witnesses specifically. He characterized much of the testimony from the four of us as "the hallway gossip we have heard here today. . . . "[5] And after repeated pressing by Congressmen Gore and Scheuer, he denied that the DOE had changed its attitude by dropping the AAAS effort. "There are some changes being made in the management of the programme," Pewitt said. "One of the changes is, we will not ask people for advice on how to put the programme plan together and turn right around and feed them money that they helped formulate the programme plan for. . . . We are not going to do social research on how Congress makes their decisions under the rubric of the carbon dioxide programme."

Pewitt did point out that the department was interested in some environmental effects, and that it had given a contract for over $700,000 to the Department of Agriculture to start looking at those environmental effects. The effects he was referring to were the CO_2 fertilization possibilities—that is, that increased CO_2 serves as a fertilizer to increase plant growth. The DOE was interested in that "what if" question, I presumed, only because CO_2 fertilization could well be a benefit to crop yields and therefore could not be used by environmentalists to attack fossil fuel energy growth. The fertilization issue is a legitimate study area, of course, but what so disturbed the four scientific witnesses was that the only impact-assessment work the department seemed willing to pursue was that which had a high likelihood of showing CO_2 buildup to be a benefit. They seemed bent on ignoring the whole other range of social and environmental impacts that might be shown to be detrimental.

Congressman Scheuer took up the questioning of the DOE witness: "However, we are concerned about the social, economic, and political implications of the destabilization that would be caused, looking into the future a generation or two, by the kind of phenomena we have heard about. . . . There are many places in the world, particularly in the developing countries, where the cost of relocating populations that are located near oceans would be painful, cruel, and nearly impossible. Those kinds of dislocations in many parts of the world can't be faced and they have potentially devastating economic and social and political implications. It is this that we think we ought to be measuring because the loss is not just crop loss, but vast human destabilization, and that goes far beyond the cost of crops or forest lands or grazing lands." Pewitt responded by saying that it was too soon for the administration to have a research programme. Scheuer then asked him for a date and a plan.

Congressman Gore pressed the issue. "I hope that the department will not cut the research . . . for analysis of the economic and social effects of this problem, and will consider the possibility in reviewing the fiscal year 1983 request of doing everything possible to accelerate the effort to narrow these areas of uncertainty. . . . Let me ask you," Gore said to Pewitt, "do you share the sense of urgency that I feel about getting a firmer grip on whether or not this problem is in fact occurring?" Pewitt replied, simply, "No." Gore demanded an explanation, to which Pewitt responded, "I think that in running a scientific research programme, we have a responsibility to the Congress and the American people to act in a fashion that is not alarmist. . . . I absolutely refuse as an official in a responsible position to engage in the type of alarmism for the American public that I have seen in these areas time and time again, and I do not think that I can responsibly encourage that sort of alarmism."

Gore responded that the Mauna Loa observatory data had clearly demonstrated increases in carbon dioxide and that it continued to show buildups. He pushed Pewitt further, "This is a rather impressive body of data that continues to accumulate. Doesn't that lead you to look at it in a different light?"

Pewitt responded defensively. "I am not an atmospheric scientist as one of the earlier witnesses alluded; I am a high energy physicist, a particle physicist. I understand false correlations; until one understands fundamental mechanisms of how things happen, they can be trapped into false correlations. Twenty-three years [referring to the length of the Mauna Loa record at the time] is not exactly a significant time frame in the weather."

Gore, a politician, not a scientist, responded on a technical point: "We have it [the CO_2 record] going back to 1958 with reliability, but the same pattern goes all the way back to 1880, and you can even see fluctuations for the Great Depression, and for the two world wars." Pewitt went on to repeat his charge about alarmism, continuing, "Nobody predicts anything to happen in less than fifty years. It is important not to waste the next decade, but it is also important not to jump off and get the American public concerned, . . . stop everything in the world, on the basis of misinformation." (This is remarkably similar to the testimony we heard seven years later by another Department of Energy official—see Chapter 2.)

Gore wouldn't let up: "I do think that the areas of uncertainty are narrowing and have been narrowing consistently over the last several years to the point that the issue is now agreed upon by the mainstream of the atmospheric sciences; that they agree that it has acquired a new percentage of certainty; if I can use that rather awkward phrase. I think that ought to lead us to reevaluate the importance of building options for our energy future, and should lead us to greatly accelerate the effort to narrow the remaining areas of uncertainty. And while we ought to avoid alarmism, we also ought to avoid a head-in-the-sand attitude at the same time."

Gore turned next to ask the new DOE CO_2 programme director, Fred Koomanoff, if he thought this was a problem of growing urgency. Koomanoff responded that there was much uncertainty, and "we will take the necessary environmental impact areas into consideration. At first, we must learn about first-order effects: the first-order effect of carbon dioxide on climate and the effect of carbon dioxide on vegetation." In other words, as director of the CO_2 programme, he was not going to support expanding economic- and social-impact assessments. Indeed, over the next seven years of his tenure in running that programme, Koomanoff largely kept that promise, funding physical science research (much of it quite good, though) almost exclusively, with some funding on biology and only small (but good) efforts in economic analyses later on.

But the fireworks were not over. Pewitt came back to the Mauna Loa CO_2 trend chart, calling it very deceptive: "It is a clever piece of chartology, in that it is intellectually accurate but can be subject to being read the wrong way." Gore was incredulous, and countered, pointing out that "if you look at a longer-range chart going back to 1880, you see virtually the same thing. . . ." Gore continued, "Dr. Pewitt, I don't want to put words in your mouth, but I got the impression that you were saying the significance of the chart should be substantially discounted."

Pewitt responded, "No. What I said is that that is chartology. It is intellectually just exactly correct. It displays 315 going to 336, but it appears to be going from 0 to very large amounts."

Gore was stunned. "It is clearly labelled 316 to . . . ."

Pewitt interrupted, "That is correct, and a person very careful in their review of this chart will see that it is true. That is the reason why I said it was intellectually correct. I mean, it is objectively a statement of fact that that is true. The impression it leaves on some people is quite a different impression. I have always had trouble playing with chartology. It appears to be going from nothing to a vast amount and that is a fact. We have a lawyer here who is nodding in agreement—it is true."

By this time most of us in the hearing room were giggling at this double talk. Gore simply read into the record the entire CO_2 emission chart from 1860 through 1980 and gave up the debate on that point. It seemed as if there was no way that this official of the Department of Energy was going to concede that there was anything other than uncertainty to the CO_2 issue. In his view, nothing more than physical and some biological (CO_2 fertilization) studies should be pursued.

Finally, Gore turned to the four scientific witnesses, asking us if we thought our presentations had been alarmist. "What do you think the proper tone of our response should be?" Gore asked. I responded: "Whether one is an alarmist or considers the CO_2 problem urgent isn't based on any scientific information. It is a value judgment. It depends upon how you personally fear potential risks versus how you personally fear the cost of mitigating them, versus your own political philosophy about whether individuals should be free to do what they want, or whether we have collective responsibility. The whole question of urgency or alarmism really is not something I think a group of experts can define.

All we can do, as experts, is try to list what the possibilities are, and, in particular, what the uncertainties are. Given that, I am happy to give you my values, which are that I think the CO_2 problem is only urgent in the sense that a lot of its solutions are also related to solutions to other pressing issues, as I said earlier, such as developing [new] crop strains, maintaining flexible energy supply options."

Congressman Scheuer wasn't through yet with Pewitt. Scheuer had been visibly annoyed by the DOE official's unusually vituperative and often incoherent testimony: "Dr. Pewitt, you have used expressions like 'navel staring' and 'alarmism' quite freely throughout your testimony. Do you consider that the panel which preceded you is engaging in navel-staring and alarmism?" Pewitt responded that the panel was impressive when it testified in its "area of expertise," but then chided us, saying that "technical experts have the responsibility to stay in their area of technical expertise when they are testifying as experts. When they are testifying as members of the public and not as experts, then I think they should clearly declare that they are not testifying as experts. We have seen that go on today." (I resented that charge since this remark came only five minutes after my testimony—quoted above—regarding personal value choices. It was as if this DOE official wasn't there or had already made up his mind about how irresponsible the scientific witnesses were.)

The hearing went on for a bit longer, with charges being traded back and forth essentially to no avail. When it was over, one of the DOE officials brushed by me, glaring and calling me an "irresponsible scientist" because I had been up there talking beyond my expertise and lying about their programme. It became clear that those of us who were concerned with serious research into the "what if" questions were going to have to wait through a long siege of nonsupport from the Department of Energy in particular and perhaps from the Reagan administration as a whole. That prediction proved to be all too true, at least for the next five years.

Fortunately, by 1986 the widely perceived antienvironmental tone of the early years of the Reagan administration began to fade. James Watt and Anne Burford were long gone—as was the Secretary of Energy. The EPA was in more competent hands, and even the Department of Energy's carbon dioxide research programme had begun to address the need for social-impact assessment in its future plans—although five years had been lost. But a very different Congress politically was elected in 1986. At the same time, the administration no longer took constant refuge in its "mandate" to cut expenditures as an excuse to avoid environmental action, and the planet continued to get warmer. Drought in the southeast in 1986 further motivated governmental interest in greenhouse questions and global warming. A number of congressional actions mandated a major study of the environmental, economic, and social impacts of climate change. This time Congress gave the money for impact assessment to the EPA, with a number of members remembering that the DOE had downplayed that kind of work when it had had the opportunity five years earlier. Poetic justice may not always be swift, but for me it was sweet: the EPA had long been viewed by many Department of Energy officials as a scientifically suspect agency, incapable of fundamental quality in research and preoccupied with the regulation of some of the DOE's clients—for instance, the energy industry. So five years after the histrionics of the AAAS/DOE congressional hearing came the irony: the Department of Energy saw Congress hand over millions of dollars to the EPA, of all agencies, for a major assessment of the potential environmental and social consequences of global warming.

In late 1988, the EPA produced a three-volume report entitled *The Potential Effects of Climate Change on the United States.*[6] The first volume was on regional climate studies, the second on national studies of various categories, and the third was a summary. The report admitted at every turn that we do not have adequate physical and biological knowledge to project by exactly how much CO_2 and other trace greenhouse gases would be increasing over time, nor could we be certain of the consequent climate changes. Nevertheless, the report asserted that it was necessary to analyse what the potential range of consequences might be for a variety of plausible scenarios—this was the same response to the "what if" question that the AAAS tried to offer to the DOE half a decade earlier. The EPA had handed various groups around the country climate-change scenarios from the Goddard Institute for Space Studies model and the Geophysical Fluid Dynamics model and asked them to suggest how the changes in climate projected by these models over the next fifty to one hundred years would affect water supplies, human health, and other impact areas. This strategy bothered some scientists, who feared that nonatmospheric researchers would take these GCM scenarios too literally. To try to allay that concern, the EPA had included a whole chapter on the problems with models.

The EPA report is enormously comprehensive. It has chapters in the national studies section that deal with water resources, sea-level rise, agriculture, forests, biological diversity, air quality, human health, urban infrastructure, and electricity demand. It used as background input the growing number of individual impact studies that had been started over the previous few years (most done without Department of Energy support), including a recent AAAS study in which the same Committee on Climate that proposed the earlier DOE project had managed to get the National Science Foundation to fund a major study of water supplies and climate change. That AAAS study, which took place in 1987 and 1988, provided major input for the EPA assessment.[7] Throughout the rest of this chapter, I will cover each of the topics on the EPA's list of impact areas, beginning with water resources and ending with electricity demand. Clearly, it takes many hundreds of pages to develop the detailed nature of impact assessments in each of these areas, and indeed the EPA reports are many hundreds of pages long, and they focus on only a half-dozen or so regions in the United States alone.

In 1987 the Australians held a major meeting, known as "Greenhouse 87," and from that came a 752-page book that described roughly similar categories of impacts for Australia alone.[8] In this chapter, I have assembled the highlights of this and a few other works and sketched out the range of potential consequences of plausible warming scenarios into the greenhouse century. The EPA report was especially helpful, in that it asked its analysts to try, where possible, to estimate costs of climate change. But the report went further, trying to estimate the costs of adapting to or trying to prevent those changes as well. While the EPA report is more comprehensive than any other to date, it is still not definitive, as the report itself makes clear. There is room for much discussion and study. Therefore, let us jump in and discuss some of the detailed "what if" issues, beginning with the very critical area of water resources.

WATER RESOURCES. Plenty of fresh water falls on the world. In amount alone, river runoff on a worldwide average would be more than adequate to handle human and environmental needs, if rainfall and runoff were fairly uniformly distributed around the world. The problem is that they are not. Rain falls where nature dictates, not where humans would prefer it. Furthermore, there are high human population densities in areas without major river runoff, which means the capacity of the water system is stressed or badly polluted. For example, more than 75% of the total land area of nearly fifty countries on four continents falls within international river basins, while globally, about 47% of all land area falls within international river basins. The Danube, for example, drains through twelve different nations. The Nile drains through nine, the Amazon seven, the Mekong six, and the Elbe and Ganges-Bramaputra both drain parts of five nations. The problem with water resources, then, is not the total amount of fresh water that falls on earth but rather the lack of uniformity in the distribution of human settlements, population centres, and national boundaries—and water resources—around the world. The health, environmental, engineering, and political problems associated with many nations sharing limited local or regional resources often give rise to conflicts over water. Therefore, climate change that might affect runoff should make many nations nervous.

The water-resource situation in the United States is similar to that of the world: there are adequate resources on a nationwide aggregate basis but many regional problems where local need exceeds supply or where high population or industrial/agriculture areas dump more pollutants into limited flow streams than the streams can easily tolerate. For example, more than 4 billion gallons (15 thousand million litres) of precipitation fall per day on the continental United States. However, nearly two-thirds of this evaporates, leaving only about 1.5 billion gallons (5.7 thousand million litres) to run off into the surface waters and groundwaters. Many of these flows are controlled through aqueducts or irrigation devices and then returned to the source after several uses or are consumed—that is, not returned to the streams. With all this manipulation of the freshwater resources of the country, the EPA estimated that from 1972 to 1985 government agencies and industries in the United States spent one-third of a trillion dollars on water pollution and control activities (over $1000 per person). In addition, the United States incurred $3 billion a year in flood damages during the past decade. In the western United States especially, water is critical not only to agriculture but to industry.

Owing to water's relative scarcity in the West, water-rights issues have become politically divisive and emotionally volatile. Typically, only about half of the 1.5 billion gallons per day of runoff in the United States is considered available for use 95% of the time. During the other 5% of the time, droughts force supplies to be well below those average figures.

The largest use of fresh water is irrigation, accounting for a little more than 40% of the fresh water withdrawn from streams. Thirty percent or so of the fresh water is consumed (removed from the system and not returned to the streams) in the United States. Irrigation accounts for only about 10% of the harvested cropland but contributes 30% of the economic value of cropland production. The reason is that many of these irrigated lands grow high-priced crops such as fruits and vegetables, which are more costly than typically unirrigated crops such as wheat or corn.[9]

Electric power plants withdrew almost as much fresh water as agriculture during the period reported (1985), but most of this fresh water was later returned to the system. Therefore, the electric power industry actually consumes less of the water than irrigation. Domestic uses in the United States account for about 10% of the total water withdrawn. Domestic water use is 25 billion gallons a day (about 100 gallons—380 litres— per person per day), up by more than 50% in the past twenty years, largely because of the increase in the number of households. There has been relatively little change in the water use of each household.

Since the enactment of the 1972 Clean Water Act, water quality has been a constant issue for debate, with contamination of the runoff of agricultural and industrial toxic wastes into streams a continuing problem. Adequate flows of fresh water must be maintained to dilute pollutants to below thresholds considered risky. Also, flows sufficient to repel the invasion of salty water into estuaries are needed to preserve downstream water quality and the aquatic life and ecosystems in lakes, streams, and rivers. Climate changes that affect runoff patterns would clearly have major implications for the management of water quality.

Finally, a major need is to reduce hazards and damages from floods. The principal method now used is to maintain adequate capture potential even for conditions of extreme weather variability. The U.S. Army Corps of Engineers has always worked to maintain flood-control structures adequate to deal with a "probable maximum flood." This issue has been politically contentious since a number of budget-weary legislators or suspicious environmentalists accused the Army Corps of overdesigning water projects by building in unnecessarily large margins of safety. If population or industrial growth in a region demands more water, then such margins of safety are used up for development and growth, exposing even more people in the future to the risks of weather variability.

The prospect of climate change that might alter runoff has led some members of the U.S. Army Corps of Engineers to respond to its critics with a rather large "I told you so." For example, three managers of the corps gave a presentation at a greenhouse-warming meeting, commenting "what makes the Corps confident that their current water management systems can absorb the anticipated near term climate changes? Many of the Corps' critics contend that the Corps projects are over designed. In some cases that is probably true. The Corps was trying to anticipate future needs without a firm basis in growth projections. In most instances, however, the design criteria merely reflected the conventional engineering approach of focusing on an extreme event or condition, for example, the critical drought period of record or the probable maximum flood or hurricane. This approach accounted for the unknown climate variability and the uncertainty that the existing data did not reflect the true range of possible events."[10] Indeed, the ultimate question of whether climate change should motivate more or bigger flood-control structures could be one of the most politically divisive issues to emerge in the greenhouse century.

As part of the AAAS study of water supply and climate change, interdisciplinary climate scientist Peter Gleick contributed a report on the vulnerability of U.S. water systems today.[11] He considered that there were five characteristics that made watersheds vulnerable to changes in climate. The first measure of vulnerability named was a relatively low water-storage capacity relative to the average amount of flow in the basin each year. River basins such as the Colorado and the Missouri, which have extensive reservoir capacities, are thus less vulnerable to flooding and drought than river systems in New England, the mid-Atlantic states, and the Great Lakes. For example, the lower Colorado River has four times more storage than the average annual flow, whereas the mid-Atlantic states have only 10% as much storage as annual river flow. Of course, the total amount of average flow in the East is much greater than in the West, which is why storage must be larger in the highly variable, semiarid western climate.

A second measure of vulnerability is high water consumption relative to the region's average annual flow. Gleick looked at the ratio between the demand for water and the supply from nature as a good measure of vulnerability. If a region uses 20% or more of the available resources, that is a warning that that watershed is vulnerable to climate change. California, on average, uses nearly 30% of its available flow each year, as does the Missouri River basin. The Rio Grande uses 64% of the flow, whereas the lower Colorado tops the vulnerable list by far with 96% of the average flow already accounted for by existing water-consumption use. This factor is a very strong indicator of the vulnerability of the Colorado River system to climatic change. In this connection, University of Arizona hydrologist Charles Stockton estimated that a 2° C temperature increase and a 10% precipitation decrease in the U.S. West would cut water supply in the upper Colorado by 40% and the Rio Grande by 76%.[12] Clearly, this scenario illustrates the vulnerability of this region to climate change as also suggested by Gleick.

Another measure of vulnerability, although not as important as the previous two, is the proportion of a region's electricity produced from hydroelectricity. Any climate change that reduces water availability in such regions could have a very serious economic impact. Any region where 25% or more of the electricity is produced from hydro power is considered a vulnerable region in Gleick's calculations. The Pacific Northwest is by far the most vulnerable, with 93% of its electric power being produced from hydroelectricity. Alaska is second with 50%, and Tennessee, the lower Colorado Basin, the Great Basin, and California all get between 25% and 30% of their power from hydroelectricity. While hydroelectricity certainly displaces some fossil fuels as an electricity-generating source, thereby cutting down on carbon dioxide and other pollution, it raises other environmental problems having to do with transmission lines, dams and diversion projects and creates a region's vulnerability to decreased flow.

The fourth category of vulnerability applies to those regions that depend heavily on groundwater for freshwater needs. The Texas Gulf, the lower Colorado, the Great Basin, the Arkansas

River, and to a lesser extent the Missouri River are all basins in which groundwater is being pumped in places faster than it is replenished naturally. If these kinds of activities continue indefinitely, the water resources in these areas would not be renewable but in fact would be mined. The mining of groundwater may be a temporary bonanza that is essential to irrigated agriculture or other economic activities today, but clearly it cannot be sustained indefinitely. The prospect of whole communities disappearing in the U.S. high plains on top of the rapidly diminishing groundwaters of the Ogallala aquifer in eastern Colorado, western Kansas, the Texas panhandle, and northeast New Mexico raises major social and economic questions for the future.[13] If climate change should decrease surface-water availability, it would also increase further the demand for groundwater, thereby causing a mining of that groundwater at even faster rates than those of today.

Gleick's final measure of vulnerability cites regions with highly variable stream flows—that is, areas where weather variability is so large that the difference in flow from year to year is very great.

Gleick examined the major hydrologic basins in the United States with respect to these five measures of vulnerability to assess the vulnerability of these regions. Only one area had shown all five vulnerability factors simultaneously—the Great Basin. California and the Missouri River basin had four vulnerability indicators, whereas the lower Colorado, the Rio Grande, the Texas Gulf area, and the Arkansas-White-Red River basins all had three. The least climatically vulnerable hydrologic basins in the U.S. were the New England and Middle Atlantic states, the Great Lakes and South Atlantic regions, and the upper Mississippi and Hawaii. The relative lack of vulnerability is in large measure due to the fairly reliable rainfall of those regions, although residents of the Northeast and South know that the relative lack of water storage can lead to serious shortages in those infrequent years when intense droughts occur—such as in 1986 or 1988.

The Pacific Northwest is primarily worried about risks of flooding, the reliability of hydroelectric power production, and water for irrigation. The principal competition for water in that region is among those seeking irrigation, hydropower, and the maintenance of fisheries habitat.

A much more vulnerable region than the Northwest is California. This is partly because of the relative variability in water supplies from year to year, but also because California, with its large population, major agricultural economy, and hot, dry summer season has very great water-consumption needs. Eighty-three percent of the total agricultural output in the state in 1982 was from irrigated land. California is the nation's leading agricultural state and also has major urban areas along the coasts and in the south. The high population density of these areas exists only because of vast engineering projects in which dams, aqueducts, and pumping stations provide the lifeblood of fresh water. Computer models driven by increased CO_2 produced climate change scenarios that were applied to California, and the EPA summarized its findings of the potential impacts of warming on California by drawing from many studies. These included a study by Peter Gleick that used three general-circulation models as well as "generic scenarios" that increased temperatures by 2° or 4° C and increased or decreased precipitation by 10% or 20%.[14] Gleick found that in those regions where water supplies are largely controlled by winter snowfall and summer snowmelt with very little rainfall, increasing the temperature would cause earlier and greater winter runoff from the mountains surrounding the Central Valley but that the increased temperatures would reduce runoff by the late spring and summer. This conclusion seemed surprisingly robust; it was relatively insensitive to switches of climate models and generic scenarios. This robustness probably would not apply to many other regions of the country or the world, but seems to be more appropriate to California because of the highly seasonal nature of the snowpack, runoff, and evaporation processes that are dominated by the state's Mediterranean–like climate.

From these CO_2 doubling scenarios and runoff estimates, the EPA concluded that the amount and reliability of the water supply from reservoirs in the Central Valley of the state would decrease, perhaps by as much as 7% to 16%. Current reservoirs would not have the room to store the heavy winter runoff and at the same time retain flood-control capacities, even if operating rules changed. Even worse, the EPA went on, is the projection of a water demand by the year 2010 that will be substantially greater than present usage. "This demand could not be reliably supplied under the current climate and resource system, and the shortage might be exacerbated under the three doubled CO_2 scenarios. The potential decrease in water deliveries could affect urban, agricultural and industrial water users in the state."[15]

Projections of the state's population growth to 2010 are typically at the level of 35 million, compared with around 27 million in 1988, an increase of some 30%. "An increase of this magnitude will significantly intensify municipal water demands, exacerbating shortfalls. Meeting the needs of the growing population in the face of the competing demands from agriculture," from industries, and from residential users will require changes in water-use efficiency, pricing, and water law, the EPA noted, "but these factors are probably impossible to project very far into the future."

The EPA suggested that several approaches could be taken to increase water deliveries in view of these perplexing demands and threats to supply from climate change. First, operating rules for reservoirs could be modified to allow increased storage in April in order to meet peak demand in midsummer. Of course, if one increases the storage in late winter, the risk increases of dangerous flooding should there be a late spring storm or intense heat wave to cause more than normal melting. The tradeoff between water supply and flood control in Northern California is a very serious policy conflict that indeed will only be exacerbated if any one of the greenhouse scenarios developed by Gleick proves true.

The second approach would be to build new water-management and storage devices. But this is expensive, raises environmental concerns over threats to wild and scenic rivers, and has been politically divisive in the past. It is possible that rather than large projects, small off-stream storage combined with improved pumping in the Sacramento Delta might help somewhat, as could artificially recharging of groundwater reservoirs during wet years.

A third approach would be to look for other sources of water, for example groundwater. Unfortunately, many areas are already pumping water at or beyond their sustainable yields, and increases could easily result in overdraft. In addition, groundwater is not usually considered to be under the jurisdiction of the state-integrated water management system, though it might need to be if water crises develop with increasing frequency in the future. Remaining options tend to be more exotic—for example, desalinization plants, cloud seeding (a highly questionable idea), and the reuse of waste water. California could also try to maintain high use of Colorado River water, which undoubtedly would lead to conflicts with Arizona and perhaps the Upper Basin states, particularly if climate change reduces the flow even beyond that which occurred since the Colorado River Compact was drawn up with erroneous allocations in 1922 (see the discussion in Chapter 1, in particular end note 4).

So far we have concentrated on water supply, but water demand can also be controlled. One way would be to allow greater flexibility in water marketing—in particular, increasing the price or cutting the subsidies that have allowed this commodity to help agriculture flourish in California with low-priced irrigation. While that might substantially reduce demand, it would be both politically divisive and economically threatening to an entire industry that depends heavily on subsidized irrigation. Water restrictions are another possibility, in particular cutting nonessential uses such as lawn watering, filling swimming pools, watering golf courses, and irrigating nonessential crops. If climate shifts do reduce summer runoff, as the models typically suggest, then these kinds of restrictions may become increasingly frequent, and by the middle of the greenhouse century essentially permanent.

Changes in runoff would affect not only agricultural, industrial, and urban water users, but also natural ecosystems. Reductions in summer flows could harm populations of birds and aquatic species that depend upon adequate water supply and quality. Existing laws protect endangered species and may be used as a legal basis for demanding increased releases of stored water to preserve these species. Obviously, conflict within governmental agencies and among industrial, agricultural, governmental, and citizens' groups concerned with the preservation of biodiversity could become acute if, as a number of models project, water resources decrease.

Finally, there is a connection between impact categories such as changes in water resources and rising sea level. The intrusion of salty water into the upper reaches of the San Francisco Bay and the Sacramento Delta could pose a major problem into the greenhouse century. This would be exacerbated by decreases of freshwater runoff and by the increased drawing of groundwater, which could cause the land to further subside. The latter would work together with sea-level rise and reduced runoff to increase the salinity of the bay. At present, water laws require that water must be released from reservoirs to improve the delta water quality at times of low flow, precisely those times when agricultural, urban, and industrial users are demanding greater protection. It is possible to devise strategies to fight the impact of saltwater intrusion. These include maintaining levees, increasing freshwater outflow, enlarging channels, constructing barriers, and digging canals around the delta's periphery. Freshwater pumping plants could be moved to less vulnerable sites or decisions could be made to abandon islands in the delta to rising waters. Property value of these islands is estimated somewhere in the neighbourhood of $2 billion, and they contain agricultural communities, highways, and other uses. The California Department of Water Resources estimated that improving the levees to protect them from flooding at the present sea level and currently assumed likelihood of extreme floods would cost about $4 billion. Clearly, that number would grow substantially if increased sea level or land subsidence from increased groundwater withdrawal is intensified. What is at stake is literally the shape of the map and the quality of life of Californians in the greenhouse century.

The Great Lakes area is another region of the United States that could be significantly affected by changes in evaporation/precipitation patterns that determine water resources. Climate models disagree to a considerable extent as to whether precipitation is likely to increase or decrease in the Great Lakes area. But the substantially higher temperatures predicted over the next fifty to one hundred years are common to all model projections for a doubling of CO_2. Several studies in Canada and the United States have suggested a drop in lake level from .50 to 2.5 metres (1.5 to 8 feet) below average historic levels.[16] A drop of only 1 metre would leave the lakes lower on average than any historic low in lake level. Lower lake levels would force shoreline communities to make adjustments over the next century. The EPA estimated that hundreds of millions of dollars would have to be spent along the Illinois shoreline alone to dredge ports and harbours and make channels deeper so existing shipping can flow. Water intake and outflow pipes would have to be relocated. On the other hand, lower lake levels could eventually expose more beaches and enhance recreational activities along the lakes – except in those areas where years of toxic runoff have accumulated in offshore sediments. A variety of nasty pollutants would then emerge from lands that used to be under the lake, threatening to expose them to people and animals, by erosion, leaching or evaporation into the air. The EPA estimated that forty-two such "hot spots" occupy bays and harbours all along the Great Lakes.

Lake-level reductions are not at all certain, given the uncertainty in model predictions of changes in precipitation. But all models agree on large temperature increases, which would affect the sea-ice coverage in winter, probably reducing the duration of that cover. This in turn would play a part in the mixing of the lake, which would influence its temperature structure and therefore its biological activity. It is difficult to estimate what such changes would imply for fisheries and other natural communities.

The United States and Canada have spent almost $7 billion on sewage treatment in the Great Lakes since 1972. By 1980, nutrient loadings into Lake Erie had been cut in half, resulting in substantial improvements in water quality. A change in climate in the surrounding agricultural regions of the lake that led to a change in farming practice would alter the runoff of nutrients or chemicals into the lake, and this series of events could alter water quality, but it is virtually impossible to predict details at this stage.

A decrease in lake level would also significantly influence shipping in the region. Channels and locks would no longer be able to take ships as fully loaded as they do today, since the bottoms of the fully loaded ships would scrape the bottoms of the channels or locks. The EPA estimated that without dredging of ports and channels, reduced cargo loads could increase shipping costs somewhere between 2% and 33%. However, reduced ice cover could lengthen the shipping season and serve to compensate somewhat for lower lake levels. If the lake states and countries were willing to make the few hundred million dollars of investment needed to expand port facilities and channels, shipping tonnages could not only be maintained but probably increased, owing to the longer shipping season. This is an example of what economist Lester Lave meant when he noted in congressional testimony that wealthier countries would be in a better position to deal with climatic change than poorer countries. The former have the resources to adapt more effectively, while the latter lack the financial capacity to mitigate the effects of climate change by infrastructure changes.

One of the greatest potential water-resources issues in the Great Lakes region involves the need to expand irrigation should increasing temperatures substantially cut average crop yields. This topic is covered later, in the discussion of agricultural impacts.

Although the southeastern United States is in general the least vulnerable to drought (1986 notwithstanding), its water-storage capacity is relatively small, since historically supply has been reliable. The Tennessee Valley reservoir system, with facilities worth over half a billion dollars, could be significantly affected if heating occurred without a compensating increase in rainfall. Lowered flows would reduce the dilution of municipal and industrial waste. The accumulation of pollution in warmer lakes and reservoirs would be exacerbated by reduced mixing from a shorter cooling season. A wetter climate would increase the risk of flooding, although it would also reduce the problems of water quality that would occur if the climate were drier. Electric power generation from hydro plants in the Tennessee Valley could also be changed by tens of percents per year—this could be a gain if the climate were wetter but a loss if it were drier or simply hotter. Estimates in the tens of millions of dollars are typical of expected impact.[17]

The Southeast has extensive coastlines, and as with the San Francisco Bay Area, the single most significant interaction in the region would be that of altered water supplies with increased sea levels. Especially in Florida and coastal Louisiana, these effects could be extremely serious. They are covered in the discussion of sea level later on.

The United States is not the only country, of course, that could be significantly affected with respect to water resources were global warming to occur at the rate and magnitude presently projected. In 1987 the Australians assembled a large group of scientists and planners to consider the implications of climate-change scenarios in various parts of Australia. Like the EPA, they developed climate scenarios based upon models and other climatology information. A. Barrie Pittock, a government climatologist, prepared the scenarios largely on the basis of U.S. computer models and the recent climatic history of Australia. His results suggested a warming of about 1° C in the northern part of Australia and about 3° C in the south along with precipitation changes largely in the increased direction except in the extreme southwest corner of western Australia near Perth.[18] The projected increase in precipitation was interpreted by a number of water managers as a potential benefit in some areas, reducing demand for irrigation water and improving the reliability of crop yields. But other Australian planners noted that in important agriculture regions in Australia salinity in the groundwater was already a very serious problem, in part brought about by increased irrigation that raised the groundwater and thereby allowed dissolved salts in the soil to reach the root zones of plants. They suggested that increased rainfall might further raise the groundwater table, making agriculture at least temporarily more difficult in these parts of Australia. But the fundamental response of Australian water planners responsible for a local or regional watershed was very similar to that of U.S. planners. For example, three water planners from western Australia pointed out that "the fundamental difficulty in establishing water resources strategies is, and for some time will continue to be, the uncertainty of expected climatic changes. For public decision making the very nature of the changes will be perceived with sufficient uncertainty to forestall decisions which would be readily accepted if expectations were more certain and the public well-informed. . . ."

The Australian report went on: "In this region the indicated effects of climate change are serious and a framework for planning and decision making needs to be established. Predictions are needed which are sufficiently firm to support general policy decisions. The first questions to be satisfied for decision makers are:

- What is the fundamental nature of the change (i.e., is it a reduction of rainfall)?
- What is the expected magnitude of the change?
- When, and at approximately what rate, will the change develop?"[19]

This list is remarkably similar to that assembled by senior water consultant Harry Schwarz, of Clark University, and his doctoral student Lee Dillard.[20] They interviewed a number of water-resource planners in the United States, asking them on behalf of the AAAS water supply and climate study what essential information the urban water planners and managers believed they needed to make decisions about climate change. People at the Indianapolis Water Company, for example, commented that for them to act they would need either "a measurable trend over at least five years or an unprecedented drought." Their actions would probably be increased drilling for more wells.

New Orleans water managers said they were most concerned about sea-level rise, but they also had concerns about low flows, which, aided by sea-level rise, would allow salinity to penetrate far up the Mississippi Delta. Increasing the level of Lake Pontchartrain by only a foot or two would reduce the efficiency of the pumps that were installed to pump out unwanted water from New Orleans, which is already below sea level. Increased rainfall would also be a threat, both from the point of view of flood protection along the Mississippi and the need to pump from the below-sea-level areas. Nevertheless, in their opinion "action would be taken if and when the Corps of Engineers recognized coming climatic changes and their local effects."

Schwarz and Dillard also interviewed a number of New York City water managers. New York City gets much of its water from large reservoirs in the upper Delaware River streams north and west of the city. In addition, there is an emergency pumping station on the Hudson River near Hyde Park. Complicated water law regulates the amount that can be withdrawn from the Delaware, since freshwater discharge is necessary to prevent encroaching salinity from damaging the lower Delaware estuary near Philadelphia. New York City officials, like New Orleans managers, worried about sea-level rise, dry hot spells in which runoff is reduced, and to a lesser extent storm drainage from big downpours. The emergency pumping station near Hyde Park on the Hudson would be a very poor backup system were sea level to rise and low river flow to occur simultaneously. Already, Schwarz and Dillard reported, saltwater wedges reached past the station at high tides and low river flows.

All the New York City managers interviewed suggested that New York could cope with climate change and that "the system is robust"—but by no means invulnerable to the kinds of changes that were being projected. One senior member of the city's environmental protection office said that no special actions could be expected "without complaints by affected people." Furthermore, before they would make costly adjustments in anticipation of climate change, they would need "near unanimity among the scientific and professional bodies." Another manager stated, "New York City is not going to be the first to act," and he did not seem alone in this view. Schwarz and Dillard interviewed water planners in Salt Lake City; Tucson, Arizona; Washington, D.C.; and Worcester, Massachusetts. In all cases, they got similar responses from local planners to the prospect of climate change: little likelihood of action without clear and specific signals of necessity—and this even though most managers conceded that the vulnerability of their water supply, flood protection, or sewage systems to the kinds of climate changes being projected were potentially serious. Perhaps the best summary of the water officials' opinions came from the western Australian planners: "History has shown that timely decisions on insidious resource management problems of this nature are difficult to achieve and tend to wait for some form of crisis before concerted societal action occurs."[21]

Before leaving the issue of climatic impacts on water supply, it is important to note briefly that the general reluctance of local planners to commit themselves to action without knowing specific regional details should not be generalized to national and international levels. As I will discuss later on, none of these managers suggested that they welcomed the prospect of rapid or substantial climatic change whose details are uncertain. Therefore, although specific cost-effective adaptations in any one watershed may be premature, given the present uncertainty about the distribution of details of climate change into the greenhouse century, national- or international-scale actions to slow down the buildup rate of greenhouse gases would buy time, allowing scientists to forecast details and planners to respond in other than crisis mode. Thus, such preventive policies at a national level would seem to be in almost everyone's interest. Meanwhile, the most general recommendation that seems justified for local water-resource planners is that they look for ways to increase the flexibility of their systems in response to all sorts of climatic variations. Not only will flexibility make adaptation to future climate change easier, but it will also buy insurance against the extreme climate variability that is inevitable whether or not global climate change proceeds as presently projected.

SEA-LEVEL RISE. Sea-level rise is undoubtedly the most dramatic and visible effect of global warming into the greenhouse century. It has inspired cartoonists to show ocean waves lapping up on the chest of the Statue of Liberty while in the background water covers half the buildings of New York City. Other artists have produced colour pictures of the bottom third of Florida's peninsula half eroded into the sea. Traditionally, perhaps no nation has been considered more vulnerable to sea-level rise than Holland, since much of this below-sea-level country seems to be chronically threatened with inundation. Despite the classical depictions, of course, there are no small boys with their fingers in the dikes, since dikes are very broad, flat structures, wide enough to support major highways and deep enough to prevent substantial storm tides from eroding the sides or breaching the top.

Not surprisingly, the Dutch have long been interested in the problem of global warming. They have been way ahead of most other nations in adapting themselves to harsh environmental conditions while building in large margins of safety to deal with unforeseen problems such as sea-level rise from global warming. Thus, while flying over the Atlantic on the Dutch national airline recently, I was gratified by an article I read in the KLM magazine, *The Holland Herald,*[22] on the country's future. The article was generously illustrated with an array of artists' conceptions of Holland's high-tech future. Elevated highways crossing vast lakes, modern structures, futuristic transportation systems, and other visions of the twenty-first century were all beautifully presented. I kept wondering if sea-level rise was going to show up in this popular treatment of the benign future. Finally, toward the end of the piece there were three maps. The first showed Holland several hundred years ago with lots of inland areas under water. The second showed how the Dutch had rolled back the sea to meet modern needs. And the third, which was side-by-side with the other two, was labelled "Back to the Future" and showed Holland flooded again as it had been hundreds of years ago. But for this map, much to my surprise, the caption read "Flooded for Recreation." There was no mention of global warming or sea-level rise from the greenhouse effect—only a suggestion that flooding would be revived for sport. I later made colour slides of these pictures, for the following week I was scheduled to give a lecture at which Gerrit Hekstra, a Dutch official in the ministry of the environment, would be in the audience. (He was to give a talk on sea-level rise himself later that day.) I thought it would be fun to tease him about how his countrymen had completely missed a principal problem for the future.

The next week in Washington, I gave the talk, got the appropriate laughs from the audience when I showed the pictures, and successfully teased the Dutch official. Later on, it was his turn, and he returned the favour. "You're right," he said. "The artist and the author of the article did not mention that the greenhouse effect and sea-level rise are reasons for flooding some land. It is true that Dutch planners project increasing the amount of open water inside of Holland, including the flooding of lands now used for agriculture. But your implicit assumption that it will be due to ocean waters rising over the dikes is not correct. There is virtually no chance that any sea-level rise that could occur in the foreseeable future would remotely cause ocean water to flow over the top of the dikes. Year-to-year changes in sea level due to local temperatures of the ocean, the passage of storms, strong on-shore winds, and other such meteorological events have already required us to build dikes some 16 to 20 metres (50 to 65 feet) above mean sea level to reduce the probability of breach to less than one event in ten thousand years."

For a typical Dutch dike, Hekstra's calculation goes as follows: "storm surge level, 5.00 metres above mean sea level; wave run-up, 9.9 metres; seiches and gust bumps, 0.35 metres; dike settlement, 0.25 metres; and finally, on top of all that, 0.25 to 0.5 metres sea-level rise for the next hundred years."[23]

"Why, then, do you think Holland will have to lose land in the next century?" I asked. "Because salt water will intrude up the Rhine and into our groundwater unless we deliberately flood part of the country with fresh water from the Rhine to prevent the contamination of our groundwater resources," Hekstra answered. Even though the sea will not come over the top of the dikes in almost any scenario we can conceive of, then, it is still likely that the Dutch will have to respond to sea-level rise, and their response will produce virtually the same result—flooding. But they will have to do the flooding deliberately and with careful timing in order to anticipate the change. Hekstra went on to point out that the Dutch have both the economic capacity and a long technological headstart that would help them adapt to sea-level rises.

But would such a resilience capacity exist in other places, such as the former Dutch colony of Indonesia?

Indonesia is a very interesting country with regard to sea level since it possesses 15% of the world's coastlines. Also, about 40% of its land surface can be considered vulnerable to a sea-level rise of as little as a metre per century, which is the typical projected rate by most scientists for the twenty-first century. In addition, its country is rich in wetland ecosystems and in the diversity of biological species that they support. However, Indonesia is a highly populated country, with more than a hundred million people living on many small islands. The islands of Java and Bali are vastly overpopulated, and the Indonesians have been trying to depopulate or transmigrate, as it is called, people from these populated islands to the more remote islands of Sumatra and Kalimantan (part of Borneo). These are heavily forested islands, covered with thick jungles, coastal swamps, and wetlands. The settlers are given agricultural lands near the coastlines as an incentive to transmigrate.

Often, local forests are chopped down to create the settlements and the lumber is sold abroad for foreign exchange. Figure 11 is Hekstra's map of Indonesia, with the shaded areas representing tidal-influence swamps and the black regions representing transmigration projects into these swamps. It is frightfully expensive, though feasible, for a rich, technically advanced nation such as Holland with a fairly small exposure to the open sea to protect iself against sea-level rise. But how can a poor country such as Indonesia with thousands of kilometres of coastline possibly protect newly transmigrated settlers who are put into lowlands that are already vulnerable to storm surges when cyclones periodically go by and whose vulnerability will substantially increase as the sea level rises? The resettlement is taking place in ecologically fragile lands that may not be environmentally sustainable even without rising sea levels. Furthermore, the resettlements are highly questionable as long-term solutions to the overpopulation of Java and Bali, even though they may be politically expedient at the moment. Where will these settlers go when the sea level rises and the likelihood of flooding increases? Can they return a generation hence to already crowded Java, to other parts of Asia, or to the United States, Europe, or their nearest sparsely populated neighbour, Australia?

Migration and the creation of environmental refugees are

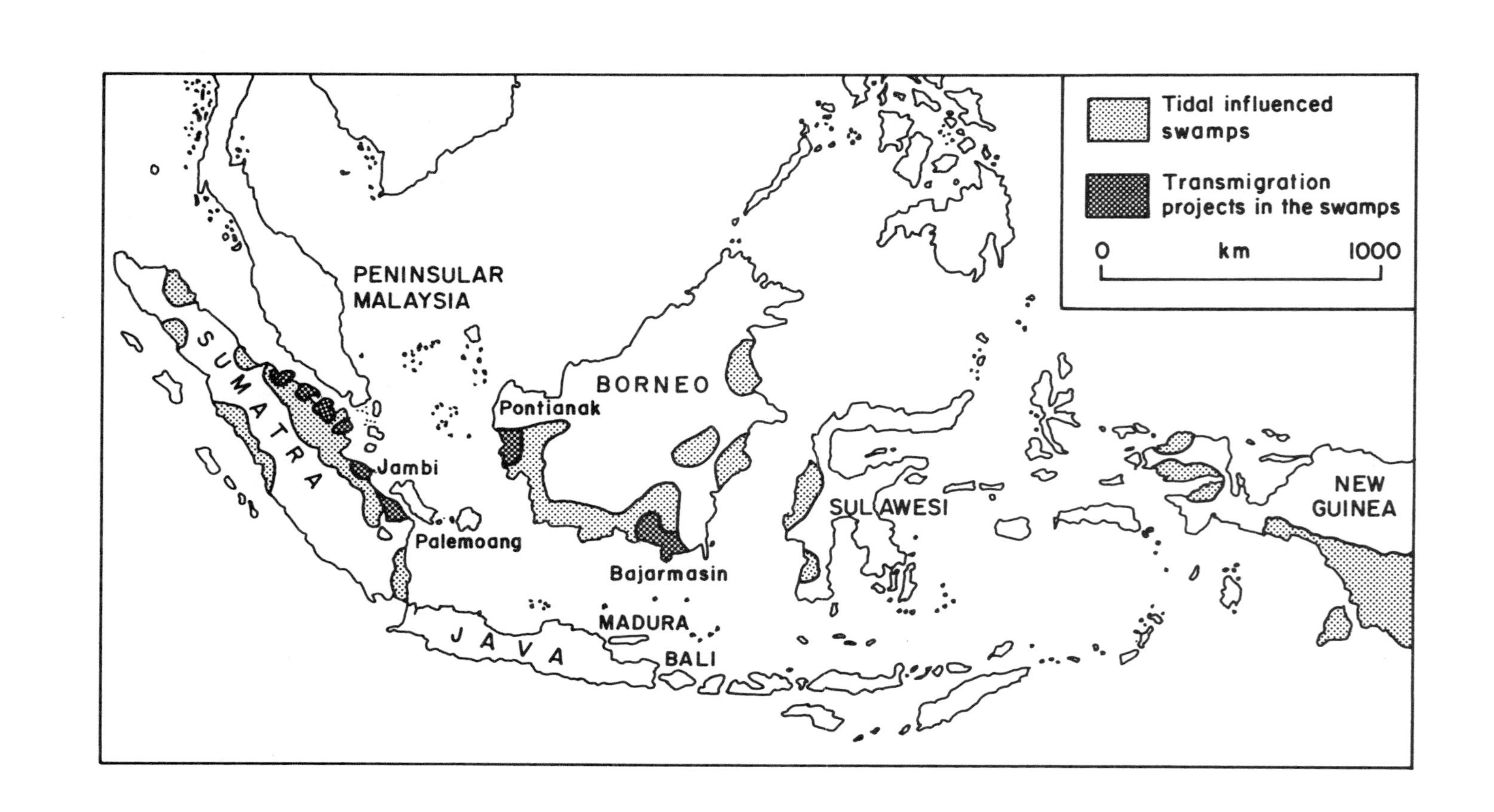
Tidal influenced swamps
Transmigration projects in the swamps
0
km
1000
PENINSULAR MALAYSIA
SUMATRA
Jambi
Palemoang
Pontianak
BORNEO
Bajarmasin
MADURA
BALI
JAVA
SULAWESI
NEW GUINEA

the kinds of strategic issues that have major implications for international peace and security. In fact, the current British ambassador to the United Nations, Sir Crispin Tickell, has been concerned about them for a number of years.[24] He has been active in trying to get the global-warming issue specifically and sustainable environmental development in general to be principal agenda items at the United Nations. "It won't be easy," he told a recent audience concerned with global warming, "to get the developed nations to put up the resources necessary to ensure environmentally sustainable development, nor will it be easy to get the developing nations to drop their suspicions that environmental sustainability is an excuse of the developed countries to prevent the developing countries from developing." He went on to argue that scientists will have to make their cases clearly and persuasively while simultaneously admitting to the uncertainties and that nations without the scientific traditions of the West will have to be convinced that they must take these urgent problems seriously.

Indonesia is not the only Asian country that has to be concerned about sea level. The Gulf of Thailand, Gerrit Hekstra pointed out, is particularly vulnerable to coastline recession, which would displace coastal villages and deprive many people of their lands and resources. The mangrove swamps that fringed this area have already been largely cleared away, while the canals that were dug to bring fresh water to rice fields are already becoming avenues for saltwater intrusion at times of low flow or when storm surges occur. "Many of the existing rice fields can become converted into brackish fish ponds," Hekstra said, "but there seems little chance that funds will be available for constructing sea walls as in the Netherlands."[25]

Australia is another country with extensive coastlines—indeed 100% of its borders are along the sea. Australia has more per capita wealth with which to adapt to the greenhouse century than Indonesia or Thailand and has already begun serious investiga-

Figure 11. Location of official transmigration projects in the tidal wetlands of Indonesia. Note that most of the settlements are in lowland areas that will become increasingly vulnerable to catastrophic flooding as the sea level rises. [Source: H. G. Walman and F. D. Fournier (eds.), "The Impact of Wetland Reclamation," Land Transformations in Agriculture Scope, *Vol. 32, (New York: Wiley and Sons, 1987):41.]*

tions into the implications of change. A number of these were discussed at the Greenhouse '87 conference mentioned earlier. At that conference, more than half a dozen contributions dealt with sea-level rises and their influence on beach erosion, shoreline-structure protection, and the loss of estuary, native habitat, and fisheries. Australia now has a population of approximately 16 million, and by 2030 most projections suggest that this population will increase by about 80%. Most of these 12 million or so additional Australians are expected to be living in and around the existing major metropolitan areas, where 85% of the current population is already concentrated. All these people will be, therefore, living close to a coastal zone. "Given the Australian penchant for coastal recreation and increasing real incomes, it is a most reasonable scenario to foresee the coastal zone from Cairns to Adelaide continuing to be our major setting for resource and environmental conflict, competition and controversy. Rising demands for accessible sites near population centres for both productive and consumptive uses will have to be met from a fixed land supply," said one Australian paper written by government ecologists.[26] These same writers pointed out that most of the species that make up the Australian seafood catch depend on coastal estuaries and tidal marshes during some or all of their life cycle. Thus, protecting them from inland water pollution and, from the other side, a sea-level rise will be important priorities for planning.

Also of concern to the Australians is the fact that their country already is half subtropical, with the northern half from time to time subjected to tropical cyclones that bring torrential rains and damaging storm tides and surges. The most severe hurricanes tend to form over the warmest waters, which are to the north of the bulk of Australia's populated southeastern region. But an increase in ocean temperature of several degrees could move the warm waters farther to the south, thereby encouraging some hurricanes to follow. This would increase the vulnerability of coastal cities such as Brisbane to periodic catastrophic storm surges. To that risk must be added that of a sea-level rise and the possibility that hurricane intensities could increase wherever hurricanes form. Whether private insurance companies will continue to insure such low-lying coastal developments is a major question.

MIT meteorologist Kerry Emanuel calculated a relationship

between sea-surface temperatures and the central low pressure of a tropical cyclone. He showed that cyclone pressures could decrease substantially with increasing sea-surface temperatures. Since the energy of a storm depends not just on the wind speed but on the wind speed multiplied by itself, increases of ocean temperatures of only a few degrees could increase the intensity of hurricanes by as much as 40% based on typical greenhouse warming projections.[27] Furthermore, an increase of sea-surface temperature of 2° C (3.6° F) could extend the cyclone belt in Australia from approximately 25° south latitude to 31° south latitude, while at the same time increasing the intensities of storms by tens of percents. For example, coastal engineer K. P. Stark from the James Cook University of Townsville, Queensland, Australia (Townsville is on the northeast coast, which is vulnerable to hurricanes) calculated that a coastal structure with a design life of fifty years made in 1987 with a floor level just above the five hundred-year return period for a rare flood would be subjected to an approximately 10% chance that a cyclone would cause flooding of its ground floor. However, for a scenario with a mean sea-level rise of 1.2 metres (4 feet) by 2030, the chance that this same structure would be flooded would go up to 27%. If, in addition, one assumed a 100% increase in cyclone frequency owing to an ocean-surface temperature rise of 2.3° C (4.1° F) by 2030, then the probability of flooding within the structure's fifty-year lifetime would be 55%. That probability goes to nearly 100% if in addition it is assumed that the cyclone intensity increases 20% on average by the year 2030. "To insure that the probability of flooding is less than 10% in the event of a scenario 3 [that is, mean sea-level rise, increased hurricane frequency, and hurricane intensity] within a fifty-year life would require a floor level at least 5.6 metres (over 18 feet) above the 1987 mean sea level, or 2.0 metres (6.5 feet) above the current design level. The cost of raising foundation levels by this order of magnitude for major tourism development projects could be many millions of dollars,"[28] Stark said.

Stark's altered probability charts for the likelihood of coastal flooding in periodic extreme storms was specific to Queensland, Australia, but the idea, of course, is general. Beaches, estuaries, and coastal structures are already vulnerable to extreme tides, storm surges, and other natural phenomena. The amount of safety margin designed into these areas is already an economic tradeoff: the damage or loss of the structure gambled against the cost of engineering extra safety margins. But greenhouse warming represents a substantial reduction in the safety margins that we had thought we had been building into structures because old designs did not account for a rising sea level or the potential for the increasing severity of travelling tropical storms. Even if the frequency of hurricanes does not increase with warming—which it may well not, since storm frequency depends upon meteorological disturbances that no one has reliably linked to an increase in global warming—there are good physical reasons to suggest that more intense storms could result. That would increase the maximum probable height of a storm tide or surge as such storms pass the coastal areas. Since coastal damage depends upon how high the ocean rises as a storm goes by, an increase in the mean sea level increases the likelihood that a certain area of coastline will be flooded even if storm frequency or intensity does not change. When these factors are added in, vulnerability is substantially multiplied, as the Australian calculation clearly demonstrated. The EPA noted that for southern U.S. coasts, "a 1-metre sea-level rise would enable a fifteen-year storm to flood many areas that today are flooded only by a hundred-year storm."[29] Hurricane Hugo in 1989 showed how vulnerable we are.

The Environmental Protection Agency estimated that "protecting developed areas against such inundation and erosion by building of bulkheads and levees, pumping sand, and raising barrier islands would cost $73–$111 billion (cumulative capital costs in 1985 dollars) for a 1-metre rise by the year 2100. . . . Developed barrier islands would likely be protected from sea-level rise because of their high property values. However, it would cost $50–$75 billion (cumulative capital costs in 1985 dollars) to elevate the beaches, houses, land, and roadways by the year 2100." If sea level rose 1 metre by the year 2100, this would drown approximately 20% to 85% of the U.S. coastal wetlands, the EPA suggested, basing the figures on studies by their own scientists and others. The ability of these wetlands to survive would then depend largely upon whether they could migrate inland or whether levees and bulkheads built to protect coastal structures would block the path to the migration. The evolution of marshlands as sea level rises can be described as follows. Five thousand years ago, when the sea level stabilized at about its current value, marshes began to be established at the margins. Marshes grew in response to five thousand years of stable sea level owing to sedimentation and peat buildup at the coastal edge. In a future scenario, an initial rapid sea-level rise would substantially reduce marshlands. Eventually, though, a few thousand years of peat accumulation at the margin could allow the marsh to expand and perhaps grow again, unless coastal structures are protected with barriers—assuming that such would be society's priority rather than protecting marshlands. This would lead to complete wetland losses. Clearly, the natural inhabitants of swamps and marshes do not flip levers in U.S. elections that help set priorities for what gets protection. Furthermore, fishing industries depend upon the survival of wetland ecosystems, so the actual cost of protecting society against environmental changes such as sea-level rise is not limited simply to the hundreds of billions of dollars involved in levees and bulkheads. It also involves the economic losses associated with the destruction of those natural systems that depend on coastal wetlands for their productivity. Currently, such losses are explicitly calculated neither in the price of engineering structures to protect coastal settlements nor in the projected cost comparisons of coal, natural gas, and solar and other renewable energy systems.

Beach erosion is another major issue facing the United States, Australia, and other coastal countries. Many studies conclude that a 30 cm (1-foot) rise in sea level causes beaches to erode from 15 to 30 metres (50 to 100 feet) from the northeast to Maryland, about 200 feet along the Carolinas, and as much as 30 to 300 metres (100 to 1,000 feet) along the Florida coast. In California the rates are closer to 60 to 120 metres (200 to 400 feet) and in Louisiana perhaps as much as several miles. That such a seemingly small increase in sea level can cause such large horizontal recession of beaches is explained schematically in Figure 12.[30] Panel A shows the present sea level and beach (shaded area labelled *A*). Panel B of the figure shows that a rise in sea level immediately results in shoreline retreat owing to inundation. However, the middle of Panel B is not a stable configuration according to most coastal experience. Rather, Panel C is the more likely situation: the raised ocean level erodes beaches and transports the sand offshore (to shaded area *Á*). Where the coastline drops steeply

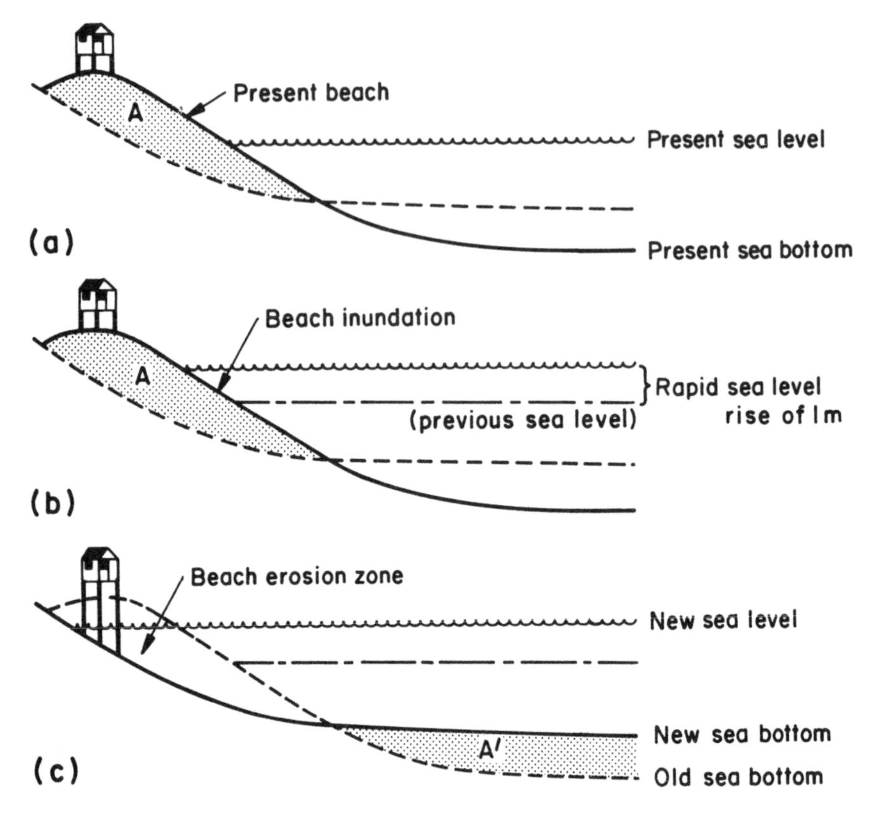

Figure 12. Beach erosion process from rapid rise of sea level by 1 metre. Present beach (A) is inundated by rapid sea level rise (panel b). However, wave erosion actually erodes entire beach region (panel c), putting sand offshore (A'), unless active steps are taken to pump sand continuously from offshore (A') onto the eroded beach site (A). [Source: Titus, J. G. 1986. Greenhouse Effect, Sea Level Rise, and Coastal Zone Management, Coastal Zone Management Journal *14:3, 147–171.]*

into the ocean, there is less loss of beach for each foot of sea-level rise, whereas in a place like Louisiana, where the slope of the land is very gradual, small rises in sea level can gobble up major fractions of bayou and coastal area. Already, because the coastline is sinking and sea level is rising, barrier islands in Louisiana are breaking up and exposing wetlands behind them to gulf waves.

To prevent this form of beach erosion, it would be necessary to dig out the sand in area Á on Figure 12 and pump it back up to try to maintain the middle profile. Stephen Leatherman, of the University of Maryland, calculated the cost of placing sand on U.S. recreational beaches, coastal barrier islands, and spits for three different scenarios of sea-level rise by the year 2100 — 50 centimetres, 100 centimetres, and 200 centimetres (1.6, 3.3, and 6.6 feet). For the nation as a whole, he calculates about a $14-billion investment for the 50-centimetre rise, a $27-billion expenditure for the 100-centimetre rise, and a $58-billion ticket for the 200-centimetre rise.[31] (For perspective, a $50-billion cost is a $200 tax for each of the approximately 250 million Americans.) Texas would require the largest investment, Louisiana the next largest, with South Carolina, Florida, North Carolina, New York, and New Jersey running close behind, all with more than a billion dollars of investment required even for the middle scenario.

How does all this translate into local decision making? Jim Titus of the EPA examined three approaches for maintaining wetland shorelines in the face of rising sea level: (1) no further development, (2) no action now but gradual abandonment later, and (3) allowing future development but with a binding agreement to force such development to revert to nature if threatened by inundation. (A fourth option is to construct engineering projects such as bulkheads, but as we saw earlier, that would protect coastal structures but even further degrade wetlands.)

The first of Titus's options, no further development, raises serious constitutional questions that have not yet been addressed in the courts. Of course, purchases of land for parks and refuges could accomplish this task without legal uncertainties, but presumably the cost of buying such lowland would be prohibitive. The second possibility — no policy today; wait to see what happens — would certainly sidestep the costs of planning for the wrong amount of sea-level rise but would take on the potential price of possible greater wetland loss than action today would entail. People usually develop coastal property on the assumption that it can be used indefinitely. Later on it would be difficult for any government — state, local, or federal — to tell property owners that within a few years they would have to abandon their land, no matter how many warnings had been issued ten, fifteen, twenty, or even fifty years earlier by their parents' or grandparents' generation. Politics generally deals with the moment, and there is little likelihood that protection of structures and wetlands would be bypassed in favour of abandoning real property to save the wetlands.

On the other hand, if there were legally binding prior contracts and if such agreements became commonplace (and if they prohibited construction of bulkheads as the sea level rises), then it is possible that conventional economics could be used to allow structures that would be expected to have a lifetime of no more than, say, twenty to fifty years. The costs of the construction and the return on the investment would then already have been calculated based on the assumed, finite lifetime. Although it might still be politically sticky to get people to abandon these structures after their official economic lifetimes, if the practice became widespread there would be much less political resistance to it.

Other state and local efforts to protect coastlines could involve choosing which areas should be protected and which should be allowed to erode. The EPA estimated that, on a nationwide basis, between $50 and $300 billion would be needed for coastal protection. It also reported that many state officials doubted their states could ever raise such money. Would communities do it? Long Beach Island in New Jersey was chosen by the EPA to illustrate the potential difficulties. "The annual cost of raising the island would average $200–$1,000 per household over the next century. Although this amount is less than one week's rent during the summer, it would more than double the property taxes, an action that is difficult for local governments to contemplate. Moreover, the island is divided among six jurisdictions, all of which would have to participate."[32]

The federal role has been most clearly defined in wetlands, with the Clean Water Act discouraging development of existing wetlands. But the act does not address current development that might have to be abandoned as the sea level rises. Sea-level rise and an increase in the probability of storm surge have yet to evoke legislative action of any sort. The U.S. Army Corps of Engineers is responsible for several major projects for rebuilding beaches and trying to curtail land loss in Louisiana, and further expansion of the Corps and their activities could protect much of the coastlines. But, as the EPA suggested, costs would run in the tens or hundreds of billions of dollars over the greenhouse century. And this is not merely a theoretical example. Louisiana is already losing wetlands at an alarming rate, and the federal government is spending tens of millions of dollars each year for protection. Coastal protection in the face of rising sea level will require investments not in the millions of dollars but in the billions. "Until someone estimates the costs and likely results of strategies with a chance of protecting a significant fraction of the wetlands in the face of rising sea level," the EPA commented, "it will be difficult for Congress to devise a long-term solution."

Congress has acted recently, passing the Upton-Jones amendment (Public Housing Act of 1988), which requires the federal government to pay for the rebuilding or relocating of houses that are about to erode into the sea. As the EPA noted, "although the cost of this provision is modest today, a sea level rise could commit the federal government to purchase the houses on all barrier islands that did not choose to hold back the sea." That might increase the likelihood that more communities would decide not to hold back the sea but to take the insurance instead. Of course, insurance rates would have to rise to reflect this risk, which would discourage construction of vulnerable houses unless they were short-term investments with potential immediate returns great enough to outweigh the likely flood damage. If agreements in advance, as suggested by Titus, could get investors in such projects to commit to abandoning structures as sea levels rose rather than arguing to protect them with ecologically damaging sea walls, then the economic use of coastlines could proceed even though costs would go up. However, the EPA noted, no assessment of the impacts of sea-level rise on any federal flood-insurance programme has yet been undertaken.

Unlike the case of water resources, for which regionally specific forecasts (for example, of how much wetter, hotter, or drier it will be and in which season) are not possible, most sea-level projections suggest a rise somewhere from .5 metres to 1.5 metres (1.6 feet to 4.9 feet) over the next century. But where do the estimates come from and how confident are scientists of them? Johannes Oerlemans of the Netherlands surveyed the scientific literature in this area, added calculations of his own, and compiled Figure 13, which shows the contribution to total sea-level rise from five categories: Antarctica, West Antarctica, Greenland, glaciers, and expansion.[33] Expansion, the largest category on the figure, simply means that a warming of the oceans will cause them physically to expand, much as warming the fluid in a thermometre causes

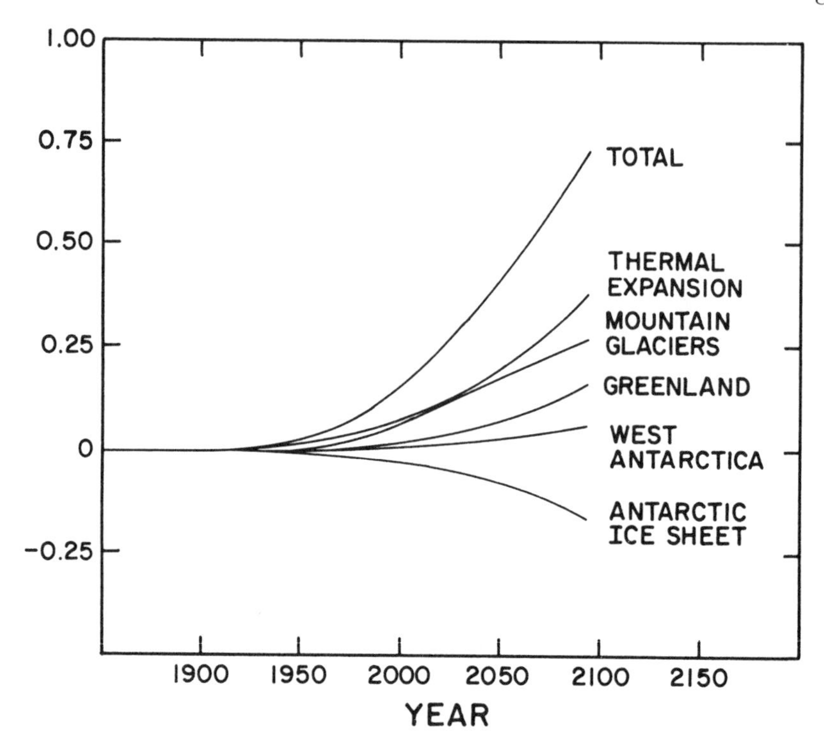

Figure 13. Projections of sea level rise in metres into the greenhouse century from various sources shown. The author suggests an uncertainty of at least 50% for each category. [Source: Oerlemans, J. A Projection of Future Sea Level, Climatic Change *(Vol. 15, pp. 151–174) 1989.]*

it to expand and thereby rise up the thermometer's walls. When the fluid of the oceans is heated, it will expand and rise up the walls of the "tube" formed by the coastal margins. That physical process is not debated, since expanding with temperature is a well-known property of sea water. However, sea water does not expand uniformly with the temperature rise; the amount of expansion depends on the average temperature of the water when heated by global warming. For example, a 100-metre-thick layer of sea water at a temperature of 25° C will expand approximately 3 centimetres per degree of warming, whereas a similar layer at a temperature of 0° C will expand only 0.5 centimetres. Therefore, if the tropical or subtropical oceans warm up first, there will be a much greater initial sea-level rise than if a comparable warming took place in the higher latitudes. To estimate how great a sea-level rise will occur from warming we have to construct regional scenarios of ocean-temperature rise, and as noted earlier, this is very difficult to do, particularly for the transient, time-evolving scenarios of greenhouse gases. Nevertheless, Tom Wigley and colleagues have attempted that kind of calculation, and their results are consistent with the line labelled "expansion" in Figure 13.[34]

Glaciers are the next factor. This does not refer to the massive ice sheets in Greenland or Antarctica, but rather to mountain glaciers in the Alps, Andes, Rockies, Himalayas, and other places. The most comprehensive study of the behaviour of mountain glaciers during the last one hundred years was made by Mark Meier, now at the University of Colorado.[35] He concluded that the bulk of glacier changes since 1850 was a long and steady retreat. He calculated that the retreat of glaciers that allowed their ice to melt and run into streams, lakes, and eventually into the ocean has raised sea level between 2 and 4 centimetres in the past hundred years. This accounts for 20% to 40% of the 10-centimetre-average sea-level rise that apparently has taken place around the world during the same time. The rest of that rise, it is usually presumed, is due to the half-degree Celsius or so of global warming that appears to have taken place during the twentieth century.

The next category on Oerlemans's graph is Greenland. Greenland is a glaciated island with a massive ice sheet that has permanent and increasing accumulations of snow on the upper and northerly portions of the ice cap and melting ice and ice flowing into the sea at the lower and southern portions. According to Oerlemans, with climatic warming, Greenland is likely to experience a decrease in ice volume, because the melting zone would move upward and northward. This would probably (no one knows for sure) cause greater losses of ice at the lower and southern margins than increases in snow accumulation at the higher northern margins. That is why it is shown as a net addition of sea level in Figure 13. Oerlemans concluded that although "the Greenland contribution to sea level change can be estimated as +0.5 millimetres per year per degree Celsius warming . . . a 50% certainty has to be accepted," because possible changes in snowfall on top are not considered, nor do his estimates reflect how iceberg formation rates might change. If entirely melted, the total volume of ice on Greenland could raise sea levels about 6 metres (20 feet).

Antarctica, a glaciated continent, has much more ice than Greenland. If all the Antarctic ice were to melt, sea level would rise about 70 metres (230 feet). One part of Antarctica that juts out into the ocean, West Antarctica, contains enough ice to raise sea levels some 5 to 8 metres; it is an ice sheet that is essentially separate from the bulk of the East Antarctic ice sheet. Most studies suggest that the West Antarctic ice sheet has the potential to disappear with several degrees climate warming, because its base is far below sea level and it might slide into the ocean. But recent estimates suggest that this would take many hundreds of years. At one point in the previous interglacial period, some 125,000 years ago, evidence from fossil beaches suggests that sea levels were about 5 metres (16 feet) higher than they are in this interglacial period. Some people have speculated that the West Antarctic ice sheet had disintegrated, since that interglacial time was slightly warmer than the current one, but that still remains uncertain. Currently, there is no appreciable melting of ice on the East Antarctic ice sheet; it is assumed that the East Antarctic ice sheet appears not to have changed its size much for many thousands of years because the ice that breaks off as icebergs into the southern oceans as the glacier flows to the edge of the continent approximately equals the amount of snow that accumulates on top of this very cold continent. In fact, almost all of Antarctica is so cold that even if temperatures were to increase by 5° or as much as 10° C, they would still be below freezing for the entire year. Therefore, the likely effect of warming on Antarctica would be to increase the amount of snow that falls on top without any additional melting. However, it is so cold there that not much snow falls (only about 15 cm—6 inches—on average), and even if the ice sheet were to increase its size and reduce sea level as shown in Figure 13, for this to have any substantial impact on sea level would probably take many centuries.[36] It is not yet known to what extent the Antarctic ice sheet is out of balance: whether it is actually growing or shrinking today. Field data and measurements of changes in ice-shelf thickness are so scarce or difficult that currently it can only be estimated that an imbalance might be as large as plus or minus 20%. Such an imbalance would correspond to a current sea-level change of about 1.2 millimetres per year. This level of uncertainty is quite high, since such a figure is large compared to the annual rate of sea-level change over the past century. Despite the uncertainty, it seems a fair bet that East Antarctica is not going to cause a major surprise in sea level over the next hundred years or so. Such surprises are more likely to be reserved for West Antarctica, which is a special circumstance.

The West Antarctic ice sheet has received intense scrutiny from scientists concerned about the greenhouse century. This ice sheet does not end at the ocean's edge, but rather has half-kilometre-thick ice shelves extending far out into the ocean. Although these shelves are floating, and if they were to melt would thus cause no further increases in sea level, they are pinned at a number of places against islands and shallow parts of the seabed. Thus, they physically restrain the flow of the ice on land, slowing its movement toward the ocean. About ten years ago, a major concern arose that a warming of the climate on the order of 5° C would be enough to raise summer temperatures to the above-freezing level in this fairly low-elevation part of Antarctica. While that would not immediately melt the ice shelves, it could cause them to break up. Such a breakup would not raise sea level any more than would the melting ice cubes in your lemonade cause a full glass of lemonade to spill over the top—a principle that Archimedes once discovered. Once the ice is in the water, it has already caused as much rise in water level as it's going to. However, if the shelf broke up, it would no longer physically restrain the ice on land from flowing more rapidly out to sea. The controversy was whether a breakup in that ice shelf could cause a rapid "surge" of the ice on land in West Antarctica, perhaps raising sea level by 5 to 8 metres (16 to 24 feet) in as little as ten to one hundred years. Today, most glaciologists would stretch that number out to five hundred years, but hardly with certainty. The number could be more like a thousand or perhaps less, given our present state of ignorance. Computer modelling of the West Antarctic ice sheet has been one method used to estimate the likelihood and timing of its breakup. William Budd, of Melbourne University, has been a leading researcher in this area and the West Antarctic contribution to sea-level rise shown in Figure 13 is based in part on his and other scientists' attempts to understand the dynamics of this very interesting ice sheet.[37]

In summary, sea-level rise seems the most probable and perhaps the most globally uniform consequence of warming projected into the next century. However, local meteorological conditions,

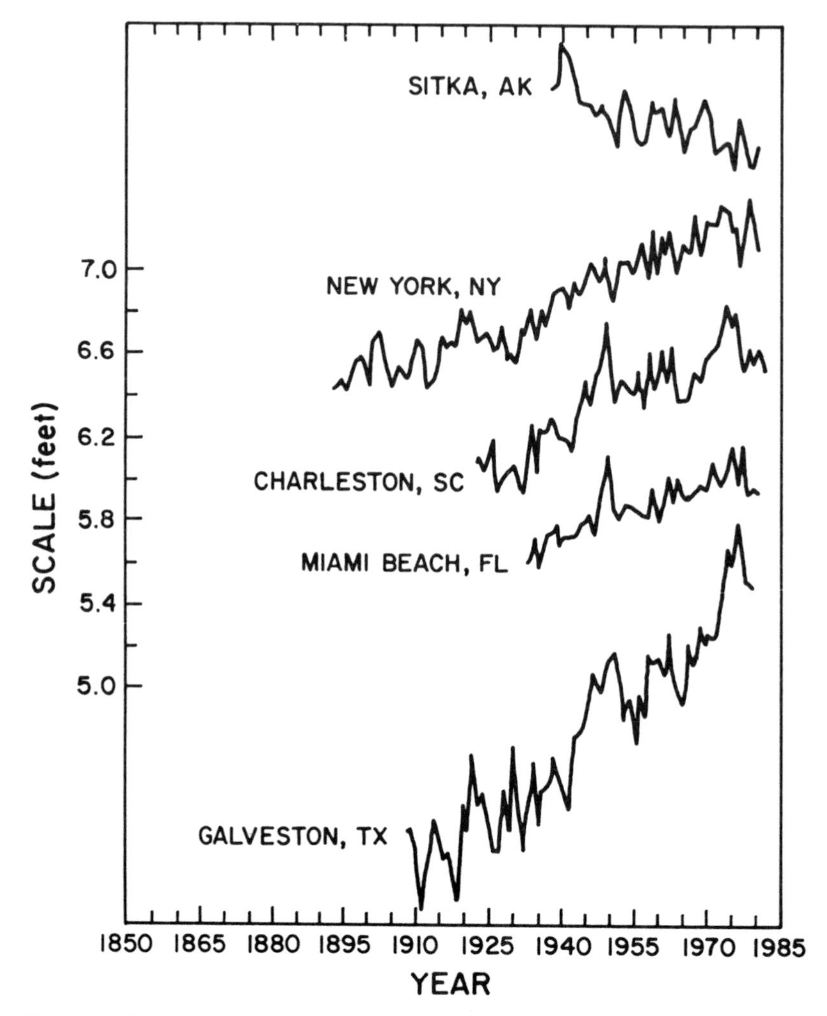

Figure 14. Changes in sea level at the locations shown. Note that local factors (such as land subsidence in Galveston, coastline sinking at New York, or coastline rising at Sitka, Alaska) cause individual locations to have substantially different rates of sea level change. Global average sea level rise over the past century is typically estimated to be about 10 centimetres (4 inches). [Source: Lyle et al., Sea Level Variations in the United States, *National Ocean Service, Rockville, Md., 1987]*

changes in ocean currents, passages of tropical cyclones, and local geological processes mean that any given location may not experience precisely the same rate of sea-level rise felt in the world average. Figure 14 shows how local sea-level changes have varied from place to place in the United States. Furthermore, very different vulnerabilities of coastal ecosystems and settlements as well as different abilities of nations or regions to invest in coastal-area protection imply that few general statements can be made other than that we need to consider immediately the implications of sea-level rise and the changing probability of storm surges for future coastal land use. It is already past time for coastal–land-use policy making that considers the high probability of the sea level continuing to increase into the next century. In my value system we simply cannot allow uncertainty over the details of global warming to prevent strategic planning that would reduce coastline vulnerability to the better-than-even chance that increased flooding will occur into the greenhouse century.

AGRICULTURE. It is hard to imagine any part of the planetary life-support system more important for our survival or more dependent on climate than agriculture. Therefore, the area of human affairs that has been studied most in the context of climate change is the potential impact of increasing carbon dioxide and other trace greenhouse gases on agricultural production. For example, English geographer Martin Parry spent many years leading a project that examined the potential impact of global climate changes on agriculture.[38] His assessment team assumed that the future climate would change according to various general-circulation-model projections. Team members recognized that regionally specific forecasts could not be trusted in detail, but nonetheless they provide at least a measure against which one could ask, What if the climate changes for agriculture? For example, if the GISS $2 \times CO_2$ climate is applied to a world map, then present locations "move" agriculturally. The climate of Finland, for example, is estimated to become similar to that of present-day northern Germany, whereas southern Saskatchewan becomes comparable to northern Nebraska. The Leningrad region becomes analogous to the western Ukraine, the climate near the central Ural mountains in the Soviet Union becomes analogous to central Norway,

Hokkaido in northern Japan becomes comparable to the more moderate climate of northern Honshu Island, and Iceland gets the climate of northeast Scotland. The Iceland/Scotland analogy, for example, might imply that hay yields in Iceland would increase by about 50% to match levels typical of northeast Scotland today. However, Parry and his colleagues have pointed out that losses to pests and diseases, which are at present limited by cold winters in Iceland, might decrease yields 10% or 15%, as in Scotland today. If crop yields were to go up as much as 50% from the improved climate, the 10% to 15% losses from diseases would have to be subtracted. Of course, an analogy does not yield a perfect climate match. Therefore, this kind of map analysis at best produces a simple scenario. What is completely clear from this discussion is that we can no longer expect agricultural practices that are currently attuned to expected climatic conditions to be maintained indefinitely. Rather, they will of necessity shift as the climate shifts.

A number of agronomists and economists have opined that of all the sectors of the global economy that depend on climate agriculture is the least likely to be vulnerable to greenhouse changes. To support this optimism, they cite the fact that crop yield (that is, production per unit of land area) has risen sharply in the last thirty years, not because of changes in climate but because of infusions of agricultural technology and capital investment. In particular, the development of new strains or varieties of crops, better able than traditional ones to take advantage of fertilizers, irrigation, pesticides, and other technological additions, has led to startling improvements in yields. A number of years ago, German philosopher Klaus Meyer-Abich, who ironically is now an environmental parliamentarian in West Germany in the Green Party, commented that he felt that the rate of projected climate change was slow enough relative to the rate at which agriculture changes to be "chalk on a white wall."[39] Indeed, in 1983 Paul Waggoner, of the Connecticut Agricultural Station, prepared a report on agricultural changes in the United States that concluded that even though a small warming and drying of the U.S. plains would indeed reduce crop yields by several percent, perhaps as much as a few tens of percent, this would be a small change over a fifty-year period relative to the expected yield increases from agricultural technology.[40] Therefore, Waggoner suggested, there was no strategic threat to the United States' capacity to feed itself from virtually any conceivable greenhouse climate-change scenario.

I objected to this conclusion at the time, not because I think there is a probable threat to our capacity to feed ourselves, but simply because the question involves not only strategic security but also economic equity. In other words, suppose crop yields were to double in the next fifty years from applications of technology—not at all clear, given the environmental and cost constraints that may unfold. Suppose also that we accept the estimate that climate change will only reduce yields by, say, 10%. The United States would indeed still be able to feed itself, and presumably would still have a surplus of grains, but it could lose billions of dollars in terms of increased production costs and decreased return on investment to farmers, since they would have to maintain the same investments in the face of lower yields. The relevant questions, then, are, How much loss will result from climate change? and Does that amount justify actions either to slow down the climate changes or to modify farming practices in advance to minimize losses? Unfortunately, analyses of this kind require a confident description of future regional climate changes, and as I have said many times in these pages, that is the most difficult part of predicting from climatic models. Therefore, the only way even to make estimates is to follow the procedure that Parry and others have used: taking scenarios from general-circulation models and asking on a crop-by-crop and region-by-region basis, What if? Let's look at some of the results.

Parry, together with geographer William Easterling, economist Pierre Crosson, and agricultural meteorologist Norman Rosenberg, examined what the GISS 2 × CO_2 scenario might mean in a variety of regions.[41] They concluded that there would be a substantial lengthening of the crop-growing season in middle and high latitudes. At first blush, that would seem to suggest increasing yields, particularly in Canada, the Soviet Union, and other areas where late spring and early autumn frosts are limiting. However, they pointed out, crops that mature early would become heat stressed by the altered conditions. In other words, farmers would no longer be able to grow spring wheat (planted in May or June and harvested in the autumn) in some areas of Canada or on the northern U.S. plains, for example, because heat stress would reduce the yields substantially. Instead, the farmers would have to plant winter wheat, as is now done in the hotter parts of the U.S. wheat belts in Oklahoma, Colorado, and Texas, and parts of southern Asia and the Soviet Union. Under the GISS 2 × CO_2 scenario, rice, for example, might increase its yield by a few percent, with a 35% increase in the number of growing-degree days (a measure of temperature over the growing season). If farmers were certain that such climatic changes would take place, then late-maturing varieties that presently cannot be grown in cooler regions for fear of late-season cold damage could be grown farther north and perhaps increase yields by 25%. Of course, this assumes similar or greater amounts of precipitation or adequate available irrigation water, which may not at all be present in all regions. The Geophysical Fluid Dynamics Laboratory's GCM run for 2 × CO_2 predicts dramatic moisture stress and temperature rise throughout the U.S. Midwest, with a 50% reduction in summer soil moisture and temperature rises of 4 or more degrees Celsius throughout the corn belt and plains states. Such a scenario would reduce corn yields dramatically, just as the heat wave and drought in 1988 reduced corn yields by 40%. (For the first time in decades, U.S. grain production in 1988 dropped below consumption.) Of course, not all parts of the world would be so reduced at the same time, which really raises the primary point that Parry, Easterling, Crosson, and Rosenberg made in a recent article: that "comparative regional advantage" would change. They point out that "even if aggregate world capacity grows equally with demand, climate change likely will alter the configuration of world comparative advantage in ways that pose both international and domestic policy issues for midlatitude governments." What that essentially means is that areas that now have excellent climates for growing and therefore relatively low production costs are blessed with the capacity to produce food more cheaply than other regions and are therefore encouraged to sell it abroad. The U.S. corn belt is one such region today. Should a climate scenario that involves a heating and drying of that region come to pass, it would not eliminate the food-production capacity but would simply make such production so expensive that its cost effectiveness would be diminished relative to areas farther north. A redistribution of comparative advantage could have major international security and economic implications. The key question is how fast and far conditions change relative to normal societal evolution.

Precipitation changes and temperature changes are not the only factors that would affect crop yields. For example, it has long been known that the carbon dioxide in the atmosphere is a sort of fertilizer that permits the growth of green plants. In the growing season, sunlight is used by the chlorophyll in leaves to take CO_2 out of the air and convert it into hydrocarbon plant tissues. Therefore, it is logical to assume that an increase in CO_2 would increase the amount of carbon taken up by plants, thereby increasing the size of the plants. Laboratory experiments have confirmed that plants do grow much bigger and more quickly when CO_2 is increased. There is debate, however, as to whether it is the leaves, stems, and roots that get bigger or the grain or fruit, but nevertheless it is commonly assumed (based on laboratory experiments and some knowledge of plant physiology) that increased CO_2 fertilization will increase the yield of plants. Philippe Martin, at the National Center for Atmospheric Research, and Norman Rosenberg and Mary McKenney at Resources for the Future estimated that not only could extra plant tissues be manufactured by increased carbon dioxide, but that the water-use efficiency of a plant could also be improved.[42] The reason for this is based on the known physiology of leaves.

In the undersides of leaves are openings called stomates. They control the amount of water and other gases that the leaf takes in and lets go. If a leaf gets so hot it can no longer get supplied with water fast enough, the stomates close and the leaf wilts. However, when increased CO_2 is present, the stomates do not remain open as long but they are more water efficient. Therefore, in the process of building plant tissues, when the CO_2 content is higher the stomates stay open for *less* time but produce *more* photosynthetic material and—because the stomates are not open as long—require *less* transpiration of water. This set of events made it seem as if more CO_2 would improve plants' tolerance of drought stress, which led Rosenberg and his colleagues to suggest that CO_2 fertilization would have a potential benefit for drought stress. However, the same laboratory experiments that show plants growing more and bigger leaves in the presence of increased CO_2 also suggest a contrary factor: even though the stomates will be closed for a larger fraction of the day and each leaf will use water more efficiently, a plant that has more total leaves may actually use *more* total water than without the CO_2 increase. Therefore, it is not at all clear what would happen to yields if water-use efficiency and greater leaf area from enhanced CO_2 fertilization and the effects of increased heating and moisture stress were present simultaneously. All we can do now is try to guess at these factors, taking them one at a time or several together in limited laboratory experiments or using models that try to represent the physiological growth stages of plants. Unfortunately, such models are based upon studies of individual plants or individual leaves; their results must be extrapolated not only to the scale of a farm field but also to the scale of a crop-reporting district or even an entire state or agricultural region. The reason for the tremendous uncertainty associated with any forecast of the specific agricultural consequences of increases in CO_2, climate changes, and so on is thus obvious. And, on top of these factors, weeds and other forms of yield-reducing environmental disturbances also change with climate and CO_2 fertilization. These variables add yet other uncertainties that have not been factored in.

Despite all this daunting complexity, the Environmental Protection Agency commissioned a number of studies to get a handle on at least the direction and potential scale of change that could be expected.[43] For a variety of reasons, the EPA studies showed that the direct effect of CO_2 fertilization did not fully compensate for the changes in climate variables for many different kinds of crops in many different states. For example, corn yields were shown to decrease dramatically, by more than 50%, in North Carolina, Tennessee, Georgia, and Mississippi, and by more than 30% in Ohio, Iowa, Nebraska, and Kansas, and by 0% to 30% in New York, Pennsylvania, Wisconsin, Indiana, and Illinois. On the other hand, Texas, Oklahoma, Missouri, Michigan, and Minnesota were shown to have increases in corn yields by something like 0% to 30%. For soybeans, yield losses were between 25% and 50% in the south central states and between 0% and 25% throughout most of the Eastern Seaboard, though improvements of 25% to 50% were obtained in Minnesota.

An obvious consequence of crop-yield decreases for rain-fed agriculture would be an increased demand for irrigated acreage. Dean Peterson, of the Utah State University, conducted a study of irrigation and climate change for the AAAS water supply and climate project.[44] He concluded that to maintain crop production at current local levels in the face of a 3° C warming, irrigation would have to increase by about 15% at thirty-nine sites distributed all over the United States. If this warming were combined with a 10% decrease in precipitation, then the increase in irrigation needed would go up to 26%, but the need for irrigation would be only 7% more if precipitation increased by 10% when the temperature warmed. In one extreme case, Peterson concluded that in southern Illinois or central Virginia, if the 3° C temperature rise were also accompanied by a 25% decrease in precipitation (a scenario closer to the Geophysical Fluid Dynamics Laboratory GCM result for 2 × CO_2), then the irrigation requirements to maintain the crop yields at present values would become comparable to areas such as New Mexico and Utah today! "This would be true for a number of combinations, for example, plus 5° C and minus 15% precipitation, which are well within the range predicted by the global climate models. Under these conditions, irrigation of most crops would become necessary." Peterson concluded: "Although our scenarios show relatively mild consequences in the east, rather modest increases in severity could trigger dramatic changes in the future of rain-fed agriculture. This hazard is particularly true for precipitation because of greater uncertainty of regional distribution."

The difficulty of such increased irrigation burdens is explained by the following numbers. In the West, irrigating only 1% of the watershed area reduces stream flow on average about 18%, Peterson calculated. In the East, however, depletions are negligible. In the future, taking irrigation water from streams and reservoirs in the amounts that might be needed in the East will cause a tangible reduction in stream flow but still much less noticeable than in the already semiarid West. Peterson concluded with the warning that people should not be misled by the numbers. Annual runoff in the United States is quite large—withdrawals are only about 25% of that runoff while consumption is 25%—but water is not as abundant as it appears. The reason is that "water of adequate quality is no longer readily available when and where needed. Especially in the West, maintenance of present irrigated areas or expansion into new areas will require costly development of surface storage and conveyance."

Daniel Dudek, of the Environmental Defense Fund, pointed out further difficulties related to the need to move agriculture and expand irrigation. He argued that agricultural adjustments to meet changing climate conditions could interfere substantially with the pathways of migratory waterfowl. "These critical breeding areas lie directly in the path of expected agricultural adjustments to a changing climate. We have made huge private and public investments in insuring the future of these waterfowl resources, our most economically valuable biological resource," Dudek argued.[45] This is an example of the conflict among land and water uses, referred to earlier, that could be exacerbated by climate change.

The conflict between agriculture and ecology would be difficult for environmentalists, no doubt, since agriculture is such an economically critical industry in the United States. Not only does it provide food for the U.S. population, but it also generates more than $40 billion in export income, which prevents the nation's already disastrous $100 billion trade imbalance from being $40 billion more severe. The food industry also employs more than 20 million people (almost 10% of the population) if processing is included with on-the-farm jobs. Farm assets were nearly $800 billion in 1985, and food and fibre accounted for more than 17% of the gross national production. The 40% reduction in corn yields in the heat and drought in 1988 led to a drought relief bill that will cost the taxpayers approximately $4 billion. With that kind of economic importance and political clout, it seems likely that the agricultural industry will be kept viable by substantial investments in such countries as the United States, Canada, and Australia, even if climatic scenarios alter the comparative advantage of various producing regions.

Perhaps that is why developed-country geographers and economists Martin Parry and colleagues are so optimistic in their conclusions about the effect of CO_2 increase on agriculture. In a paper they prepared for a conference at Resources for the Future in 1988, they summarized the best research available in the following way: "Assuming that climate changes for the worse in the Midwest [meaning warmer and drier], it is possible to conclude from the above research that yields would decrease on average by 3–18%. How important such yield decreases might be largely depends on the yardstick that is used. If the yardstick is the ability of production to keep pace with world food demand, then there appears to be little reason for concern." However, they went on, "even if climate change reduces yields by double the largest of these figures (up to 36%), production capacity would still be adequate to meet demand. If our projections are correct, technological improvements in yielding ability alone will increase production sufficiently to offset increasing world demand and any climate induced yield decreases." Fortunately, they tempered this rather optimistic strategic view of food production with the following caveat: "However, if the yardstick is the economic health of the farmer and the regional economy, impacts may be significant."

The opinion of agricultural specialists that somehow changes in technology and the adaptability of farmers will occur at equal or faster rates than changes of climate is pervasive among developed-country specialists. It is interesting to contrast this view to that of three Indian researchers, S. K. Sinha, N. H. Rao, and M. S. Swaminathan (who was the director of the International Rice Research Institute in the Philippines and is the president of the International Union for the Conservation of Nature and Natural Resources). These very experienced agricultural scientists addressed the question of food security in a changing global climate for the Toronto Conference on the Changing Atmosphere in June 1988. They tried to put into perspective the differences in agricultural importance between developed and developing countries:

> In most developing countries, land and water based occupations consisting of crop husbandry, animal husbandry, fisheries and forestry are the major sources of employment and income in rural areas. In this context, agriculture assumes a more significant role in the development of national and global food and nutrition security systems than being just a source of food. Therefore, in predominantly agricultural countries, importing food will have the additional consequence of enhancing rural unemployment, when this is done to compensate for inadequate national attention to agricultural development. Thus, food security has to be viewed in the contexts of food production, job creation, and income generation. An additional issue of overriding importance, if we are to ensure that today's progress is not at the expense of tomorrow's prospects, is that of conservation of the ecological base for sustained agricultural production.[46]

They go on to point out that the biggest difference between developed and developing countries is the technological character of rich country agriculture, in particular fossil fuel subsidies in the form of tractors, fertilizers, irrigation, and pesticides. If the energy input of the entire agricultural system is accounted for—including land preparation, on-the-farm work, transport to processing plants, and sale in supermarkets with transport, refrigeration, and lighting included—then "more commercial energy was being spent than the harvested solar energy through crops. Thus, it would be virtually true to say that ultimately in developed countries, fossil fuels serve as food." In developing countries, the situation is quite the opposite on most farms, although some larger farms use energy-intensive modern methods. In developing countries, more than 50% of the population continues to work on farms. A Western-style agriculture would displace many of these people to city slums.

One of the greatest problems related to producing adequate food on a sustainable basis today is the destruction of the resource base through soil erosion, water logging of soils, and stress on the natural predators of insects and pests. As environmental pressures cause this resource to dwindle, the flexibility to adapt to changing climate may be reduced in developing countries, precisely those places with the fewest financial resources. One Canadian development expert noted that "small farmers are held responsible for environmental destruction as if they had a choice of resources to depend on for their livelihood, when they really don't. In the context of basic survival, today's needs tend to overshadow consideration for the environmental future. It is poverty that is responsible for the destruction of natural resources, not the poor."[47]

It is against this background that the Indian agriculturalists conclude, in rather stark contrast to the optimism of the Western analysts, that "projected population growth rates and the ensuing food demands, even in the current global climate, would make it difficult to provide for human sustenance and food security in the twenty-first century. Africa and South America are the most vulnerable regions from the point of view of food security. The food demands of these regions can be met by increasing the cultivated area. Asia is next in order of vulnerability." With regard to climate change, they caution that current calculations suggest that the production of rice would be relatively unaffected, whereas wheat may decline a little in the midlatitudes and corn would increase from expanded area and wetter subtropics. But, they caution, such "recent reports on the favourable impact of the climate change on agricultural production should not lead to complacency. The available evidence in support of this is inconclusive and is not based on the complex dynamics of the interactions between agricultural production processes and the environment." For example, while most general-circulation models predict that warmer and moister conditions in the tropics and longer frost-free periods in the midlatitudes could, by themselves, increase crop yields, such conditions are "highly conducive to crop pests. By far," the Indian specialists warn, "the most predictable effect of climate change is that it will cause significant increases in pest populations. Thus, some of the advantages of present temperate regions may be cancelled by increased temperature and uncertain precipitation."

To me, these divergent opinions and vastly differing forecasts of agricultural benefits and catastrophes suggest that, in agricultural terms, rapid climate change is a global gamble. It is true that, just as a water-supply manager in a specific district is hard pressed to know how to modify planning because of the uncertainties in regional climate projections, an agricultural official responsible for a crop district is hard pressed to plan for climate change except by looking for ways to make the system more flexible and adaptable. However, as with the water-supply case, it is doubtful that most farmers would welcome the gamble associated with the prospect of rapid climate changes of the magnitudes that present predictions suggest are plausible. Therefore, the relative inability to know how to react yet at the local level should not be misconstrued as a lack of need to act at national and international levels to slow down the rates of change and buy time. This strategy will allow scientists to improve their forecasts so that sounder planning can be made at the local level. Extra time will also allow the development of agricultural infrastructure, testing of new seeds, and improvement of irrigation systems. With these efforts made, humanity will be better able to adapt as climate change inevitably unfolds than if it came unknown and in a giant rush.

NATURAL ECOLOGICAL SYSTEMS. In the United States, about 65% of all the forested areas are productive commercial lands. Although they represent only about 10% of the world's forest area, they supply nearly a quarter of the world's wood products. Nearly 2 million people (nearly 1% of the U.S. population) are employed in timber industries, and the total value of timber products is nearly $50 billion for slightly under 5% of the nation's gross national product. While these numbers may seem staggeringly large, they are much smaller than the agricultural equivalents mentioned earlier. However, it is my belief that climate threats to forests and other natural ecosystems from a rapid buildup of greenhouse gases are a vastly more serious problem in the long term than the potential threats to agriculture. It is not the dollar value that motivates my concern, but comparative rates of adaption. The rates at which agriculture can adapt to change are much faster, as many economists correctly point out, than are those at which natural systems can respond.

For example, the following factors are all relevant to the growth and distribution of trees: temperature, precipitation, CO_2 concentration, light, nutrient availability, chemical environment, disturbances (for example, diseases, hurricanes, bushfires, or human land use), and the combination of these factors. In Chapter 3 I pointed out that trees "moved" thousands of kilometres in response to global temperature changes on the order of 5° C occurring over many thousands of years. At the rate of climate change of a degree or so per thousand years, forest species were able to move and "keep up" with the changing climate. However, as Thompson Webb and other members of an interdisciplinary project[48] discovered in their extensive analyses of fossil pollen in lakes around North America, intact forest and animal ecosystems did not simply move north. The species moved, but they changed their relative abundances and thus the habitats changed. (Perhaps this explains some animal extinctions that occurred as the ice age ended.)

In order to predict the future, it is necessary to build models of forest growth and productivity. We can validate these models by checking them against the actual forest changes that occurred from the last ice age to the present. The simplest of these models predicts only what kinds of trees would exist based upon the summer and winter temperatures and the annual amount of precipitation. Webb, Patrick Bartlein of the University of Oregon, and John Kutzbach from the University of Wisconsin used six species of plants, tracking their abundance every 3,000 years from

18,000 years ago to the present.[49] They then used the NCAR general-circulation model to simulate how the trees should have moved given that model's prediction of the climate in each of those 3,000 years. While the correspondences were not perfect, they were remarkably close, providing some satisfying degree of general validation for both the forests and the climate models.

Foresters have much more sophisticated models of forest growth than the simple temperature and precipitation method just mentioned. Daniel Botkin, of the University of California at Santa Barbara, developed such a model back in the 1970s. It has been extended by many investigators, such as Hank Shugart at the University of Virginia, Allen Solomon at Oak Ridge, Margaret Davis at the University of Minnesota, and John Pastor at the University of Minnesota at Duluth.[50] These investigators modified the model to improve it for predicting different forest species so it could be applied to various regions. Scenarios of climate change, usually from atmospheric-circulation models, are then fed into these forest models in order to ask our favourite question: What if the climate changes from the perspective of the trees?

The EPA collected a number of these studies with the following general conclusions. In the Great Lakes region, for example, significant forest decline and species change becomes evident within thirty to sixty years. In the U.S. Southeast, forest declines become evident slightly later in the moist parts, with declines occurring even sooner in the drier western portions. Existing forests would not shift intact, the EPA noted, but would change in composition. Tree species now in the South could well find a new home in northern states, hardwoods presently restricted to the northern U.S. and southern Canada could well encroach into the boreal forest area, and substantial shrinkage in tundra would take place with boreal species moving northward. This would, some Arctic ecologists fear, squeeze out some Arctic birds and animals that would be pushed literally into the Arctic Sea into extinction. Therefore, as with agriculture, a substantial redistribution of natural resources is implied by typical general-circulation model scenarios applied to forest-ecosystem models. But the situation is not that simple. As forest ecologists John Pastor and W. Post point out, on soils where moisture is easily retained and there is not a drastic decrease in precipitation, current mixed spruce, fir, and northern hardwood forests in Ontario and Minnesota could be replaced with more productive northern hardwoods. However, on soils that hold less water or where there is less overall rainfall, spruce, fir, and hardwood forests would be replaced "by a stunted pine-oak forest of much lower carbon storage." The modellers went on to say, though, that they did not include direct effects of CO_2 on the water efficiency or photosynthetic capacity of the trees, nor did they account for pests or other disturbances such as fire, which could become a substantial agent for ecological change during the transition.

Indeed, it is the transition that has most forest ecologists concerned. In these forest models, it is assumed that species can flourish when the climate conditions are appropriate for them. However, if there are no seeds in the neighbourhood, then they cannot grow. In other words, if climate changes rapidly enough, many species may simply go extinct because they are unable to migrate quickly enough to the new location appropriate for them in the greenhouse century. This possibility has raised a major philosophical debate in the biological conservation community. In the EPA report, for example, the question is asked, Should national forests be left to decline as a natural process, whereby we lose aesthetic values in parks, water yields from watersheds, and highly productive timber crops? Or should silvicultural forest techniques, such as thinning, weed control, and fertilization, be employed to "save them?" The EPA tried to estimate how much it would cost to plant species that might be going extinct because of an inability to move rapidly to a new location: "To keep pace with the rapid climatic changes projected, the U.S. reforestation effort conceivably would need to be doubled or tripled in size. In recent years, about 800,000 hectares (about 2 million acres) per year (approximately 700+ million seedlings) have been reforested in the United States. Costs range from $200 to $700 per hectare ($80 to $280 per acre), depending on species, site preparation, plantation density, and planting method. Using $500 per hectare ($200 per acre) as a [mean value], the total annual expenditure is near $400 million [about $2 per U.S. citizen]."[51] The EPA went on to point out that at that rate it would take about a hundred years to reforest 40% of the U.S. forest land, assuming no mistakes were made and that the trees planted would in fact grow as the climate evolved throughout the greenhouse century. Who should pay is a major question, of course, since this concerns landowners, forest users, consumers, and all taxpayers. That question, clearly, will spark a public policy debate that will require an informed electorate.

Since forests are habitats that preserve the biological diversity of the planet, the EPA was suggesting what was later called the practice of "restoration ecology," by no means an uncontroversial doctrine among conservation biologists. In October 1988, the World Wildlife Fund and the Conservation Foundation put on a meeting run by Rob Peters, of the foundation, and Tom Lovejoy, of the Smithsonian. Peters was one of the first people to raise the question of how climate change might affect the preservation of biological diversity, since the reserves set aside to protect the diversity of species on the planet could be substantially upset if the climate changed rapidly, forcing species out of the reserves into surrounding lands that were no longer natural.[52] At the October 1988 meeting, biologist Jerry Franklin, of the University of Washington, pointed out that if we simply allow rapid human disturbance to the environment in the form of global climate change to render species extinct rather than transplanting them to new habitats in an effort to save them, we will be partly responsible for the extinctions ourselves. "Those of us who are into 'the natural,'" said Franklin, "are not going to like what we will have to do, which is damage control to minimize the amount of loss. We will have to become ecological engineers, managing natural areas."[53] That touched off quite a debate from the floor, with Berkeley physicist-turned-biologist John Harte, for example, pointing out that planting species in new areas that we suspect might preserve them could indeed save them, but at what expense to the inhabitants of the engineered area? Indeed, the more rapidly the climate changes the more intensely will these terrible ethical and moral dilemmas within ecology be exacerbated.

Let's analyse a bit further the relationship between biological diversity and climate change in natural systems. The diversity of living things on the planet is clearly a legacy that we must preserve to the extent possible. This I consider a simple moral imperative, and one that questions our right to drive other life forms to the irreversible state of extinction. But morality aside, there are practical reasons to keep species around for future potential human uses (exotic plants could hold the key to cures for cancer, for example).[54] The bulk of these species live in forests, in particular, tropical forests.

In 1967, Robert MacArthur and Edward O. Wilson developed an important theory known as island biogeography.[55] This suggests that the number of species on an island is determined by the immigration of species from the mainland and the extinction of species on the island. First, the theory predicts that larger islands will have more species and larger populations than smaller islands. Chance disturbances such as droughts, fires, floods, or landslides are therefore less likely to exterminate populations on large islands than on smaller ones. Second, islands close to a source of potential immigrants would have a greater rate of immigration and a closer distribution of species than more distant islands. The point here is how this theory is applied today to the practical problem of conserving biological diversity. For example, tropical forests are being cut and burned at the alarming rate of one Tennessee per year. At a recent meeting, Tennessee Senator Albert Gore, Jr., said that while some may not be alarmed by this fact, if we were invaded from space by aliens with football-field-sized feet that were so hot that they burned the forest every time they touched it and they kept stomping at a rate of one footprint a second, that would undoubtedly get people's attention. That, he asserted, is the rate at which Brazilians are currently cutting down their own forests. The relevance of this to island biogeography is that conservationists have convinced national governments, such as the Brazilians, to maintain reserves for the protection of the genetic diversity. But, as Rob Peters and Joan Darling argued a number of years ago, if the climate changes, then a park of a given size maintained to protect a certain number of species will no longer be able to do its job. That is, if the temperature and rainfall patterns change, then species preserved in the park will attempt to migrate—assuming that the rate of change isn't so fast that they can't make it. Let's assume for the moment that they can migrate. Where would they migrate to if the park is surrounded by deforested land occupied by cows and crops? In other words, the effective size of the park or refugia will effectively be substantially shrunk by climate change. The conferees at the October 1988 meeting concluded that more and larger parks would be needed in order to deal with the added stress of rapid human-induced climate changes. In addition, migration corridors—so-called greenways—among parks seem essential to protect these ecological islands from being artificially shrunk by rapidly changing climate.

Paul Ehrlich and population biology colleagues from Stanford University addressed this topic recently in an article in *Natural History* discussing the island biogeography of the U.S. Great Basin:

> Since the Earth as a whole is being rapidly converted into a system of habitat islands surrounded by a sea of human disturbance, the Great Basin can be viewed as a model for the global conservation of biological diversity. As island biogeographic theory predicts, the process of fragmentation with the associated shrinking habitat, increases the isolation of those areas and raises the extinction rate while lowering the immigration rate. Thus, the faunas of the Basin islands equilibrate with fewer species. Precisely the same thing is happening throughout the entire planet.[56]

Rapid climate change can also threaten habitat change through a disturbance such as fire. During the transition to a climate that would force a redistribution of forest species, trees that were no longer suitable for the new climate might begin to die. If the climate change included a summer that was substantially drier and hotter than normal, as climatic models suggest is quite plausible in the United States, then the dying trees subjected to that kind of heat stress would be much more vulnerable to fire. In other words, the combination of old, dying trees and climate change, which increases proneness to wildfire, could serve as the agent for a rapid transition to new sets of evolving habitats. Such combinations of factors, while highly likely to occur in reality in hard-to-predict ways, are not accounted for in most forest models, except as random terms that would only resemble reality by luck. The message for the forests of the future, with their recreational, aesthetic, and economic importance, is very similar to that for managed agricultural systems: the faster the climate changes, the more difficult it will be to predict the consequences and the greater are the margins of safety we will need to build into reserve systems and other measures of protecting these resources now.

HUMAN HEALTH. One of the least studied and potentially most interesting impacts of rapidly changing climate are those on human health. Human health could be influenced by weather and climate changes in a number of ways. First, the incidence of cardiovascular and respiratory disease is already highly correlated with extreme climatic conditions. For example, influenza tends to occur in the wintertime, and heat stress and diseases borne by mosquitoes or ticks, which are suppressed in the winter, tend to strike in the summer. Therefore, a change in temperature or moisture conditions could change both the direct physiological impacts on people and disease factors such as insects.

An EPA-commissioned study by L. S. Kalkstein and colleagues, from the University of Delaware, suggests that total summertime heat-related mortality in 15 United States cities would grow from a current estimate of nearly 1,200 deaths to almost 7,500 deaths in a CO_2-doubled world.[57] The elderly, aged 65 and over, represent about 70% of each of these figures. Since the current percentage of elderly in the U.S. population is increasing, the number of mortalities estimated in the future could be larger than this, because Kalkstein assumed that the age structure of the population would remain consistent over time. But he encountered an interesting phenomenon known as acclimatization and related it to the concept of the threshold temperature, the maximum temperature above which sharp increases in mortality occur from summer heat waves, as hospital records suggest. This threshold temperature is much higher in southern cities such as New Orleans or Washington than in more northerly cities such as New York or relatively cool places such as San Francisco. In other words, in New York City temperatures would have to rise consistently into the high nineties (Fahrenheit) to result in a clear increase in mortality, whereas in Washington or New Orleans they would have to rise to more than 100° (38° C) to produce the same result, and in San Francisco they would only need to be sustained above the mid eighties to create a substantial increase in death rate. Thus, in order to predict the number of excess deaths related to the intensified heat waves we might expect with global warming, it would be necessary to decide whether the population were or were not acclimatized. This factor, in turn, could depend on how extensive were the air-conditioning and other artificial climate-control mechanisms in operation. The greater the number of climate-controlled hours a person spends in a cooler-than-outside environment, the less likely he or she is to be fully acclimatized.

These interrelationships raise the possibility that in an extended super heat wave sometime in the greenhouse century that also caused a simultaneous brownout, reducing electric power and cutting air-conditioning, the population might be more vulnerable than it would have been without air-conditioning and therefore better acclimatized. Such a possibility makes estimates of future deaths from heat stress more speculative, but nonetheless Kalkstein was able to estimate that the number of deaths per season in New York City would increase from about 320 per summer currently to more than 1,700 without acclimatization but only 23 might die with full acclimatization. To extend the figures, Kalkstein estimated, on the assumption of no acclimatization, that the 1,150 deaths in 15 U.S. cities currently attributable each summer to physiological distress from heat waves would increase to about 7,500 in a CO_2-doubled world. With acclimatization, his $2 \times CO_2$ figure is substantially smaller—around 2,200—but still nearly double the present rate. Kalkstein also estimated the number of winter-related deaths currently at about 243 and suggested that these would drop by half or less, depending on the degree of acclimatization, because of the reduction in winter cold stress that would be expected with CO_2 increases.

The respiratory distress, for example, that is related to temperature is also related to levels of air pollution. One well-known observation is that temperature increase generally correlates with the amount of ozone created in the lower atmosphere around cities. For example, the EPA estimates that a 4° C temperature increase in the San Francisco Bay Area would increase ozone concentrations about 20%. The summer of 1988 provides direct evidence that weather can create pollution episodes in the United States. There has been progress in reducing emissions over the last decade, but the extended stagnation periods plus high temperatures of this intense summer caused ozone levels in seventy-six cities across the United States to exceed national standards by at least 25%. Of course, the extent to which the 1988 summer is an appropriate analogue for the future is not clear, but it is well known that heat waves are correlated with pollution buildups, which then work synergistically with heat stress to create physiological distress in humans.

One of the more interesting potential categories of impact

would be changes in diseases brought by insects. Sleeping sickness in Africa currently "protects" vast tracts of land from human habitation because humans who live in these arid regions generally bring herds of animals, which are vulnerable to the disease. Andrew Dobson, an ecologist from the University of Rochester, studied the way various diseases are restricted in their ranges by temperature or other meteorological phenomena. He showed that it is possible that increased temperatures in central Africa would actually spread encephalitis-bearing flies out of their current range into areas currently free of the disease, while simultaneously freeing relatively ecologically fragile lands for human habitation—an ecological concern.[58] Another potential interaction can be seen in the need of mosquitoes, which can bear diseases of all sorts, for standing water to breed. If increased temperatures or decreased moisture availability enhances the need for irrigation in many regions, this may result in greater amounts of standing water in fields, potentially increasing local mosquito populations. Such speculations are clearly hard to back up with facts, but they are nonetheless examples of the kinds of surprises that may yet unfold if climate changes rapidly.

Increasing use of chlorofluorocarbons, assuming they are not entirely banned by international agreement in the next few years, is very likely to continue to decrease the amount of stratospheric ozone, thereby increasing the ultraviolet light reaching the earth's surface and increasing the incidence of skin cancer in humans. There may be an interaction between climate change and increased UV on human health in the following sense. If warmer climates lengthen the recreation season, during which people are more scantily clad or spend extra time at the beach, then it is possible that increasing temperature could work with decreasing ozone to lead to substantial increases in skin cancer based on increased ultraviolet exposure, especially for those light-skinned members of the population who do not protect themselves properly with sunscreens.

Finally, plant diseases moving out of their currently restricted zones owing to winter chill could be an indirect potential health threat from climate change associated with health factors. Crop diseases going further poleward into croplands could reduce yields. If such yield reductions were to occur in those countries where nutritional stress was already serious, populations could become more vulnerable to severe morbidity or mortality from normally preventable diseases. It is through this circuitous pathway that nonhuman disease factors that affect crops or livestock could lead to human health degradation.

More than almost any category discussed so far, the impacts of rapidly changing climate on human health are clearly poorly predicted. Economists such as Tom Schelling of Harvard have pointed out that people get off aeroplanes in Los Angeles in December having just left from New York—which involves a "climate change" substantially greater than New Yorkers at home would experience from the greenhouse effect over the next fifty years. Just how adaptable and acclimatized we are is still a matter of debate, but the statistical studies and some of the physiological information available suggest that this relatively neglected area of research certainly needs more careful and serious attention in the future.

URBAN INFRASTRUCTURE. The next category of impact assessment deals with how urban development over the next thirty to fifty years might be influenced by changes in climate. Earlier I noted that coastal cities could be most severely impacted by sea-level rise, which could affect these cities' water supplies and demands as well as the need to substantially alter expensive infrastructure such as storm drainage. The dredging of lakes and elevating of port facilities on coastlines are other examples of expenses mentioned earlier. But different aspects of urban infrastructure also are important. For example, U.S. urban drainage systems are currently worth more than $60 billion, streets almost $500 billion, mass transit more than $30 billion, and private electric power utilities more than $250 billion. All these urban infrastructural elements could be affected by climate change. The heat wave in 1988 suggested some examples. According to the EPA,[59] 100° F weather distorted railroad tracks, forcing Amtrak to cut speeds from 200 to 128 km per hour (125 to 80 mph) between Washington and Philadelphia. Further, these temperatures might have been factors in a train crash in which 160 people were hurt in Chicago on a Chicago/Seattle run. In Manhattan, heat amplified the effects of long-standing leaks in more than 250 km (150 mi) of steam pipes, which caused asphalt to soften. As cars and trucks passed over the soft asphalt, thousands of bumps formed on city streets requiring extensive repairs by the city, to say nothing of the damage done to private and public vehicles. In Washington, D.C., expansion joints along a 21-km (13-mi) section of Interstate Highway 66 bubbled in the heat wave.

The EPA looked at three cities in the United States with regard to potential impacts of climate change on their infrastructure. For Miami, they suggested that global climate change could cause as much as half a billion dollars in needed capital investments in the next century, primarily owing to a sea-level rise. Many bridges would have to be raised to ensure adequate clearances and to reduce vulnerability to storm surge in a hurricane. A rising underground water table could become quite serious and might require substantial construction of canals and levees. The effects would be exacerbated if population pressure continued to push Miami into further investment needs. Cleveland, on the other hand, was judged not necessarily vulnerable to negative impacts related to climate change. The EPA predicted that heating costs would be reduced by millions of dollars in the wintertime, although it saw air-conditioning as probably raising costs two to three times more than heating would reduce them. However, the city could save somewhere under $5 million in snow and ice control, $500,000 in reduced frost damage to roads, and presumably another $500,000 or so with respect to road construction and maintenance. All in all, given the range of uncertainties, the EPA estimated the annual capital risk Cleveland might have to face in dealing with CO_2 doubling would be somewhere around plus or minus $1.5 million. New York City, on the other hand, would be more likely to experience negative impacts from climate change, primarily as a result of sea-level rise. Climate change could require an investment of more than a quarter of a billion dollars to improve the reliability of water supplies in order to cope with potential heat waves. Also, increased air-conditioning use could raise peak demand 10% to 20% in New York City, and perhaps more than that in the southeastern parts of the country. This could raise electricity demand by a large enough margin to require that utilities consider building more than a quarter of a trillion-dollars worth of infrastructure ($1,000 per person in the United States) to deliver the needed energy. Of course, this estimate assumes no substantial improvements in energy efficiency, which could easily offset some or even all the additional energy needed to meet air-conditioning demands. Of course, if new fossil-fuel-driven power plants were built, they would just feed back on themselves by adding more greenhouse gases, thereby further exacerbating the problem. New solar or nuclear electric power plants would not contribute further to the greenhouse effect, except to the extent that fossil energy was used to build or decommission them.

One potential problem related to global warming is a change in the seasonal energy demand, probably reducing the winter needs for heating and substantially increasing summer needs for air-conditioning. This suggests that additional capital investment to give plants peak demand capacity part of the year would yield a poor return on investment, since the plants would be running at near full capacity for only a small part of the year. One way to mitigate this problem would be to increase power sharing across a grid around the United States, an option that obviously needs further study.

Increased electrical demand at times of intense heat waves would mean increased local air pollution, which is often exacerbated at these warm times as well by the meteorological conditions. This synergism could create substantial health problems that might necessitate forced power cuts for air pollution control reasons. Again, these are speculative possibilities, but they suggest the sorts of future uncertainties we may be facing as the climate changes.

CARBON DIOXIDE EMISSION OFFSETS. A number of people have suggested that one way to deal with the carbon dioxide problem is to offset any and every injection of CO_2 into the atmosphere. For example, one New England utility has agreed to plant trees in tropical areas in direct proportion to the amount of CO_2 that plant injects. Thus, if a typical tree will soak up about a ton of carbon dioxide in twenty or thirty years, then it is possible to calculate the area to be planted that would offset the current 5 billion tons of carbon in the form of carbon dioxide the world injects annually and from there the possible 10 billion tons that might be injected each year in fifty years. Economist Roger Sedjo and forester Allen Solomon calculated that reducing the current rate of tropical deforestation plus substantially increasing reforestation by planting fast-growing trees could indeed provide an offset that would act as a global carbon sink.[60] They calculated that sequestering approximately half the 5.5 billion tons of carbon currently injected as CO_2 annually would require an area that is 50% larger than the total closed forested area of the United States, or about 10% of the world's total forest area. Some people have suggested planting trees over an area the size of Australia as a CO_2 offset, but others have argued that this action would not be sufficient to do the job, since the trees would only temporarily soak up about half the CO_2 that is currently injected. Of course, in a generation or so, we would have to do something with those trees; burning them would only put that CO_2 back in the air. However, one should never say *only* with regard to an issue of delay, for delaying the greenhouse gas buildup and stopping "only" half the buildup of one greenhouse gas would by no means be a trivial achievement. After all, it is the rate of buildup that poses major threats and potential surprises. The initial costs of such a tree plantation project, Sedjo and Solomon estimate, would be about $250 billion ($1,000 per person in the United States), perhaps much more. However, the forests later on could be harvested for economic return, so exactly how prohibitive these costs are requires additional thought.

Other ways to offset the CO_2 injected by any process would not necessarily require the development of new lands; rather, they could entail further intensification of existing land. For example, faster growing species could be planted in the normal harvest cycle of forests that would take CO_2 out more quickly. Also, deforested lands might not be burned; rather, they might be allowed to keep the carbon in their wood for a long time. The wood could be converted to timber or some other more long-lived form rather than being burned or allowed to rot, as is done in most tropical areas, where it then gives its CO_2 back in a matter of a decade. One fanciful suggestion is to dispose of chopped-down trees in abandoned mine shafts, where decomposition would be slowed and the return of their CO_2 back into the air retarded. This principle could be realized, of course, in above-ground storage as well, with large piles of logs covered in plastic or kept in sheds. This would make them recoverable for further economic uses somewhere down the line, when it is to be hoped we would have a better understanding of how to slow down greenhouse-gas buildup so that we could use the trees. We should not dismiss any solution too quickly just because it is small or seemingly wild, but clearly any suggestion of this type will require much careful economic and environmental consideration before the political establishments around the world will give it serious support.

If this chapter has revealed any one answer to the general question, What if the climate changes?, it is that *we insult the environment more rapidly than we can understand the consequences of our actions.* To be sure, political scientists and economists are correct when they say that on a regional basis positive changes accompany negative ones. However, the concept of "winners" and "losers" is highly questionable in this context, partly because of the great uncertainty, but also because any local region that might experience climate improvement with respect to one activity, such as corn growing or tree harvesting, could well find itself a net loser if the overall national economic condition were degraded or if national borders were pressed by immigrating neighbors whose own climate was deteriorating. We can estimate to a rough order of magnitude and direction the kinds of costs or benefits that could be associated with specific scenarios of climate change. We know these scenarios are plausible, but at present it is very difficult to assign a probability to any of them. Therefore, the best we can do is assess the magnitude of the problem, recognize the size of the potential issues, and then decide whether or not to risk those kinds of alternative futures without considering some investments to slow down the rate of change.

The tough question is whether the rates of climate change and the natural responses to the change will be faster than society's own rate of change. That is, will we be rebuilding our cities anyway at a rate that outstrips a sea-level rise? Will our developing technology enhance agriculture at a rate as fast or faster than that of a climate change? These issues lead to substantial debate among economists and environmentalists over whether adaptation or prevention should be our primary policy response. However, with natural systems such as forests, where it often takes hundreds of years for a stable ecological condition to be reached, these concepts of rapid human economic change have no meaning. Moreover, although I personally believe that the adaptability of human society is very great, that greatness is primarily concentrated in the developed countries, which simply have more resources than the developing countries. Furthermore, all countries will face greater risks and uncertainties as climate changes accelerate, for the faster the changes unfold, the less ability we will have to forecast the details and the greater will be the likelihood that we will simply not know how to adjust. Furthermore, what we do in the transition phase, in which we implement whatever adaptive actions we choose, could very well differ substantially from the actions needed over the long term. After all, matching our planting or water-supply strategies to a moving climatic target will be a lot harder than deciding how to respond with confidence to changes we can anticipate accurately. Communicating this incredibly complex issue to a public that must send the policy signals to its leaders is not a trivial task, as the next chapter will show.

7

Mediarology

Was the "Summer of '88" a real example of climate change in the making or simply a media event? I am constantly asked that question in some form by journalists, business leaders, politicians, and people I happen to sit next to on aeroplanes, and I heard it asked of others literally hundreds of times since the summer of 1988. The daily newspaper and broadcast stories on the heat, drought, forest fires, and super hurricane Gilbert were overwhelming factors in bringing climate-change issues to the public consciousness. The term *greenhouse effect,* which had been so difficult to squeeze into the public vocabulary over the preceding fifteen years, had entered the media jargon even before 50% of Yellowstone National Park was blackened by tremendous uncontrollable fires in late summer. How did a relatively obscure scientific concept, such as the trapping of thermal infrared radiant energy in the lower atmosphere—that is, the greenhouse effect—move from the professionals to the press so rapidly?

The rise in public consciousness engendered by the Summer of '88 actually began in the early seventies, when stories on human impact on the global climate first started appearing. Of course, much of the talk then was about cooling from human-generated dust, but the greenhouse effect still got substantial play. Neither theory, though, gathered much serious public attention. In 1976 I testified to Congress in support of the creation of a National

Climate Program that was to consolidate fragmented government-agency efforts into a coherent national programme.[1] Part of my testimony dealt with the potential of global warming from increasing carbon dioxide. "Are you saying," asked one congressman rhetorically, "that there might be some global environmental limits to the use of our nearly unlimited supplies of cheap coal?" While I acknowledged the alacrity with which he picked up this subtlety, privately I was stunned to realize that people of his stature had not heard of the greenhouse effect. And even up to 1988, this was still possible!

Since 1976, testimony on the issue has taken place before Congress, in parliamentary proceedings in Europe or Australia, at the World Climate Conference in 1979 (sponsored by the United Nations agency the World Meteorological Organization), and at numerous national and international meetings. And, gratifyingly, media coverage of the issue, at least that by the well-established science writers, has become increasingly sophisticated. Of course, even these journalists had to make some compromises to get the next century's news into print today. Many, for example, needed to call me or other scientists in the field on the heels of an unusual heat wave, a drought, a hurricane, even a cold wave, or some other "weather peg" on which they could hang the greenhouse issue. After a hurricane, for instance, a journalist might provide a few obligatory paragraphs about the millions of dollars' worth of destruction and the 150-mile-an-hour winds before slipping gradually into a decent feature article about the overall prospect of climate change and its implications for hurricane intensity, the environment, and society. Many good science writers would call for a quote or two about the weather peg, as an excuse to write an article about the more important longer-term issues that are difficult to get in to the news media without some dramatic and immediate hook. With the hundreds of such stories plus the government forums that had occurred since 1970, by 1987 I assumed that the greenhouse effect was pretty much of a household term. As things turned out, there was a long way to go.

Throughout 1987, the climate committee of the AAAS sponsored a series of meetings. First a special panel was convened to analyse and then to write a book about how the United States might cope with the greenhouse effect in the area of water resources. I was on it (its conclusions were mentioned several times in Chapter 6). One of our panelists was a distinguished political scientist, and her initial contribution suggested that climate change was unlikely to become a part of the national, let alone international, political agenda. The issues involved simply changed too slowly; they were long-term issues with relatively little immediacy, no focused constituency, no single identifiable villain, and no clear and visible proof that could be trumpeted in terms familiar to most people. In other words, to get attention our issue had to be "soon, serious and certain."[2] Without these characteristics, we were told, no issue could be high on the agenda. Surveys and other professional judgements bolstered this hard-nosed conclusion born of political experience.

In May 1988, I attended a meeting in Washington sponsored by the economist-oriented think tank Resources for the Future (RFF). In one debate session, several economists argued that the government would never express concern about, let alone take action on, the greenhouse effect until scientists could provide detailed forecasts of the winners and the losers and the wheres and the whens. Another tack would be to create popular movies or television shows with known media stars who would embrace the subject sufficiently to push it through the public's rather narrow sensors. I wasn't so cynical and expressed the opinion that real-world events would have much more impact on the public consciousness than a few clever scientists or cute stars hyping the issue on the box. A long heat wave, complete with power cuts in the Northeast, or a month of bushfire smoke streaming across California, I speculated, would do more to raise public consciousness than any of us could possibly do through high-profile public relations. What happened in the few months after the RFF meeting was beyond my wildest expectations of how events could, through the media, totally transform the public awareness.

THE SUMMER OF '88. May and June in the United States saw rare and anomalous weather. Normally, the jet stream, a five-mile-high river of air, steers storms directly across the midsection of the United States as it meanders from an average position roughly around southern California across the Midwest into New England. But that spring, the jet stream split in two: a weak branch went into southern California and across northern Mexico (keeping cool Pacific air in southern California) and a strong branch looped up into Canada, diverting the storms away from the U.S. breadbasket. Given the high intensity of sunlight in May and June and the relative absence of clouds and rain, drought and heat began to build at an alarming rate across the country. Since northern California, many parts of the West, and the Missouri River basin were already drier than normal when the spring of 1988 began, this rare climate anomaly (a persistently split springtime jet stream) moved toward catastrophe. Newly planted crops withered, Mississippi River levels dropped so low that thousands of barges were halted until the Army Corps of Engineers could dredge narrow channels, and salt water moved up the Mississippi toward New Orleans. The hundreds of greenhouse effect stories that had been dribbling out of the media for years seemed suddenly to combine, producing a spontaneous media conclusion all over America: The greenhouse effect has finally arrived!

No longer was a weather peg needed to get a climate-change story on the front page; such a story was the lead almost every day for weeks in every form of media. Cracked earth, withering plants, stranded barges, and record high 100° + F (38° + C) temperatures flashing on outdoor bank building thermometers were commonplace visual images in the media accounts. An international gathering in Toronto at the end of June attracted so many reporters that extra press rooms had to be added to handle the hordes of descending journalists. To be fair, the Canadian organizers had already built unique elements into this conference: a mixture of scientists, government-agency civil servants, elected political officials, diplomats, and even a few prime ministers. The purpose was to discuss policy implications of global warming. Initially, there were hopes that there might be some press interest since the subject had been increasingly covered in the past few years. But by late June 1988, the almost daily bombardment of climate news had turned this conference into a media mecca.

A U.S. Senate energy committee hearing on global warming a couple of weeks earlier had gathered even more momentum. I had been invited to testify, but declined because of a schedule conflict.[3] In doing so, I missed one of the most effective speeches in raising the public's consciousness about the greenhouse effect, by Dr. James Hansen of NASA's Goddard Institute for Space Studies in New York.[4]

Jim Hansen, an atmospheric physicist, spent most of his early career in the rather arcane business of calculating how light (or other forms of radiant energy) was propagated, scattered, and absorbed by planetary atmospheres containing gases and particles. I spent 1970–1972 at the Goddard Institute for Space Studies as a visitor and postdoctoral fellow, and had almost daily lunches with Jim, where I discussed the new emerging field of human impacts on climate. He, in turn, would tell me of complicated mathematical techniques for making highly accurate radiative-transfer calculations.[5]

Many years after I had left this small, quiet research outpost of NASA, both Jim's interest in climate and his administrative responsibility had grown, and he had assembled a multidisciplinary group of atmospheric modellers, radiative-transfer specialists, geologists, biologists, and others to work on the greenhouse effect. This was GISS, a small satellite branch of NASA's large Goddard Space Flight Center in Greenbelt, Maryland. GISS was constantly struggling for money. It had received some early funding help from the Environmental Protection Agency but never had much prominence or support within NASA. But by the mideighties Jim had testified a number of times before Congress, and several juicy quotes from him had made it into the national media. He began to develop the reputation as an ardent believer in the seriousness of the greenhouse effect, and had calculations from his institute team to back up his strong words.

Then, one fateful day in June 1988 (the 23rd—a day selected months earlier by Senator Tim Wirth, since it was the anniversary of the hottest day ever recorded in D.C.), Hansen appeared before the Senate Energy Committee, the television lights glaring and the pens of the world's journalists poised. That day he gave the greenhouse-effect buffs a big shot in the arm but himself a load of trouble. He was "99%" sure that the warming of the 1980s evident in his calculations was not a chance event, he said. It is time we stop the "waffling" around about uncertainty and started taking the greenhouse effect pretty seriously, he went on. Immediately that "99%" was everywhere. Journalists loved it. Environmentalists were ecstatic. Many meteorologists were upset. Jim appeared on a dozen or more national television programmes, was quoted in a front-page story in the *New York Times,* and even showed up on David Brinkley's Sunday television programme sporting a large pair of dice on which he painted some faces to represent warmer years.

A lightning rod attracts lightning. In science, if you enhance your visibility through the media, you're almost certainly going to stir the passions of some of the quiet workers who have virtually no opportunity for such notoriety. A few of those workers—people who had been toiling to resolve the uncertainty bit by bit without the world watching—began to look for ways to shoot down Jim Hansen's now famous "99%" assertion as to the likelihood that the greenhouse effect had been detected in the climatic record. Natural scientific scepticism, aversion to public discussion of still uncertain science, a misunderstanding of the policy-making process and the media, and yes, plain jealousy all contributed to the scrutiny Hansen was about to receive from different quarters. Newspaper accounts soon appeared quoting opposing scientists. Patrick Michaels, a University of Virginia climatologist, found that a thermometer on the island of St. Helena, in the South Atlantic, had been moved down a mountain slope in the 1970s, thereby giving the false impression of a recent rapid climatic warming. In some stories, this data error got equal billing with the thousands of valid thermometers Hansen and Lebedeff had averaged. (Later on, Hansen made a correction and concluded that this error changed his global record by less than a tenth of a degree.)

In order to validate one of the GISS model predictions, Hansen had worked with his colleague Sergei Lebedeff to try to reconstruct a record of the earth's temperatures over the past hundred years. This issue is not straightforward. Not only must researchers assemble millions of bits of data from thousands of widely scattered thermometers around the globe, but they must rely on people to read them, and that means human error. Further, some thermometers, as noted, have had cities grow up around them, while others have been moved out to airports, and up or down mountains. Although Hansen and Lebedeff did attempt to correct for these variables, the computed planetary temperature changes over the last century contained significant elements of uncertainty. (In late summer 1988, an entire meeting was devoted to critical analysis of the Hansen-Lebedeff curve.)

Furthermore, when Jim stated that there was only a 1% chance that the record year of 1987 could have been so warm by accident, he implied that every year of the climatic record was independent from every other year. In other words, each year of climate has no memory of what has come before. Analyses of long-term climatic trends suggest that this assumption is not necessarily a good one, and that trends can persist for decades if one accounts for the temperature memory in the oceans.[6] This meant that the statistical method the GISS scientists used was, technically speaking, questionable for drawing a quantitative conclusion such as the "99%" confidence one. Despite these qualifications, I fully agree with Jim that it is much more likely than not that the greenhouse gas buildups in this century have contributed to the observed global warming. It is simply impossible to assign formal probabilities without making intuitive assumptions.

Incidentally, Phil Jones and Tom Wigley in England produced a similar but not quite as dramatic graph of the global temperature trend that agrees the world has warmed up by more than 0.5° C (1° F) this century. Jones and Wigley used different techniques to average their thermometers, and included many more measurements from the oceans as well. Both records were compared in Figure 4. The English group's warming is about 0.6° C and the Hansen record about 0.8° C over the same 100-year time period. Considering the difference in techniques and the additional records, this is good agreement. Furthermore, Tom Karl of the U.S. government's National Climate Data Center in Asheville, North Carolina, compared these records with a very detailed study that made use of thousands of additional thermometers in the United States for this century that helped eliminate the biases owing to urban heating. What he found was that both the GISS and the CRU records captured the warming in the U.S. stations up to the middle of the twentieth century and also captured a slight cooling thereafter. (The slight cooling in the United States in the last several decades proves nothing about the greenhouse effect since the greenhouse effect is a global, not a local, phenomenon.) Karl found that for the United States, Jones and Wigley had a century-long trend with about 0.15° C too much warming,[7] whereas the Hansen record had about 0.38° C too much warming. If you apply these correction factors from the United States to the global trends of the English or NASA scientists, both would come out with a global trend value of nearly 0.5° C warming over the past 100 years. And that correction makes the radical assumption that the entire world would experience as severe an urban-heat-island bias as the United States, which I think is unlikely. At a recent meeting at the National Academy of Sciences,

Tom Karl assured the assembled group that he was not trying to suggest that there had been no global warming during this century, nor that the magnitude of change projected by these famous trend curves in Figure 4 are incorrect. He argued that the simple averaging of thermometers, with their observer errors and other complicating factors, is a very tricky business and to get precise quantitative answers is not easy. Unfortunately, that message has often been lost in the media, obscured by a highly polarized account of Hansen versus his doubting or cautious critics.

Personally, I think Jim would have been better off not using the "99%" figure since the problem with the urban heat islands and the technical arguments that are possible over the assumption that each year was independent from the others gave ammunition to his detractors, regardless of whether they were motivated by an honest search for the truth, political attempts to damn Hansen's credibility, or simple jealousy. In December 1988, Jim Hansen and I appeared simultaneously at a press conference held in Washington at a well-attended meeting sponsored by the Climate Institute. At the end, one of the reporters asked Jim the inevitable question: "Dr. Hansen, after all the criticism you have received, do you still stand by the 99% statement?" I presumed that whatever Jim said I would be getting the question next. Jim answered that he never should have said it, but since he believed in the statement, he still stood by it. Then it was my turn. I argued that I wouldn't have used the 99% number because it simply had no meaning for me. I noted that there were too many assumptions we could not verify (such as the independence of one year's temperature from the next) and things we didn't know (how to correct for urban bias effect) to be able to assign meaningful probabilities. Nevertheless, I said I completely agreed with Jim that there was a very good chance that the world had warmed up at least half a degree Celsius in the past century and that it was very likely as well that much of this was caused by the buildup of greenhouse gases. The fact that I believed the enhanced greenhouse effect was already present in the observed records was not the same as proving it, which was why I preferred to use verbal rather than numerical descriptions of its likelihood. Another reporter then asked, "But isn't the greenhouse effect just a theory? Why do you sound so confident?" Yes, it is a theory, I said, but so is gravity. If you wish to revalidate that theory, then please stick your neck into my guillotine and we'll conduct another test. I don't think the greenhouse effect is quite yet as solid a theory as gravity, I continued, but the point is not whether the greenhouse effect is a "theory" but rather whether we can validate the theory. I told him that I believed the greenhouse effect to be a well-validated theory, based not on the performance of the planet in the last hundred years but on other evidence such as the climates of Mars, Earth, and Venus, the workings of the seasonal cycle, hundreds of physical measurements of the properties of atmospheric gases, and millions of spacecraft measurements of the energy balance of the earth.

Two days later, when I came home I found a headline from the local newspaper posted on the bulletin board: "Boulder scientist says 'greenhouse effect' not just a theory." This Associated Press story correctly quoted my wisecrack about the guillotine, but did not contain the extra few sentences about the importance of *validation,* and not the word *theory.* Two of my colleagues told me they thought I had been excessive in that statement, until I explained what I had said in total. I could only wonder how many others would never bother to ask. Such experiences are typical of those involved in the public discussion of complicated issues in which polarized opinions abound. Scientists who go public with their work, particularly if it could potentially affect policy, often meet with a mixed reception from their own colleagues. Let me recount some of my earliest experiences with this problem.

In the early 1970s, the northern hemisphere appeared to have been cooling at an alarming rate. There was frequent talk of a new ice age. Books and documentaries appeared, hypothesizing a snowblitz or sporting titles such as *The Cooling.* Even the CIA got into the act, sponsoring several meetings and writing a controversial report warning of threats to American security from the potential collapse of Third World governments in the wake of climate change. (I described one such meeting at the White House in 1974 in *The Genesis Strategy.*) I believed then (and now, too) that climatic variability could be a major security threat to nations, particularly those chronically short of food and without adequate financial and agricultural reserves to deal with the bouts of bad weather that nature might select. Some scientists had an additional concern that the climate could be altered as an inadvertent by-product of economic and population growth. Industrial and agricultural dust and smoke drifting in the skies were being blamed for blocking out sunlight, thereby cooling the climate—a view made popular by Reid Bryson of the University of Wisconsin. Such logic often seemed plausible to those who looked skyward through polluted air at a dimmed, orange sun. But by 1972 a number of us began to think it wasn't only the visible junk in the sky we should worry about, but also an ever-growing burden of invisible gases, which might turn out to be the more sinister threat to the climate.

In Baltimore in December 1972 I gave a talk on the issue of human weather control to the annual convention of the American Association for the Advancement of Science (AAAS). AAAS meetings are internationally known because they bring together research scientists and policy makers to discuss the societal implications of new knowledge. In addition, AAAS meetings attract hundreds of journalists; by contrast, most ordinary scientific meetings are unlikely to attract even half a dozen media people. For me, an upstart twenty-seven-year-old postdoctoral fellow, my invitation to speak at the meeting was an exciting opportunity to share the podium with scientific leaders typically twice my age. After speaking for half an hour or so on how various kinds of human activities could change the climate, I concluded that, unfortunately, only a relatively few people were aware of the possibilities. I then quipped: "Nowadays, everybody is doing something with the weather, but nobody is talking about it."

At the front of the audience a distinguished-looking gentleman was taking notes: he turned out to be the doyen of all science writers, Walter Sullivan of the *New York Times*. Since journalists love one-liners—especially if they boil down complicated issues into a quick phrase, or reflect or create controversy, the next day's *New York Times* featured a story on weather control that closed with my reverse Mark Twain quip.[8] From then on, for better and for worse, my opinions were no longer my own property.

Meanwhile, in Boulder, Colorado, at the National Center for Atmospheric Research (NCAR), most scientists were busy doing what the vast majority of good scientists do well: working quietly and carefully to advance knowledge by learning small bits about their discipline or inventing new tools or techniques to help facilitate such learning. Scientists toil for years in some specialty trying to uncover some of nature's mysteries. Then they publish their findings. Over time, those publications contribute to the most precious intangible a scientist ever owns—his or her scientific reputation. The unwritten rules in science decree that recognition is supposed to be based on years of careful work backed up by scores of publications appearing in the most strictly peer-reviewed scientific journals dealing with narrowly defined topics. Published deeds that stand the tests of time are supposed to build one's recognition, not clever phrases that capture the public's—or worse, the media's—attention. These unprinted rules were still in full force in the early 1970s. Therefore, it shouldn't have been surprising that when I returned to Colorado I found a clipping of the *New York Times* story prominently displayed on the door of the weather map room (the most conspicuous spot imaginable). Stamped on it in large letters, right next to the quote from me, was an anonymous peer review, unusually clear and to the point: "BULLSHIT." Over the next few weeks, things went even further downhill.

NCAR employs a "press intelligence service," which is a company that reads all the newspapers and magazines readily available. Any time the service finds the name National Center for Atmospheric Research, it clips the article and sends it to NCAR. NCAR's public information office then compiles a month's worth of clippings, photocopies them, and distributes them to hundreds of scientists around the country. It was interesting to watch what happened in the wake of Walter Sullivan's well-balanced article. Within days of its original appearance, better newspapers picked up the story with Sullivan's byline and the *New York Times* copyright. Some dailies shortened the article and nearly all of them had different headlines. (Headlines are rarely written by the journalist who writes the article, I later learned, as on many occasions journalists have called me to apologize in advance for the sensational or inaccurate headlines that their papers were about to put over their story.) As the Sullivan article continued to work its way into smaller and smaller markets, parts got cut out and eventually some papers dropped the *New York Times* copyright and even the Sullivan byline! The worst offender, I recall, was a short condensation under a local reporter's byline, with a headline about "water" modification rather than "weather" modification. That article also appeared on the map room door—

this time with two friendly peer comments written in bright blue ink. What happened to the Sullivan story as it journeyed from city to city and eventually into the journalistic backwaters of America was akin to what happens in the children's game of 'Chinese Whispers': a group of kids sit in a circle; one whispers a word or phrase in the ear of his neighbour who then whispers it to his neighbour and so on, until the last child repeats what he thinks his neighbour just told him. How often have children giggled wildly when the one who originated the phrase tells out loud what he actually said. But this was no game; this involved informing the public.

I learned several lessons from this experience. First, that no one should believe all the details of newspaper stories—hardly an original discovery. However, one can usually trust the broader outline of such stories, such as what the issues are, and, if they are controversial, why, and who the people are who are involved with these issues. Secondly, I decided there are probably two safe ways for a scientist to deal with the press: not at all, or a lot. If you deal with the media at all and choose the latter way, you minimize the risk of damaging your reputation from a few bad stories since there is a greater chance for more good ones as well. Ultimately, however, I resolved to write a book about climate change myself rather than expect the media to carry my message about this complex, uncertainty-riddled topic to the public.

My first book, *The Genesis Strategy: Climate and Global Survival*, came off the presses and was announced and distributed in February 1976—appropriately enough, at an AAAS meeting in Boston. (By then, the core of science journalists who routinely attend this annual meeting had been exposed to several years of climate talk and were now much more sophisticated about climate issues. The quality and number of stories written by them—not by general assignment writers unfamiliar with scientific content and methods—dramatically increased.) The book hit about the same time that a number of nasty climatic events were unfolding that helped to increase public exposure to issues of climate change. But the greenhouse effect hardly became a household phrase. Indeed, in *The Genesis Strategy* I focused more on the need to plan for large (but precedented) natural variability of climate. However, the book also clearly argues that climate change from human activities may pose an ultimate limit to growth in the long term and that therefore we must attempt to control environmental pollution.

In the decade that followed, scientific consensus grew for the theory that the injection of invisible greenhouse gases was likely to cause unambiguous global-scale warming by around 1990 and create unprecedented climatic change sometime in the twenty-first century. Assessments by teams of scientists from the U.S. National Academy of Sciences, the United Nations, the U.S. Congress, the Canadian Meteorological Society, the Australian government, the European community, business, and other institutions increasingly took place to discuss the climate issues; and increasingly, news of global atmospheric change was moving up from the back pages of the science section of newspapers and magazines, particularly when bad weather (like heat waves) struck. But until 1988, acute awareness of the greenhouse effect and global warming was still largely confined to academic, congressional, environmental, or industrial professionals.

Despite very sound physical principles that drove a few dozen scientists around the world to boldly predict ten to fifteen years ago (when most people thought the earth was cooling) that a global warming trend would become clear by 1990, the issue was not raised to prominence in the public consciousness or on the political agenda. For one thing, the public had long been subjected to doom-laden placard carriers warning of various disasters that never came to pass. Moreover, global atmospheric change had few dramatic, familiar metaphors—at least until 1986 when the ozone hole mysteriously appeared over Antarctica, threatening additional skin cancers and untold ecological damage. Then, in 1988, nature did more for the notoriety of global warming in fifteen weeks than any of us or the sympathetic journalists and politicians were able to do in the previous fifteen years.

Journalists have a tradition, quite appropriate I believe in political reporting, of providing balance in a story. This tradition stems from the need to give both sides of the political spectrum equal exposure. However, balance is not a very good doctrine when applied to science reporting. It does not help the public to understand the nature of complex technical questions to balance an extreme position of a scientist or advocate at one end of the spectrum against an extreme position of a scientist or advocate on the other end. Technical issues often have more than two sides, which means that polarized reporting can create a false dichotomy. The public, and the politicians who must ultimately make policy, need to know not only what the members of the community think, but also what the broad spectrum of responsible and knowledgeable opinion is on an issue. If only the irreconcilable debates of implacable expert enemies are reported, the typical public reaction (and probably those of politicians as well) will be, "Well, if the experts don't know what's going on, how can I decide?" The next reaction would probably be, "You folks go back and study some more, and when you have more certainty come and tell us so we can decide how to act." Instead, knowledge of a consensus over how much we already know or a rough estimate of how long it will take to learn a great deal more could very well lead to public and political reactions quite different from those created by a noisy, angry, and dichotomized debate.[9]

In the June 1988 Toronto meeting, for instance, I was asked to hold a press conference. A reporter from a U.S. newsweekly kept pressing me on whether I differed from Jim Hansen and his 99% statement. I said that I felt the 1980s, having had several record warm years, provided increasingly strong circumstantial evidence that a greenhouse signal was emerging. I even said that I believed the issue would be settled over the next decade or two, but that as a scientist I'd have to admit that some uncertainty still remained. "Then you disagree with Hansen," said the reporter. I then said, "I'm not going to get into a false dichotomy debate with Jim Hansen when we both agree that the physical basis of the greenhouse effect is very strong, that greenhouse gases have increased, and that global warming is occurring on the planet. True, I choose to state this problem differently than he has, but focusing only on that difference is not going to give the right impression if it is presented in the name of 'balance' as an opinion opposite to Jim." Several reporters asked other questions. Then, toward the end, my persistent friend from the newsweekly said, "I take it, then, that you don't agree with Hansen that the greenhouse effect has been detected with 99% certainty."

Despite a few such frustrations, the Toronto meeting had a major impact. It got excellent press coverage in the United States, Canada, and other parts of the world. The participants called for radical and specific policies to reduce carbon dioxide emissions by 20% in the year 2005. That stirred quite an internal debate (described later) but was effective in capturing press attention. On the closing day of the plenary session, the meeting chair (the then Canadian ambassador to the United Nations, Stephen Lewis) asked me for my view of how the meeting's strong conclusions would be received in the United States. There were basically two categories of conclusions he wanted me to address: (1) reducing emissions of greenhouse gases through energy-conservation programmes or the development of nonfossil energy systems; and (2) recognizing the special problems of Third World countries who desperately lacked economic development and needed extra help if they were to develop with less polluting technologies. To answer his question, I considered the probable U.S. reaction of five separate groups: the scientists, the bureaucrats, the politicians, the media, and the public. I offer my response below as an overview of current thinking among these groups in the United States.

Most scientists probably appreciate the attention their fields are getting, but many consider statements suggesting concrete policies premature as long as major uncertainties remain. Unfortunately, too many scientists still regard a decision to react to possible change in the face of uncertainty as a scientific judgement, when in fact it is a value-question, requiring that we weigh the risk of unabated change against the risk of possible wasted investment.

Most bureaucrats' reactions are also relatively easy to identify and describe. The bureaucratic culture thrives on information. Those who live and work in it subsist on a daily diet of cost-benefit studies, flowchart assessments, and so forth. While environmental protection officials would largely welcome the possibility of more regulatory activities that would expand their influence, they would probably be uneasy with the conference's recommendation for action until further cost-benefit studies showed the impact of the proposed specific reductions in greenhouse gas emission on various countries and various economic sectors.

Politicians, on the other hand, care much less about information than about perception—the perceptions of their constituents. If they feel the public is alarmed about an issue such as climate change, they will scramble over each other to look as if they are making hard decisions and lending leadership to implement concrete actions for dealing with the situation.

The public's perceptions, of course, are primarily shaped by the next category, the media, which thrive on the four Ds: drama, disaster, debate, and dichotomy. The summer of 1988 – not even a month old at the end of June, when the Toronto meeting took place – had already provided enough of these to bring the media on board the climate problem wholeheartedly. Keeping their interest up without overstating the relatively weak connection between the greenhouse effect and one season of weather was no easy chore.

What, then, was to be the public reaction? The public, of course, has long been subjected to placard-carrying advocates warning of one impending disaster or another, the forestalling of which supposedly demand immediate attention – and resources. Although I thought people had all heard of global warming, I had long feared what my political science colleagues on the AAAS water-supply panel had documented: that a problem like climate change simply lacked sufficient familiarity and credibility to capture the public's attention, let alone its imagination. Moreover, faced with a steady diet of dichotomy and debate in the media – with one scientist saying warming while others said cooling – it was not surprising that it had been difficult to mobilize public attention on the issue over the past decade. But the Summer of '88 had reversed all that. In the United States, public perception of the problem was becoming so acute, I told the Toronto assembly, that U.S. politicians would soon be proposing dramatic actions, such as the 20% CO_2 emissions reductions being proposed in Toronto (indeed, before the end of 1988, several major bills based in part on the Toronto recommendations had been introduced in the Congress). However, there was a risk of severe credibility loss for climatology if nature rolled a cold, wet summer or two soon, and this was quite possible given all the faces on the climate dice, even if they are loaded.

So far, my answers to Ambassador Lewis's question had been mostly positive: I thought the United States would be able to react positively to the conference's recommendations – as long as we didn't get too much cold, wet weather in the next few years. But I had so far dealt only with the first category of conclusions – U.S. emissions reduction. It was much tougher to be optimistic about the second: encouraging developed countries to help developing countries recognize and meet their special development needs in an environmentally sustainable way.

Only a few studies have been done to help the bureaucrats determine cost-effective ways in which Third World countries could develop in a reasonably pollution-free manner. The media have primarily focused on climate emergencies at home, and with the exception of floods in Bangladesh and hurricane destruction in the Caribbean and Mexico, in 1988 there was little coverage of the every-day crises in food, health, shelter, and employment in many Third World countries. Can we ask China not to develop the abundant coal resources she is counting on to improve her economy (as we ourselves did in the early days of the Industrial Revolution) simply because the increased use of this highly polluting stuff will make our world warmer? We could only ask the Chinese (or the Indians, or the Brazilians, who are chopping down their forests) to change their development plans if we provided alternative, more efficient, and less polluting technologies or helped those populous, debt-ridden countries develop less energy-intensive economies. But it will be a political struggle even to pursue these objectives faster at home, and we will have no moral authority to advise the Third World to make sacrifices until we make some at home. Before we can expect our political leadership to agree to transfer substantial resources to the Chinese, we will need to generate a great deal more media attention and work to foster public understanding of the global development issue. Therefore, I concluded, while we may have experienced an unprecedented jump in climate-change consciousness that might encourage action to curb some of our own emissions, we are not yet at the level of perception and understanding needed to address the greenhouse-effect problem on the global scale.

As the summer of 1988 wore on, more hearings were held, more crops withered, and by late August fires had become the new media event. Yellowstone National Park, perhaps America's most important symbol of nature protected from the ravages of civilization, was ablaze. Day after day, flames over Yellowstone were seen on televisions in living rooms across America. Firefighters, old monuments, buildings, fleeing wildlife, and obscuring smoke drifted across the tube. Before long, senators, governors, and cabinet secretaries appeared at the scene with their sleeves rolled up looking at charts while smoke billowed behind them. We must question the park service's policy of "let burn," they typically remarked, and work harder to put these fires out before they spread. The media dutifully covered their seemingly grave concern and a number of good feature stories appeared on the issue,[10] but unfortunately there was little widespread reporting of the underlying basis of the conflagration. The great Yellowstone fire of 1988 was as much a manmade phenomenon as one of nature. True, intense drought dried the woods and underbrush to such a tinder that any spark, whether from lightning or an irresponsible campfire, could ignite it. Strong winds would blow such fires out of control, even if hundreds or thousands of firefighters were waiting to contain a small plot that was deliberately being allowed to burn. What lay behind all this, however, was the fact that Smokey the Bear had been too successful.

The fairly recent let-burn policy of the park service was based on the ecological principle that natural environments had evolved with fire, and that fire was part of—in fact, beneficial to—the health of forests. Some seeds cannot even be released from their pine cones until heated. The ash from burning underbrush, dead leaves, and trunks fertilizes the soil, generally reducing its acidity. Acid soil is both bad for many species of trees and promotes the growth of moss, which further acidifies the soil. Eventually, enough moss can kill trees, turning the land into bogs.[11] Fire is a disturbance that allows ecosystems to remain dynamic. Thus, fire is one of nature's ways of rejuvenating forests by turning over the nutrient stock and maintaining the vigour of the ecosystems. However, when people move into forested areas and build houses or recreation centres, they lose sight of long-term ecological considerations. Their primary goal is to protect their property. That fact, coupled with the mid-twentieth-century view that human beings know better than nature, helped create the fire-suppression mentality that has hovered over forest lands in this century. We must not forget either that timber is a big business and that burned trees cannot be logged and turned into lumber.

All these pressures were combined with one of the most successful public relations campaigns ever mounted to prevent the natural rate of burning. Many of us remember the stern-faced Smokey the Bear pointing his finger at us from out of the small black-and-white television sets of the 1950s and 1960s and saying, "Remember, only you can prevent forest fires." Unfortunately,

Smokey forgot to say one thing: if fires don't occur often enough, then dead and accumulating plant matter will increase the fuel loading on the surface of the woods to such an extent that when the inevitable drought occurs not only will a fire occur but it will be one of such intense heat and scale that it could actually damage the ecosystem. This, apparently, is what happened in Yellowstone. Fuel levels built up over decades of intense fire suppression. Such a catastrophic fire was inevitable, a number of foresters said at a recent meeting. It was only a matter of time. Drought and heat are bound to occur sometime—and they did in the summer of 1988. The foresters were hardly surprised when these conditions conspired. Nor were they surprised, they said rather cynically, that when the fire hit the politicians went straight to the press to warn us sternly that we must work harder to suppress fires so this would never happen again. Such short-term political expediency, uncritically reported through the media, is likely to drive us even further in the wrong direction based on a misinterpretation of the underlying cause of the intensity of the Yellowstone disaster. Of course, too much political attention to short-term solutions is hardly news. A few anecdotes may make the short-term bias of society more tangible.

In July 1980, in the middle of a heat wave in the southcentral states, I received an exciting invitation from NBC News: fly to Washington and appear live on "Meet the Press." The Sunday morning on which the programme was to be aired was not a feasible date for me, so I asked if we could prerecord it. An hour later brought the affirmative response. That evening, I had dinner with the moderator; the programme was to be taped later that night. It was a pleasant evening, enhanced by the elegant style of wining and dining. We discussed the media, and I told the moderator of the frustration of having several lengthy interviews of mine cut or discarded altogether unless there was a strong weather peg. He sympathized and recounted similar stories of his own. Has that ever happened with "Meet the Press," I asked. "No," he replied. "It would take an assassination of the president to bump this programme," he said. Off we went to the studio to tape the show.

The taping went well, the producer expressed satisfaction with the programme, the director said the tape was technically good, and we all left happy. The programme was then advertised in the newspapers for the upcoming Sunday. However, that Saturday night the phone rang. It was the producer. "Remember what I said about an assassination? Well, it's essentially happened. Four Democratic congressmen have just announced a 'dump Carter' movement, and my executive producer insists that we put them on live this Sunday morning. We're going to put your programme on the following week."

What then happened to the first-ever embargoed "Meet the Press"? A week later, they weren't sure whether they were going to run it on the next Sunday—the Democratic Convention was coming up. Two weeks later, a major flood struck in drought-ravaged Texas. NBC decided it couldn't do a weather programme that makes no mention of this highly visible weather event—once again they failed to run the programme, and so it went.

One final anecdote illustrates our society's inevitable emphasis on short-term priorities. In July 1988, Democratic presidential candidate Michael Dukakis visited Colorado. Senator Tim Wirth called to invite several Colorado environmental specialists to join him for a 6 A.M. expedition to the Rocky Mountain National Park, where the candidate was scheduled to tour the park and give an address outlining his environmental views. The senator had arranged a breakfast briefing where a number of us would have a few minutes to explain our particular viewpoints to Dukakis. At the breakfast, three of us briefed the candidate: one on wilderness, another on urban air pollution and transportation problems in the West, and me, on global warming. I had six minutes at the end to deal with the next sixty years.

After breakfast, two buses loaded up, carrying an entourage of some 100 campaign staff, national press people, secret service agents, and state officials in a police-escorted drive through the Rocky Mountain National Park to the beautiful Bear Lake location. At one point along the narrow pathway around the lake, the Colorado senator began explaining to the candidate how an eroded gully could not be repaired as long as the current administration continued to sequester funds from park receipts rather than spend them on trail repair. Some seventy-five members of the press elbowed past each other in an effort to zero in on this conversation, with technicians, producers, camera operators, boom holders, and others all racing to the scene—trampling flowers, stomping trees, cracking branches, and scaring wildlife, all to hear a rather short but important political discussion on preserving the park environment!

In the scenic beauty of the national park, the candidate nicely summarized what he had heard that morning and added several additional interesting twists of his own—proposing, for example, an environmental summit between the U.S. president and Soviet leader Gorbachev to deal with the greenhouse effect, ozone, and acid rain. After his short speech, the scores of press began to question the candidate. There were nine questions, of which only one was even on the environment, to draw attention to which was the announced purpose for the trip to Rocky Mountain National Park. And that question came from a European journalist, not an American. The Americans asked questions about the candidate's political rivals (especially Jesse Jackson) and Dukakis's views on taxes. The commercial broadcast networks as well as the cable news services all covered the event, but the twenty-second sound bites that made the evening news programmes primarily dealt with the issues of taxes or Dukakis versus Jackson. I heard virtually no discussion of the environment, except in one passing reference to that as the purpose of the trip.

The relevance of these stories to the greenhouse-effect issue is obvious: short-term, immediate issues always take precedence in the public view over long-term, slowly building questions. Indeed, in more despairing moments I often wonder if the combination of short-term electoral politics with the high ratings value of short-term news stories will prove to be an insurmountable obstacle that will virtually banish the possibility of fostering long-term social consciousness and sparking political action on global warming. The situation is not always that bleak, however, for sometimes nature rolls snake eyes, and all of a sudden the strengths of the media become an advantage thanks to a dramatic short-term event. I refer to the Summer of '88. This media event did more for climate consciousness in fifteen weeks than all the "Nightlines," "Novas," *Newsweeks,* I and other scientists were able to achieve in the preceding fifteen years. Whether this was enough to precipitate remedial action remains to be seen. And, overly dramatic coverage of the Summer of '88 and overstrong association with the greenhouse effect inevitably created a backlash.

The hype and hoopla of the Summer of '88, overblown as it sometimes was, inevitably created an angry response from some scientists. Several Op-Ed pieces appeared from scientists, decrying the irresponsibility of the media hypsters and public scientists for misleading the public and politicians about the seriousness or immediacy of the greenhouse effect. For example, a young statistician at the Woods Hole Institute of Oceanography, Andrew Solow, had an Op-Ed piece in the *New York Times*—"Greenhouse Effects: Hot Air in Lieu of Evidence." The piece was sarcastic in style (certainly not typical of scientific literature). Solow argued that "existing data show no evidence of the greenhouse effect. Many people will be surprised to hear that this is more or less the view expressed in scientific journals, where articles are subject to peer review. Unsubstantiated or misleading statements appear in such journals only when the review process fails. Congressional testimony and interviews in the press are not subject to peer review, and that is how unsubstantiated and misleading statements come to dominate public discussions."[12] I was, frankly, annoyed at that latter remark, since I dislike innuendo and sweeping generalities. I prefer critics to state forthrightly who is the perpetrator of such misleading statements, and who use some quotations to let us judge for ourselves whether they are in fact misleading. Without such specifics this kind of criticism simply sets up an aunt Sally to knock down. Later on Solow attacked those of us (and I plead guilty to his charge) who argue for policy response to the greenhouse effect: "Some will say that if we wait until we are sure about climate change, it will be too late to do anything about it. This argument applies equally to an invasion of aliens from space. More seriously, this argument neglects the costs of overreaction now." Articles such as this, and others (like the Op-Ed piece by S. Fred Singer that appeared in the *Wall Street Journal* and that of Patrick Michaels in the *Washington Post*)[13] created a substantial wake of media articles that essentially put the greenhouse effect on trial in their headlines.

On the morning of February 21, 1989 two Congressional hearings were held in which I was asked to testify about the latest scientific controversies challenging the existence of the greenhouse effect (the controversies were a direct result of the Op-Ed pieces and the media response to them). I noted, quoting Solow's crack about aliens from space, that "of course, what Solow neglects to make explicit is that waiting for more evidence before acting is his value position. He also fails (which is surprising for a statistician) to suggest anything about the different probabilities between an invasion of aliens from space and rapid future climate change. In my estimation, space invaders are no more likely than the collision of the earth with a large asteroid—which would be a catastrophe! The odds of such a collision are presently assessed at something like one chance in 10,000,000 each year. But the odds of rapid, potentially unprecedented—even catastrophic—climate change are certainly in the first decimal place of probability, and I believe a likelihood of several degrees warming by fifty years is a better-than-even bet. In any case, the cost of overreaction is a legitimate issue but so is the cost of underreaction. The policy process is not advanced when such issues are ridiculed with hollow rhetoric. Instead, scientists should be asked to provide what they are technically competent to offer: estimates of specific consequences of greenhouse gas buildups and their likelihoods of occurrence. Any statements beyond that are the personal opinions of those scientists. Although I believe that scientists, like all citizens, are entitled to opinions on how to deal with those probabilities and consequences, they must always be scrupulous to point out that such opinions are personal value judgments."[14]

The real *cause célèbre* of the hearings was not the Solow piece but rather a much lengthier and more reasoned critique from Patrick Michaels, who appeared simultaneously with me in front of Indiana Congressman Phil Sharp's House Subcommittee on Energy and Power. In the *Washington Post,* Michaels had charged James Hansen in particular, but other media scientists by implication, with misleading the public on the nature of the greenhouse evidence. In particular, he cited Jerome Namias of the Scripps Institution of Oceanography, "the dean of American climatologists," as "saying there is no way one can scientifically defend any statement linking causation of last summer's drought to the greenhouse effect. Then in the December 23 issue of *Science,* Kevin Trenberth of the National Center for Atmospheric Research and his co-authors convincingly argued that the drought of '88 was caused by warm ocean temperatures in the tropical Pacific, which have since dropped to near record low values." Michaels went on to attack Hansen's June 1988 congressional testimony and cited instead the recent work of Tom Karl, "who arguably knows more about regional climate variations than anyone in the world," to the effect that "there may have been no global warming to speak of during the last century." Pat concluded somewhat heroically: "Karl's findings surprised none of us who merely toil with the data. But it should be a major shock to those who are using those figures for policy purposes. Is it irresponsible to point this out in public?"

At the February congressional hearings I read out loud some of these quotes and tried to put them in perspective. With regard to the Namias/Trenberth view that the drought was caused by factors other than the greenhouse effect, I cited a recent letter to the *New York Times* from Jim Hansen: "As I testified to the Senate during the 1988 heat wave, the greenhouse effect cannot be blamed for specific drought, but it alters the probabilities. Our climate model, tested by simulations of climate on other planets and past climates on earth, indicates that the greenhouse effect is now becoming large enough to compete with natural variability." It was fun for me then to read a passage from Trenberth and colleagues' *Science* article: "Climate simulations indicate that a doubling of carbon dioxide concentrations could increase the frequency of summer droughts over North America. Thus, the greenhouse effect may tilt the balance such that the conditions for droughts and heat waves are more likely, but it cannot be blamed for an individual drought." That statement is clearly reasonable and, contrary to the impression in the media set off by the Op-Ed pieces, is quite consistent with those of Jim Hansen, myself, and most other scientists that I'm aware of who speak on these issues.[15] "Trial-by-media of the greenhouse effect was thus a nonscientific issue from the very beginning," I told the Congress. "Nevertheless many Op-Ed pieces continue to appear from scientists and others criticizing the 'hysteria' being generated by some (usually unnamed) public scientists on this issue."

Next I turned to Tom Karl and the urban heat island issue by noting that on several occasions Tom had personally told me of his frustration with frequent media phone calls asking him to give the "other side" of the greenhouse effect when in fact he does not represent another side of the greenhouse effect, but is simply trying to maintain quantitative reliability in the world's temperature record. He told Richard Kerr of *Science* magazine in an interview that "the long-term global warming is something on the order of 0.4° C during the past century. Is the bias 0.05° or 0.2°? The chances that it is the same size as the warming are pretty remote. It's a matter of adjusting the rate of rise, not questioning the rise itself."[16] In other words, a man cited so often as the destroyer of the warming trend actually believes in it, and is simply trying to get the numbers right as any good scientist should do.

Finally, I discussed with the congressmen two rapid-fire articles that had appeared in the *New York Times* and had gathered much attention. Both were written by the well-respected and careful science writer, Philip Shabecoff. The first one reported on a 100-year record of annual temperatures and rainfall in the United States prepared by NOAA scientists, including Tom Karl. Well-balanced, the article quoted the authors as saying this limited record did not imply that the greenhouse effect was a false issue. Nevertheless, the headline (not written by the science writer) did the damage: "U.S. Data Since 1895 Failed To Show Warming Trend." Shabecoff's next article appeared a week later, on February 4, 1989. It was entitled "Global Warmth in '88 Is Found To Set a Record." In it Phil discusses the latest conclusions from Phil Jones and Tom Wigley at the University of East Anglia: that 1988 is the new record warm year, followed in order by 1987, 1983, 1981, 1980, and 1986.[17] While urban heating could perhaps help to explain some of the century-long warming trend, it obviously could no more explain the rapid warming of the 1980s than the greenhouse effect could explain the rapid onset of drought in 1988 compared to 1987.

The congressional representatives could not understand how global warmth could be setting a record when the United States was showing no warming trend. I answered by analogy, showing two slides of the United States on which the states that were won by Democrats in the 1976 and 1984 presidential elections were coloured in (I coloured in the Democratic states because it was less work). In the 1976 election, in which Jimmy Carter beat Gerald Ford in a close contest, roughly half the states were won by the Democratic victor. It would be foolish, I argued, to try to guess who won the 1976 election by simply picking two states at random. On the other hand, the 1984 election in which Ronald Reagan won by a landslide—with only the District of Columbia and Minnesota going to the Democratic challenger—is a different story. It would be relatively easy to guess the winner of that election by looking at only a limited sample of the country. The relevance of this to global warming, I went on, is that the United States contains less than 2% of the area of the world. Therefore, for a very small climate trend—as with a close election—it is ludicrous to expect to make large-scale inferences from a very small sample. On the other hand, if the warming trend were 2° C instead of the current 0.5° C, then you would expect—as with a landslide election—that even a small sampling (of states or of the world area represented by the United States) could give you a fairly accurate indication of larger-scale trends. In other words, the larger the climatic signal the less likely that small regions like the United States would be exceptions. That metaphor certainly seemed clear to the congressmen in the room.

I concluded my testimony by pointing out that scientific concern for unprecedented climate change in the next century is not based on the performance of the planet in the past century. Indeed, at the level of global temperature change to date (about 0.5° C), the noise of natural climate variability is simply too large to be able to clearly detect a greenhouse effect signal of sufficient clarity to verify precisely how sensitive the climate has been to increasing greenhouse gases. (But even if not particularly relevant for detecting greenhouse signals, the detailed patterns of climate variability this century are worth careful analysis to determine whether they were due to unaccounted-for influences like human dust or solar activity, or alternatively to some internal manifestation of the complicated dynamics of a chaotic climatic system.)[18] Rather, the case for rapid future climate change is based on the millions of satellite and laboratory observations that validate heat trapping as a theory essentially beyond any reasonable doubt. If we couple that validation with the certainty that global greenhouse gases have increased substantially and are very likely to continue if economic and population growth maintains its expected trends, we see a substantial probability of growing climate change; and in my opinion it is a better-than-even-money bet that large global change is likely. In their Op-Ed pieces, Pat Michaels and Fred Singer also acknowledged that they were not questioning the heat-trapping properties of greenhouse gases, but were focusing instead on uncertainties having to do with the nature of twentieth-century climate variability. I agree that these are legitimate scientific issues and that until they are resolved there will be substantial questions about the detailed nature of climatic change. But these uncertainties do not call into question the relatively high probability for unprecedented climate change into the next century. Neither can the world's thermometers resolve the greenhouse-signal-detection debate without another ten to twenty years of data. And, I repeated to the Congress, waiting for that extra degree of certainty is not a risk-free proposition, since it could force us to adapt to a much larger dose of change than if we took preventive actions now.

The role of the media in questioning the greenhouse effect because of the largely irrelevant weather phenomena that created the drought of 1988 was, of course, a two-edged sword. On the one hand, because of media attention in 1988, the issue is now firmly in the front of public consciousness. On the other hand, a debate in 1989 and 1990 that is largely irrelevant to the scientific validity of the greenhouse effect was turned into a trial-by-media of a phenomenon whose existence is proved essentially beyond doubt. My fear is that a cold and wet summer in the next few years, which certainly could occur since even loaded climatic dice have many faces, will be misinterpreted as evidence of the nonexistence of a greenhouse effect.[19]

There is, of course, a generic underlying concern reflected in these remarks: that democracy works only to the extent that the citizenry is informed about the complex nature of the issues their leaders must deal with. To be sure, without the media, people would be much more poorly informed than they are. But if the most popular media largely stick with the four Ds—drama, debate, disaster, and dichotomy—the public will be much more poorly informed than it could be and the policy-making process will be that much poorer.

I do not assign to journalism sole responsibility for the public's general lack of understanding of the many complex issues of the day. This is more a symptom of the problem than a cause. The media executives produce a product that their experience tells them will either sell to the public or attract advertisers. At the root of my worry is the need for all of us, the citizens, to develop an intense interest in how our world works and how we might improve it. We need to be more sceptical about what we hear without being too cynical. We need to learn to question while still recognizing that uncertainty is not automatically a reason for inaction. The problems with most press coverage, I believe, reflect the demands of their audiences for entertainment or brevity more than a conscious conspiracy of journalists to reduce the complexity of modern life to slogans that will fit on window stickers. We need to build curiosity and interest about the future at home, and to teach young people how to think and question in schools. Teachers (or any experts) shouldn't be portrayed as the owners of truth, but rather as guides to the vast fields of knowledge and as models of how to observe, question, and synthesize ideas. If the public is receptive, the media organizations—whose rewards are usually driven by audience sizes—will respond quickly to the tastes of the public, just as politicians respond to the polls they constantly examine.

OZONE DEPLETION. It would be inappropriate to leave the discussion of the Summer of '88 without mentioning another issue that went a long way toward raising public consciousness about global atmospheric problems. In this case, the media played an important and mostly responsible role, partly because the high degree of polarization they reported honestly reflected differing perceptions and values of two rather diametrically opposed debating camps. The problem I refer to is ozone depletion and, more recently, the ozone hole.[20]

Ozone is a trace gas in the atmosphere critical to life on earth. It consists of a molecule of three oxygen atoms (O_3) and is very chemically reactive in the earth's atmosphere. Too much ozone in the lower atmosphere helps to create an eye-burning, lung-damaging, and plant-harming component of urban smog, whereas too little ozone in the stratosphere overhead (16 to 48 kilometres, or 10 to 30 miles) allows extra ultraviolet (UV) radiation from the sun to reach the earth's surface. Ozone has filtered out such UV for perhaps as much as a billon years since the buildup of oxygen in the atmosphere. UV damages DNA molecules and is known to reduce the effectiveness of the immune system and cause skin cancer in humans. Each 1% reduction in the total amount of ozone increases UV reaching the surface by about 2% or more, which, in turn, is thought to increase the risk of skin cancer, particularly in fair skinned people or those who spend considerable time in the sun.[21] UV may also be harmful to ecosystems at many levels in the food chain. Some think that enhanced UV reaching the ocean surface could reduce the productivity of phytoplankton (which make DMS, see Chapters 4 and 5) upon which krill (small shrimplike creatures) feed.[22] These, in turn, are the food upon which larger creatures such as whales are ultimately dependent.

Pollutants produced by human activities eventually make their way up to the stratosphere, and several of them, including chemicals manufactured by major industries, could be remarkably effective in reducing the ozone shield. Unfortunately, the most direct way to prevent this would be to put those manufacturing enterprises out of business. Thus, ozone depletion in the stratosphere has all of the elements of a good news story: dread consequences, dramatic potential confrontations between environmentalists and industrialists, and dichotomous debates about what is known, what is speculated, and what should be done. Because the ozone-depletion controversy often arises along with the greenhouse-effect issue (for example, both got equal headline space on the October 19, 1987 cover of *Time* magazine) and because the ozone issue can serve as an analogy to climate change with respect to policy making and the media, it seems worthwhile to spend a few pages now briefly reviewing the debate.

Ozone has been known for many decades to exist in the stratosphere. Its occurrence was explained for quite awhile by a simple set of chemical reactions that predicted its existence and its approximate amount at each height in the atmosphere. But the amounts of O_3 this first theory predicted were much too large. Thus, scientists searched for additional natural phenomena that might aid in the destruction of some ozone to bring the simple theory in line with what was known to exist overhead. One idea proposed "catalytic reactions," whereby repeated encounters of ozone molecules with a single molecule of some catalyst could destroy thousands of ozone molecules. Thus, seemingly minuscule amounts of a trace gas could have a substantial impact on ozone. By 1970 it was suggested that hydrogen oxide generated from high-altitude water vapour could be part of such a catalytic cycle. This soon led to a major controversy since proof of this theory would mean that water-vapour emissions from a proposed fleet of high-altitude supersonic transports (SSTs) might pose a threat to the ozone layer. Industrial and environmental chemists began to square off over the plausibility of this hypothesis almost two decades ago. But concern about the potential damage from water-vapour emissions was quickly replaced when Paul Crutzen in Europe and Harold Johnston at Berkeley independently discovered that oxides of nitrogen in the stratosphere can destroy ozone by very active catalytic reactions. These oxides of nitrogen can exist naturally, being produced by the action of cosmic rays hitting molecules of nitrogen in the air or by the injection into the stratosphere of relatively stable, biologically produced gases such as nitrous oxide (common laughing gas). Over a decade or so, this gas percolates slowly through the lower atmosphere to the stratosphere. Some nitrogen oxides are also an inevitable by-product of combustion since high-temperature burning in the presence of air (with 78% of the volume of air occupied by nitrogen) inevitably results in the production of nitrogen oxides. The obvious source for combustion at stratospheric elevations would be a fleet of high-flying SSTs.

In the early 1970s, a major political debate took place about whether the federal government should sponsor the development of an American SST, which would be primarily centred at the Boeing Corporation in Seattle. The debate led to an ironic reversal of the normal ideological polarization. Liberal members of Congress, who typically vote for spending programmes that increase employment, generally opposed SST federal development on the environmental grounds that the exhaust emissions of a large fleet of these aircraft could damage ozone and enhance ultraviolet levels at the earth's surface. Normally conservative senators and representatives, ever mindful of federal budgetary expansion, nonetheless largely supported the venture. Even without environmental considerations the vote would have been close, but most analysts think that the presence of the ozone-depletion complication tipped the balance slightly in favour of the ultimate vote that killed the U.S. project. The Anglo-French Concorde SST project proceeded despite the ozone debate. In fact, many British scientists and industrialists were hostile to U.S. environmentalists for years over this issue.

It is useful to recall the political context in the early 1970s. The United States had been embroiled in a half dozen years of a stalemated and unpopular war that had led to tremendous internal dissension. A war-weary nation built consensus quickly and moved with remarkable speed to take political action on concerns over deteriorating environmental conditions that were poisoning lakes and damaging the health of communities. But understanding the tangible air and water pollution crises that were visible—and sometimes painful—to the naked eyes of those in polluted areas was one thing; comprehending an invisible, hypothetical threat to the global ozone layer was another, requiring a stunning transition. The media carried the debate and it was well known at the time. However, the publicity surrounding the SST/ozone issue would prove to be merely a warm-up for a much larger controversy: the hypothesized ozone-destroying properties of the chlorofluorocarbons (CFCs), a major group of industrial chemicals used as refrigerants, foam-blowing agents, propellants for sprays cans, and in other applications that required a very inert gas.

In 1974, Mario Molina and F. Sherwood Rowland, two chemists from the University of California at Irvine, made a landmark discovery.[23] They theorized that CFCs, because they are so inert in the lower atmosphere, would not react with other atmospheric chemicals and thus would eventually reach the stratosphere. There they would diffuse above the ozone layer and be subjected to increased levels of the sun's ultraviolet light, which would break them up into their constituent molecules—chlorine, fluorine, and carbon (*ergo* the name chlorofluorocarbons). The most common CFC is known as Freon, Dupont's trade name for the chemical. Molina and Rowland went further, suggesting that chlorine, for example, would combine with oxygen to form a chlorine-oxide molecule in the stratosphere that could, through catalytic reactions, eat up many thousands of times its own weight in ozone. Their early estimates suggested that industrial production capacity over the next several decades could eventually lead to a destruction of as much as 20% of the ozone layer.

Battle lines were quickly drawn. The industry labelled the theory as speculative without a shred of real-world evidence. Environmentalists countered by saying that to tolerate such a frivolity as spray cans was unthinkable in the face of even the slightest chance that their products could damage the integrity of the global ozone layer. Charges and countercharges were traded in front of congressional committees, on the pages of newspapers, and in the broadcast media. At one stage, some of my colleagues at NCAR reported that a preliminary chemical analysis of stratospheric air suggested some of the reaction products expected in the Rowland-Molina theory hadn't been observed. Industry immediately jumped on the news, publishing full-page ads in the

New York Times and elsewhere calling the industry's sober judgment justified and the environmental concern premature. That this finding was given so much weight was ironic since several weeks later the scientists involved discovered a mistake in their work. Its correction led to measurements that actually supported a CFC-ozone destruction theory.

The media, of course, drank it all in and for awhile in 1974 it was hard to find a programme or read a newspaper that did not mention the controversy. The administration, deeply embroiled in the Watergate scandal, wasted little energy defending industry or taking major sides on the issue. Many members of the federal government were thus free to reach their own judgments as to the issue's relative importance and the appropriateness of dramatic regulations that environmentalists were calling for. Despite the fact that the chemistry predicting the amounts of ozone destruction was extremely complicated, rendering any estimate highly tentative at best, the United States nonetheless chose to regulate the industry. In 1976, a Food and Drug Administration official said, "It's a simple case of negligible benefit measured against possible catastrophic risk for both individual citizens and for society. Our course of action seems clear beyond doubt."[24]

Many national and international assessment groups were convened and quite a number of scientists, economists, and others spent weeks criss-crossing the skies (mostly in lower-flying subsonic aircraft) on their way to meetings on the science and policy implications of the CFC-ozone controversy. The National Academy of Sciences convened a committee on the impacts of stratospheric change (CISC). The initial publication of the CISC report was delayed because some of it was leaked to the press, which only served to heighten interest. This in turn made the academy committee, which consisted of thirteen physical scientists, more cautious than ever—they didn't want to be burned by unforeseen scientific developments that could either kill the theory or rapidly accelerate it. As a result, the CISC report (while strongly supporting the hypothesis that CFC emissions could deplete the ozone layer) included a curious and, I believe, inappropriate recommendation *against* the decision to "regulate at this time." Instead, the committee suggested that two more years be devoted to studies aimed at reducing uncertainties in the ozone-depletion theory.

The committee completely disregarded the difficult position

this recommendation would put regulatory agencies in. Regulatory processes can be lengthy, in the order of years to perhaps a decade, so a delay of two years before even starting the process could result in a regulatory delay much longer than even the committee might have expected or approved. Furthermore, the industry immediately cited the prestige of the Academy to imply that there was no need for emission controls, certainly not for the two-year period. Russell Peterson, former Republican governor of Delaware, was then the chairman of the President's Council on Environmental Quality. He complained that the academy committee was out of line in deciding how much uncertainty was enough for policy and thereby assuming the role of policy maker. Instead, he argued, it should have stuck to the scientific issues, leaving the value judgments of whether or not current uncertainties justified regulatory action to those responsible for those judgments—government officials. The CISC's recommendation underscores once again the need for experts to carefully avoid mixing facts and values, an all-too-common pitfall when dealing with complex technical issues.

Nevertheless, by 1977 the Environmental Protection Agency, the Food and Drug Administration, and the Consumer Product and Safety Commission jointly announced U.S. emission controls prohibiting nonessential uses of CFCs in aerosol spray cans. This was the first time a substance suspected of causing global harm had been regulated before the effects had been demonstrated fully. (The earlier ban on atmospheric testing of nuclear weapons was implemented well after widespread contamination from radioactive fallout had occurred and evidence already existed that radioactivity could cause health damage.) Thus, a precedent was set that would prove important in the future for the protection of the ozone layer and that could have major implications as well for global control of other atmospheric gases, such as carbon dioxide.

By late 1982, about twenty countries had taken some form of action to control CFC emissions. Several factors helped make this regulatory action possible: largely pro-environmental public attitudes in the 1970s, strong media attention to the issue, the presence of a strong regulatory authority, limited influence of CFC production on the national economy, and a reasonable availability of substitutes, including hand pumps for many aerosol applications. But not all countries faced the CFC ban with the enthusiasm the United States showed. The Soviet Union, Great Britain, and Japan were much more sceptical. For example, in 1979 I attended the first World Climate Conference in Geneva. It was a major international event featuring such important issues as the greenhouse effect, drought, soil erosion, and the role of climate change in global development. Sir Basil Mason, then director-general of the British Meteorological Office and the major British meteorological figure, gave a keynote address. He noted that in the early 1970s initial predictions of stratospheric ozone decrease from supersonic transports were as large as 10% or 20%, but that now, in 1979, improved chemistry in the models showed that a fleet of SSTs—Concordes, which fly at lower altitudes than the hypothesized U.S. SSTs, as he neglected to mention—would, according to the latest chemical calculations, actually *increase* the ozone a couple of percent. "A rather expensive way to maintain the ozone layer," he joked. The atmosphere is resilient, he said; it will make fools out of those who do not understand its complexity. This lesson, which he drew from the SST-ozone controversy, he then applied to the greenhouse effect, suggesting that perhaps nature has some strong negative feedback mechanisms that will make this problem go away too.[25]

I was annoyed. I squirmed in my seat as questioner after questioner was recognized from the podium and the time for questions was frittered away. Finally, with seconds left, I was recognized. "Professor Mason," I said, "I agree that there is much uncertainty in the chemistry of the stratosphere and the climate response to CO_2. Indeed, perhaps our calculations are way off, as you showed for the SSTs. But time probably didn't permit you to show that the very same chemistry that made the SST-ozone depletions rise from an early 1970s estimate of a 10% decrease to a late 1970s estimate of about a 4% increase also made the CFC estimates of ozone depletion from CFC buildup grow from an 8% drop predicted around 1977 to nearly a 20% decrease by 1979 calculations. I agree that uncertainty means that the numbers we project could be wrong. But the sword of uncertainty cuts two ways. The lesson isn't necessarily that the atmosphere is resilient, but that when we modify it with disturbances that are comparable in magnitude to the chemical composition or energy flows in nature we are bound to experience some significant changes."

Happily, Sir Basil and I have since had several cordial interactions. Unfortunately, that is more than could be said for some members of the British entourage, for, as discussed later, some U.K. representatives strongly oppose the notion of environmental regulation of industry before absolute scientific certainty has been established.

By the early 1980s, the CFC-ozone problem faded from public view, probably because the initial regulatory action gave the impression that the problem had been solved. Furthermore, the Reagan administration's anti-environmental stance in its earliest years hardly encouraged regulators or environmentalists to try to expand the chlorofluorocarbon ban beyond spray cans or to pressure the other countries that were doing very little. The loss of American leadership in environmental issues in the early 1980s was noted recently at a *Time* magazine environment conference, where a number of international political, industry, and scientific leaders came together to discuss the revival of the environment as an issue in 1988. At that conference, French environmental official Brice Lalonde remarked, "Through the late 1970s, lots of things we learned about the environment came from the United States. And [in the] late seventies, it stops, and the lead [switched to] Scandinavia, Germany, and the Netherlands." To this Tennessee Democrat Senator Albert Gore quickly responded, "January of 1981, to be precise."[26]

Despite the fading of the U.S. environmental leadership in the early 1980s, there was a small movement toward international action to protect the ozone layer from human disruption. In fact, under the auspices of the United Nations Environment Programme, a convention on a global framework for protection of the ozone layer was organized. Still, the wheels of international diplomatic progress turned very slowly relative to the tastes of environmental activists. For endless years, it seemed to them, the chlorofluorocarbon industry had been growing rapidly, emissions had been and still were increasing, and the ozone layer was falling increasingly under threat. Then, in 1985 nature finally stood up and took centre stage—with much help from the media. A gaping hole opened up in the ozone layer over Antarctica, providing impetus to what I believe is a major swing of the political pendulum back toward environmental concern. Ironically, publicity associated with the ozone hole arose and the enhanced pressure to regulate emissions of CFCs was exerted while the Reagan administration was preoccupied with the Iran-contra scandal. Once again, middle-level government officials at regulatory agencies and other places, many of whom had virtually been in hiding during the early years of the administration's anti-regulation crusade, reasserted themselves and began working quietly to protect the global atmosphere.

It's ironic that with all the modern satellite technology in the possession of NASA and other U.S. agencies, the opening of a hole in the ozone over the South Pole went undetected for years. It wasn't that the satellites were failing; rather, the scientists who wrote the computer programs that diagnosed vast volumes of satellite data were specifically asking the computers to reject measurements that diverged sharply from normal conditions. In other words, every time a high or low value came in, the computer program rejected it and called it to no one's attention. Incredibly, the phenomenon went undetected by the high-technology branch of science for nearly a decade. Instead, it was discovered by British scientists plotting by hand their own records of how much UV radiation was reaching the earth's surface at their station on the coast of Antarctica.[27] They detected a steady decrease in the southern springtime amount of ozone from the mid-1970s to the mid-1980s, a phenomenon that no one expected and no one could explain. Immediately, U.S. space scientists reprogrammed their data-analysis programs to allow all values, and there in beautiful living colours for all to see (ideal for television) were maps showing a deep hole in the ozone growing in intensity and length over the Antarctic continent and nearby oceans.

Cries for regulation came up again from the environmental community, predictably met by industry denials that the theory had been conclusively demonstrated. Environmentalists claimed that the ozone hole proved they had been right a decade ago, that you don't fool around with Mother Nature or you get nasty surprises. Industry retorted that *it* was right a decade ago when it charged that the atmospheric chemists really hadn't understood what they were doing or they would have been able to predict the ozone hole. In any case, industry argued, the scientists still had no idea whether the hole was caused by CFC buildup or was entirely natural.

In the mid-1980s, a major set of scientific expeditions to the

Antarctic were mounted to find out what was going on. This time, camera crews from all over the world accompanied the scientists, and a press conference led by Boulder NOAA scientist Susan Solomon was even held on the south polar continent. The first year's data were suggestive that chlorofluorocarbons were connected to the ozone hole but were not conclusive. But by 1988 the chemical fingerprints measured in the cold Antarctic stratosphere as the sun rose over that frozen continent clearly showed the presence of chlorine compounds concurrent with a decrease in ozone. The reason the hole occurred there, catching everybody by surprise, was that the stratosphere over Antarctica is nature's coldest spot. There it is so cold that sulphuric and nitric acids exist in the atmosphere in a frozen state as thin clouds, and the ice particles of these clouds are ideal surfaces for the catalytic reactions that destroy ozone. No one had predicted the hole because people were using "homogeneous chemistry" in their calculations – that is, chemistry based on reactions among gases without solid particles in the mix.

Amazingly, even before the CFC-ozone-hole link had been strongly established, a stunning political-environmental achievement was won: in the fall of 1987, the Montreal Protocol on Substances That Deplete the Ozone Layer was approved. This protocol established international regulations on chemicals that can both destroy the stratospheric ozone and exacerbate the greenhouse effect – since CFCs are projected to account for as much as 25% of the greenhouse effect's increase in the next century if they are not controlled. Against these potential dangers, the negotiators had weighed the social, environmental, and economic costs of regulating chemicals upon which the livelihoods of many workers and investors rested. Activities in food processing, plastics, transportation, electronics, cosmetics, fire prevention, and health care would all be affected. The resulting regulations were based on state-of-the-art scientific theory with the acknowledgement that major, let alone certain, evidence of both ozone depletion and ecological or human damages from ozone depletion or climatic change was not in hand. Although existence of the ozone hole had been known and widely publicized prior to and during the Montreal deliberations, very few – except Sherry Rowland, who first pointed out the CFC-ozone connection – were willing to claim that there was any clear connection between the ozone hole and chlorofluorocarbons. That connection would only be brought out many months after the protocol had been signed. As noted by the chief U.S. negotiator, Ambassador Richard Benedick, "The accord was . . . unique in the astonishing rapidity with which it was achieved, considering the complexities involved: thirteen years from the first scientific hypothesis in 1974, and only nine months of actual diplomatic negotiations in four formal sessions beginning in December 1986."

How did Richard Benedick, a career diplomat in the U.S. State Department, become the mild-mannered hero of the ozone treaty? I first met Ambassador Benedick when we appeared on a joint panel discussing atmospheric change before the members of the fifty-eight universities forming the nonprofit corporation that runs my laboratory. Benedick was just beginning his involvement with environmental treaties when this meeting took place. The primary thrust of his brief comments to the audience was that we scientists should not overstate our knowledge or minimize our uncertainties lest we lose the credibility that is so vital to achieving pollutant-control policy action. I recall thinking that this was yet another Reagan administration official using platitudes about uncertainty as an excuse to delay concrete pollution-control actions. How wrong that snap assessment would ultimately be.

For the next year, Richard Benedick spent considerable time with scientists and industry and environmental advocates, all of whom would influence him greatly. In particular, Robert Watson, an important official in NASA's attempt to understand the stratosphere in general and the ozone hole in particular, had many opportunities to explain to Benedick the science behind the belief of many world-class atmospheric scientists that regulating CFCs was not premature, despite the remaining uncertainties. Others, such as environmentalists Michael Oppenheimer of the Environmental Defense Fund and Rafe Pomerance of the World Resources Institute, also spoke repeatedly with Ambassador Benedick, pointing out some of the potential risks of delaying action. Indeed, at one stage, Watson, Benedick, and a few other Americans travelled to numerous foreign capitals to provide scientific and other information about the issue. Their tireless efforts to discuss the questions with the officials and industry representatives of various countries helped to change many minds about the importance of pressing on with an international treaty. At one stage, the British government complained to the U.S. State Department that it was encouraging U.S. environmentalists to lobby inappropriately against British foreign policy. Indeed, Ambassador Benedick has provided a thorough and eye-opening account of the complex and politically intricate nature of the negotiation process in a monograph.[28] Here I will briefly touch on some of the highlights of that year of negotiations.

The chief antagonists in the negotiations were the United States and the European Economic Community (EEC). Despite their political alliances and common economic and historical connections, the United States and the EEC, which together accounted for more than 80% of the world's output of CFCs, differed substantially about the issue of regulation. For example, after the United States banned chlorofluorocarbons as spray-can propellants in 1978, the EEC grabbed a much greater share of the market. However, European public opinion cut considerably the growth rate in domestic chlorofluorocarbon consumption in Europe, so EEC producers adjusted by becoming the principal nations to export CFCs to other parts of the world. By 1985, the year before the negotiations began, the EEC countries dominated world production for two principal CFCs, whereas the U.S. production was a third lower than it had been ten years earlier and its share of the market was about half that of the Europeans. The United States resented the Europeans' general refusal to impose the kind of controls we had had in the 1970s. On the other hand, the Europeans, especially the British and French, had not forgotten their anger over the 1970–1971 Concord SST controversy and were suspicious that this was simply a ploy of the U.S.—using an environmental scare to weaken a European industrial market superiority. Furthermore, they suspected that if we were so gung-ho on regulating CFCs, we must have ready substitutes waiting in the wings, which meant we would profit from regulation. So, in Richard Benedick's words, the European industry had as its "primary objective to preserve their expanded markets and to avoid the costs of switching to alternative products for as long as possible." The Europeans thought the Americans "had been panicked into enacting the aerosol ban," continued Benedick, "and therefore had only themselves to blame for any market losses. Indeed, they hoped that delaying agreement on international controls would provoke impatient environmentalists in the United States to demand a second round of unilateral regulations, which could further consolidate Europe's competitive edge."

However, as the negotiations wore on, it became clear that the European block was by no means united. Germany, a principal manufacturer of CFCs, had a strong domestic environmental constituency and was not nearly as implacably opposed to regulations as Britain or France. Furthermore, Belgium, the Netherlands, and the Scandinavian countries had already been leaning toward strong environmental control. Greece, Spain, Ireland, and Portugal, on the other hand, virtually ignored the entire proceeding. Europeans have long tried to smooth over historical animosities, and the European common market has gone a long way toward that goal. Therefore, the EEC members agreed to achieve internal consensus before and during negotiations, which, as Ambassador Benedick put it, "tended to make it a difficult and inflexible negotiating partner." The central executive of the EEC demanded to be the only spokesman for all members at formal negotiations but was frustrated when individual European nations ignored this idea.

Bitter behind-the-scenes fights were apparent to more than the EEC negotiators. The "Council of Ministers"—that is, the environmental ministers of the EEC countries who had the authority to determine a negotiating position—only met (and still only meet) twice a year. Therefore, there were times when EEC negotiators, while waiting for the ministers to meet in their next scheduled get-together, simply could not negotiate directly. This situation, along with the demand of EEC unanimity, appeared for a time to be an insurmountable obstacle to rapid treaty approval. By chance, the EEC presidency, which automatically changes hands every half year from one member country to another, proved to be one of the most important steps leading to the eventual reversal of the EEC hard line on the chlorofluorocarbon regulations. A British delegate had been president up until January 1987, when a Belgian replaced the British official, at which point the past, present, and future presidents all met together in closed-door meetings to determine the EEC position. But when the presidency revolved again in July, the British had rotated out of this inner circle, which now consisted of representatives from Belgium, Denmark, and Germany—the EEC countries that favoured stringent controls. As Richard Benedick noted, "It is interesting to speculate how much this serendipitous constellation, in the right place at the right time, influenced the final EEC acceptance in Montreal of considerably stronger measures than it had originally favoured." In the final chapter of the protocol, which called for 50% reductions in CFC production by the turn of the century, most EEC countries, including Germany, France, Italy, the Netherlands, Belgium, and Denmark, sent high-level representatives to sign the treaty. There was one exception: the United Kingdom, whose representative was a mid-level environmental official.

The impression so far is that the United States led the fight for strong regulations with little or no internal dissent. That would be a misapprehension, for there was indeed substantial internal administration opposition, reportedly from the departments of commerce and interior, to any international regulations dealing with environmental questions marked by large uncertainty. But by the time these forces became organized, the publicity surrounding progress at Montreal had reached Congress and the public. According to Richard Benedick, secret orders from President Reagan to continue the process and see it completed overwhelmed rear-guard action from some administration conservatives. This action was consistent with the substantial reversal in anti-environmental attitudes of the Reagan administration in its final few years, a trend that I hope will grow in the Bush administration, although at the time of writing this in mid-1990 most environmentalists are disappointed.

Even though the United States had signed the treaty and was the principal protagonist in its eventual adoption, almost immediately after the 1987 Antarctic expedition, there was substantial complaint from the environmental community that a 50% CFC ban simply would not be enough if the ozone hole was to be checked. The treaty does permit a reexamination of the appropriateness of the ban, with the possibility of further strengthening it, depending upon new scientific information. Indeed, disturbing new scientific information surfaced in the months following the Montreal signing. Antarctic ozone expeditions had led to a widespread consensus that the ozone hole was caused by chlorofluorocarbons. Furthermore, a number of estimates have suggested that we need at least a 90% ban—a percentage originally proposed by U.S. environmentalists but rejected by the other countries participating in the protocol negotiations as too radical—if we are to prevent the ozone hole from expanding further. Even more disturbing was the conclusion of a U.S.-led scientific committee that not only was a springtime ozone hole in Antarctica proof of the seriousness of chlorofluorocarbon buildup, but also that the best calculation of *global* ozone trends suggested a depletion of several percent in the past decade.[29] When taken together with the growing consensus that the ozone hole is a creation of CFC buildup, this finding will result in substantial pressure on most nations of the world to ratify the present Montreal Protocol, and will substantially strengthen the protocol's regulatory grip on these potentially dangerous chemical emissions. (October 1988 saw some amelioration in the ozone hole, but this apparently was caused by a periodic atmospheric variation that kept the stratosphere over Antarctica relatively warm and thus not as packed with clouds as in 1986 or 1987 or 1989.)

Does this Montreal process, whose complexity has been only briefly outlined here, lend any hope for a global treaty on emissions of greenhouse gases, especially CO_2? Opinions are divided on this issue, with sceptics pointing out that half the chlorofluorocarbon uses are frivolous—in spray cans or foam cups, for example. Moreover, substitutes should be readily available for these applications—and indeed it is rumoured that Dupont is about to manufacture some. Nevertheless, it may not be that easy to get an agreement for a total phasing-out of CFCs on a global basis, despite the fact that this single action would substantially reduce two global atmospheric threats, ozone depletion and global warming. For example, at the *Time* magazine meeting mentioned earlier, when World Resources Institute president Gus Speth and U.S. Senator Albert Gore suggested a phasing-out of chlorofluorocarbons by 95% at the end of the century, the Soviet representative, somewhat reluctantly, added a cautionary note: "I agree completely with Senator Gore that CFC production must be banned," said Vladimir Sakharov, "but I'm afraid that for us that it's not feasible in the near future. I'm afraid very much that it would be extremely difficult, if not impossible, to go further than the Montreal Protocol in the near future. I wouldn't specify what does it mean in this case 'near future,' but five years."

"Why?" Gore asked.

"Because for us it's not only a political question but an economic question. Our economy is not as flexible as yours . . . and when we discuss in our country the Montreal Protocol, we had very hard times with the industry, with the chemical people."

Gore asked if the initial Soviet resistance to the Montreal Protocol was due to the Soviets' desire to build semiconductors, to which Sakharov replied, "like practically in every country, spray cans, refrigeration and semiconductors as well. And, you know, we have to change mentality in many things. Even in this particular building [NCAR] where we are now, downstairs they are using CFCs for cooling of semiconductors [he was referring to NCAR's supercomputer] and it is done by people who know very well the dangers of it."

Gore acknowledged that the resistance to further bans was legitimate, "but that doesn't mean it shouldn't be proposed and pushed. And maybe with *glasnost* it will be a groundswell of support," Gore said.

"Absolutely. Because, me, myself, I am an environmentalist," said Vladimir Sakharov, "so I'm quite on your side. And we're going to push our industry." Sakharov went on to say that if the Soviets were to achieve environmental improvement, they would need to replace their ageing technology with new technologies, presumably from the West.

Even though a phasing-out of CFCs would be the single biggest step toward stabilizing the global atmosphere, this exchange shows that it may not be simple to achieve, despite the desire on all sides and excellent media coverage of the event—after all, this meeting itself was held by a major media representative, *Time* magazine, which made the transcript just quoted.

Sceptics would argue that substitutes for fossil fuel burning are even more difficult to introduce than those for CFCs, and that a "law of the atmosphere" will be much tougher to achieve. Furthermore, the pessimists contend, the consequences of ozone depletion are clearly dangerous and seem uniformly negative—that is, serious, soon, and certain—, whereas the climate changes associated with the greenhouse effect cannot all as easily be proved bad.

But Ambassador Benedick is less pessimistic. He feels that many lessons from Montreal are immediately applicable to the process of developing a law of the atmosphere. He cites four important components. First, it is important to build scientific consensus and to impress this consensus on government policy makers through close collaboration in which each side comes to understand the working context of the other. This collaboration, he asserts, "contributed to the irresistible logic of the American position on ozone, and greatly strengthened the persuasiveness of U.S. negotiators; in contrast, the European commission based its tactics largely on self-serving contentions and data provided by industry, which, ultimately, proved less convincing in the international arena" than the less biased materials provided by independent scientists. Second, the political will of nations can only follow if public opinion is adequately informed. The role of individual scientists at national academies in building a credible case, and in helping to translate and disseminate information through the media, is critical. International organizations such as the United Nations Environment Programme (UNEP) have been engaged in major educational efforts. UNEP's director, Moustafa Tolba, and the head of its New York office, Noel Brown, were generous in their efforts to promote the issue. Similar efforts will be needed if greenhouse gas emissions are to become publicly familiar, Benedick believes. (Of course, no public education programme can easily make climate change appear as obvious and dangerous as skin cancer, as the pessimists have pointed out.) Third, the success in the ozone protocol suggests that a treaty on the greenhouse effect may require a multiple-step process. Initially, a framework convention would be called, primarily as an expression of various nations' commitment to act. That would be administratively (not necessarily politically) easy, for a Vienna Convention has already been signed. It set up the framework for the subsequent Montreal Protocol. Konrad von Moltke, of the Conservation Foundation, has suggested that all we need to do is add the words "greenhouse gases" after the phrases in the Vienna convention on controlling ozone-depleting substances. This would be much easier than making any new conventions, he has argued.[30] This updated convention would then be followed by individual protocols with specific actions.

All along, international scientific assessments of the highest credibility would be needed. However, Benedick cautions, climate change, unlike ozone, has many aspects—energy production and demand, population, deforestation, agriculture, coastal zones, biological diversity, and so on—and he agrees with the pessimists that it will be difficult to address them all in a protocol treaty. It is for this reason that he recommends a step-by-step approach.

Finally, the most important component of the Montreal Protocol is its dynamic and flexible capacity to modify itself as new scientific, economic, or technical assessments are made. The treaty has provisions both for calling emergency meetings in the case of fast-breaking developments and for differentiating the regulatory impact on developed and developing countries. In short, the treaty's sensitivity to special situations ultimately brought about the compromise necessary to get it approved. The media, on balance, did a fine job of carrying the technical ozone debate to the public, but they have spent much less ink or air time on these complicated diplomatic manoeuverings.

The greenhouse effect and ozone depletion teach the same lesson: we cannot continue to use the atmosphere as a sewer without expecting substantial and potentially irreversible global environmental disruption. The problem scientists face in motivating people to make immediate economic investments to minimize this disruption involves the tough tradeoff between the need to give clear and dramatic statements versus the ethical requirement to tell the whole story, complete with both uncertainties and caveats. This is the double ethical bind I described in the Preface. On September 7, 1988, with the Summer of '88 still fully in American consciousness, the ABC news programme "Nightline" broadcast a segment dedicated to the greenhouse effect. I was contacted as a possible guest but was later told my views were "too moderate." Some of the exchange between "Nightline" moderator Ted Koppel and the environmental activist Michael Oppenheimer, of the Environmental Defense Fund, helps to make this dilemma quite explicit:

KOPPEL: Dr. Oppenheimer, I'd love to be able to say to you that I think the American public can get energized over some perceived threat forty years down the road, but I don't believe it. Do you?

OPPENHEIMER: Well, I think that they can. This summer has provided us with a vivid example of the kinds of changes that are in store for us if we don't move to limit the greenhouse gases—record heat, record drought, record smog levels, forest fires putting our forests up in flames. It doesn't matter whether the summer was due to the greenhouse effect, and in fact, we'll never know whether it was or wasn't. . . .

KOPPEL: Well, forgive me, but I think it does matter. If—if it was not caused, if it is simply a coincidence . . . that's one thing. If it is the consequence of the greenhouse effect and you can draw some kind of correlation there, then presumably we can present evidence to the American public and say, "Look, it may happen in forty years, it may happen in sixty years, but folks, there's disaster that far down the road and there are some really bad things that are going to happen ten years from now, or five years from now."

Koppel then introduced Alan Hecht, Director of the National Climate Program office, who has long been concerned with issues of climate change. However, true to the need for dichotomy, Hecht was brought on as a "balance" to Oppenheimer and indeed pointed out substantially more uncertainties than Oppenheimer had. Despite that, a few minutes later in the programme, Koppel came back to the magnitude of the risk.

KOPPEL: Dr. Oppenheimer . . . I mean—there are not many opportunities when someone with your background and expertise, or Dr. Hecht, has a chance to talk to several million Americans at the same time and say, "Hey, dummies, wake up. What we are doing here is causing a problem." Now you can either say it in such a way that people sort of doze off while you're saying it, or you can say it in such a way as to convey a sense of alarm.

Even the normally outspoken Mike Oppenheimer was unable (to his credit) to overcome his scientist's instinct and take Koppel's invitation as an opportunity to deliver a "certain disaster if we don't act" speech. He led off his reply with a caveat.

This exchange once again reflects the difficulties scientists have in trying to communicate the seriousness they feel about a rapidly increasing buildup of atmospheric pollution. There is a tension between the scientific culture of caution and reticence and the media's penchant for the drama, dread, and debate that keeps the show lively and the audience tuned in. I often close public lectures with a slide of a licence plate from Missouri displaying the slogan "Show-Me State." While this is certainly a practical piece of folk wisdom, in applying it to environmental matters we will be forced to experiment with our own planet if we must prove potential threats beyond a reasonable doubt before we take action. To me, a rational society anticipates problems, just as a rational person not shackled with poverty buys insurance. How much insurance is appropriate for any person, enterprise, nation, or world is itself a difficult question, one that is addressed in the next chapter. In this discussion of the media's role in dealing with hypothetical or long-term threats, I want to end by suggesting that there is a growing mismatch between the complex nature of reality and the way such problems are usually reported in the popular media or perceived by the public. Self-governance demands an informed public and a knowledgeable political leadership. If information is distorted by an overemphasis of extreme opposing views, then the policy-making process will not be rooted in the level of understanding appropriate to the reality of the issues. Scientists must learn to communicate clearly without fear of using familiar language and metaphors. (At a minimum, I wish they would be less critical of those who do!) For example, when a scientist or statistician suggests there is "no evidence" of a greenhouse gas-induced signal in twentieth century temperature records because of his implicit assumption that "evidence" begins only after 95% or 99% statistical significance is established, this is not a responsible public statement. In my opinion, few non-statisticians can be expected to know the assumptions professionals implicitly make. Any specialist making public comments is obliged to use language familiar to the audience and to anticipate their level of technical comprehension before making pronouncements, particularly on controversial issues or those with public policy overtones. The media, on the other hand, must come to report the spectrum of scientific opinions without concentrating on the most contentious and extreme opposing views, interesting as these may be. Mainstream views may not be as exciting, sell as many newspapers, or increase advertising revenue on television, but if the media is to carry out its essential role as the independent guardian of truth in a democracy, then it needs to give us more of the *whole* truth, not simply a selection of the most dramatic or visually fascinating bits.

8

Coping with the Greenhouse Century

Is a growing greenhouse effect inevitable, or is there something we can do to reverse it, stabilize it, or at least adapt to it? Those questions are rarely raised in purely scientific debates but are among those I am asked most frequently at public forums around the world. At this writing, the event freshest in my mind is the ten-day Greenhouse '88 event in Australia, designed to raise the consciousness of the Australian public. Because the emphasis there was public awareness, not scientific evaluation, a quick summary of the effort to explain publicly the greenhouse effect will prove highly relevant to the question, Can we cope?

GREENHOUSE DOWN UNDER. Australia's total population is only 60% of California's, but the country is roughly the size of the continental United States. As inhabitants of an island continent with nearly the whole population living in coastal cities, Australians are very sea-level conscious, and because a large proportion of their nation is arid or semiarid, they are similarly concerned about temperature increases and rainfall decreases. Perhaps for these reasons, Australia was the first country to sponsor a nationwide media event on the greenhouse effect.

The facilitators of this saturation coverage of global warming were a small, dedicated group centred in Melbourne. They are part of a compact minibureaucracy (only a million-dollar operation) known as the Commission for the Future (CFF). As I understand it, CFF was the brainchild of Barry Jones, Australia's federal minister for science, customs, and small business. Jones struck me as a middle-aged *enfant terrible,* a former quiz kid whose enormous energy, intelligence, and verbal talents had been tapped by the political system. Although voluble and feisty, Jones is a rare breed as far as government officials go. His perceptiveness and honesty are widely admired, even though these same characteristics seem to keep him from gaining major political power.[1]

In 1987 the Commission for the Future had, with Australia's science branch, the Commonwealth of Scientific and Industrial Research Organization (CSIRO), conducted a major scientific conference that addressed how specific scenarios of climatic change in Australia—presumably wetter in the centre and north, drier in the southwest, and increased sea level and heat waves everywhere—would impact on Australian farming, water supplies, ecosystems, coastal development, public health, and so on. That scientific meeting netted a very impressive collection of thoughtful essays[2] by professional planners to the effect that many sectors of Australian interests were indeed vulnerable to climate change, that some sectors could benefit from some types of changes, and, most problematically, that few planners could make many immediate decisions about how to change their own planning activities specifically without knowing more about where and when it would be wetter and drier (that story is by now familiar and will be echoed in many other places, as we saw in Chapter 6). That meeting, Greenhouse '87, got a fair degree of publicity Down Under. Then a warm winter in 1988 (which happened during the northern summer of '88) combined with a very dry October in Sydney (Australia's largest city, population nearly 4 million) to fuel intense media interest in the first week of November when Greenhouse '88 was to take place.

My involvement began in January of 1988 when Philip Noyce of the commission invited me to be the keynote speaker on the first day of a three-day meeting in Melbourne. This was not to be a scientific meeting like Greenhouse '87, he explained, but rather a media event. The opening-night show in Melbourne would be linked by satellite to nine other major Australian cities. The purpose was to focus on policy responses to the greenhouse effect, not to debate once more what detailed changes might take place. "Why me?" I asked Phil, mentioning a number of Australian scientists who were both competent in the technical issue and articulate. "We Australians seem to pay more attention to foreigners, particularly Americans," he quipped, "and we need someone who is both scientifically credible and experienced in dealing with politicians and the media." The event sounded unique, and Australia seemed a great place for a family vacation (even though it didn't turn out that way), so I agreed.

Sometime in mid-October, the phone started to ring. Australian Broadcasting Corporation (ABC) radio journalists were calling to get advance statements on the issue. A few days before the departure, my itinerary arrived. I was stunned. From the day I landed in Sydney, it was nonstop media, government, or public events. It looked like a political campaign. A dozen interviews in one day plus flights from Melbourne to Canberra (to brief the national Parliament), to Sydney (to do a press conference), and back to Melbourne (to talk to water planners, and so on) were all crammed into twenty-four hours! This was typical of the pace for five days before the national broadcast on Thursday evening. After that, I had to fly to Perth to participate in still more meetings, press conferences, and public discussions. I had to insist on getting three hours "off" to give a technical talk to my scientific colleagues at CSIRO near Melbourne, a visit I had to make if my sometimes sceptical scientific friends weren't to think I had moved out of science completely and over into the political arena.

I don't disapprove of media events; nor am I uncomfortable with them. But I was used to dealing with the media all at once following a presentation to a public but largely professional audience. The Greenhouse '88 week was to be composed of dozens of ten-minute interviews by radio and television stations and reporters from localities all over Australia. Moreover, I suspected they would all want to know what the local implications were going to be, something I couldn't say much about confidently for the United States, let alone Australia. Nevertheless, a commitment is a commitment, so off I went.

The week began in the beautiful city of Sydney. As "luck" would have it, on my second day in late October (seasonally equivalent to the northern spring month of April), strong desert winds from the north heated Sydney to near record levels of about 37° C

(100° F). This was quite a day for beginning the media blitz on the greenhouse effect. I was pleasantly surprised to find that most of the journalists I dealt with were reasonably well informed about the issues. Earlier, the Commission for the Future had distributed a copy of my congressional testimony on global warming along with some statements by other scientists. Most of the reporters had read them, and their questions were largely relevant and appropriate: "How do we know who is telling the truth over this controversy?" "What does global climate change mean for New South Wales?" "Exactly how much will the sea level rise?" "How much energy conservation can we afford?" "Is the greenhouse effect only an excuse for more nuclear power?" "Is there anything Australia can do by itself?" "How do we get Third World countries to take this seriously?" and not least, the ubiquitous question: "What is it the average person can do?"

Part of my schedule involved doing a number of "talk-back" shows, what we call talk shows. These I anticipated with some trepidation. For example, one of them was on ABC's "country radio"—a nationwide farm broadcast network in which my interview followed a programme known as "Earthworm." I was fearing the worst: someone calling in to say that God gave us the earth to exploit as we could to multiply and prosper. All that worry was unfounded. The people who called in were by and large well informed, polite, and sincere. Their questions roughly paralleled those of the professional press but were usually less contentious. One obviously elderly lady asked me if it was all right for her to finish using up the can of insect spray she had had on hand for the nasty wasps that had got into her house and were bothering her. She had actually read the label, discovered that the spray contained chlorofluorocarbons, and concluded it would be unfair to destroy the ozone layer simply to avoid a few stings. (I wonder how many of us have such fine ethics and would be willing to make the sacrifice?) I assured her that indeed the propellant she named both could reduce ozone and increase the greenhouse effect, but that in any case it would escape from the can at some point on its own, so she didn't have to feel guilty about protecting herself with the remaining contents of the can (but that she should cover the food first). However, I told her, and I hoped others were listening, there were many products in Australia that contain nonchlorofluorocarbon propellants, and

I hoped she would read the labels and purchase accordingly.

"What are some of the things *I* can do myself," said one man who called in. "I mean, not the grandiose stuff you were talking about like negotiating with the Chinese about their inefficient coal use, but simple things in my own life?"

"Well," I said, stringing together a set of sentences I was to repeat at least four dozen times in the next six days, "when you went out this morning did you remember to turn off your light?"

"No, I don't think I did," was the reply.

"Well, try to remember to do that next time," I said. "I know it sounds preachy—it's like telling people that their vote could make a difference, even though everyone knows nobody wins elections by one vote. Still, a number of elections would have been different if only one more person in each precinct had voted for the loser. Likewise, when a billion people conserve energy a thousand times each in a year, the savings add up. Moreover, doing so sets the right tone for a lot of the small lifestyle changes we need to make if we are going to be effective."

The radio host asked for more ideas. I took his lead. "For example, if you want to buy a refrigerator and you find two models that look identical with similar features but one costs $300 and the other costs $325, which one are you likely to buy?"

"Probably the cheaper one," was the natural response.

"Do you have labelling in Australia for energy efficiency?" I asked. The answer was yes. "So do we in the United States, and the kind of lifestyle changes we're asking for aren't really so difficult in this context. For example, if you read the label, you might find that the $325 refrigerator uses slightly less energy than the $300 refrigerator. Perhaps it costs more money because it's better insulated. All we need to do is a little mental arithmetic—"

"On your solar-powered calculator, of course," quipped the ABC moderator.

"What we need to do," I said, "is calculate from the information given on the label just how much we might save on our electric bill each year from the better-insulated refrigerator. Let's say it's $5 a year. That means in five years, we've gotten our $25 back, and for the ten- to fifteen-year lifetime that is typical of refrigerators, we will have actually come out ahead. At the same time, we will have helped the environment, since using less energy means less pollution."

"What else?" a caller asked.

"We all need to replace our cars," I went on, "and the next time we do that, let's read the labels for petrol mileage. Is it really so important to have the biggest and the fastest? Why not do something that's good for the environment and our pocketbooks at the same time? Get a more energy-efficient car. There's one more lifestyle alteration that is important for all of us," I concluded. "At the next election or town meeting or in the next letter you write to the editor dealing with the political process, ask the politician, 'What are you going to do to improve the economic outlook for next year, but in an environmentally sustainable way?' Politicians are remarkably good at reacting to the perceptions of their constituents, and if we are going to influence political leaders to be creative and work for long-term solutions to help us use energy efficiently at home, to provide alternatives to the Chinese for their planned inefficient use of dirty coal, and to help the Indonesians and Brazilians reverse their rapid deforestation, the politicians have got to know we want them to." That point typically evoked chuckles from the Australians (who often call their politicians polis), since "poli bashing" seems to be a national sport. After giving that little speech for probably the tenth time, one listener called in to ask why I hadn't suggested planting trees to take up CO_2 and to help draw down the groundwater that is salinating our soils and wrecking our crops.

"I'm afraid," I cautioned, "that in order to absorb most of the CO_2 people are likely to inject into the atmosphere over the next several decades, we would have to plant fast-growing forest species in an area the size of Australia, something that clearly wouldn't be easy economically or politically. Moreover, it would be foolish to think that we could entirely offset the changes to the greenhouse effect by reforestation alone. But 2° C warming is not as bad as 5° C warming, and if it is taking place in 150 years, that is not nearly as risky as its taking place in 50 years. So, planting trees here, saving energy there, helping the Chinese become a bit more efficient, and substituting natural gas for coal where possible may each be only a nickel and a dime, but if we nickel and dime long enough, soon we'll get to a half dollar." I was pleased that a listening audience was already beginning to create its own solutions without having to be led by professionals.

That aspect was the most uplifting part of the trip. For example, in the final phase, I flew out to Perth, the frontier of Australia in the state of Western Australia. This beautiful, shining city near the Indian Ocean in the far lower corner of Australia (about as geographically far away from home as I could get and not be in the middle of the South Indian Ocean) was the site of a follow-up meeting to the Thursday-night national broadcast. The Western Australians invited me to comment on the work of six groups of local citizens who had put together what they believed to be solutions to the greenhouse effect. The context was localized to their state of Western Australia (a very arid place except for its extreme southwestern tip, with a total population of about 1.5 million and a size equivalent to that of five or six western U.S. states combined). The meeting was chaired by a man from the state environmental protection authority with the appropriately ironic name, Barry Carbon.

The very first presentation was by a woman who was a tree farmer. She said that the problem with any large-scale efforts to plant trees and preserve the environment in Western Australia was that there were too many interest groups—within the government, the environmental community, and industry. These groups needed coordination, she pointed out, so they didn't duplicate each other and waste efforts. That recommendation was as close to the mark as it could be and equally applicable to Washington, D.C., London, or Peking. "We need demonstration farms to educate farmers on how it can be done," she pointed out, "since they are unlikely to take our word for it without seeing the precise advantages of reforestation." That kind of wisdom was put into practice by the U.S. Soil Conservation Service after the soil-erosion disasters in the dust bowl days of the 1930s, and it helped provide the agricultural extension effort, whose practical demonstrations became the credible driving force behind agricultural productivity.

She went on. "We need tax incentives and must recognize that some of our effort will not be without risk—thus we need insurance for failed experiments. For example, fencing is more expensive than tree planting, and we need a little bit of state help to encourage fence building to keep the animals out of the sapling areas." Finally, commenting on the debate over the welfare state in Australia, her group recommended waiving the "job-seeking" provision for welfare recipients if they do a certain amount of volunteer tree planting. "That way," she said, "we can do good work for the state and at the same time keep these people productively employed by giving them something to do that will lead to self-respect."

The next group report was on water resources, and it picked up where the agriculture and forestry group left off. Add environmental education to the state school curriculum, demonstrate how water conservation and reuse can be safely and economically implemented, integrate interbasin management and build flexibility into management systems to deal with the uncertainty in future climate—these were the kinds of recommendations this group made. They even went so far as to coin an acronym, WATER. This could have two possible meanings: either We All Take Extra Responsibility, or if we don't, We Accept Total External Regulation. Two more team reports were presented; after each, I took questions. One high school teacher said, "I'm not sure what to teach my students any more since you told us you can't be sure precisely what will happen in Western Australia." "Teach them how to question, and not always to look for *the* answers from experts," I said, "and don't feel bad just because we don't know the details about future change. It's most important that the students understand the range of possible consequences so they can choose how they wish to take risks with their own planet."

The energy and industry group argued that political polarization is the enemy of action. "I'm tired of the greenies versus the gainers," said the presenter. "All the people do is get stiff-backed and ego involved, and they cling to ideology." This excellent statement led to a policy discussion dealing with the contentious issue of taxes on polluting fuels. I took the opportunity to state my position. "If we tax fuels in proportion to the amount of pollution they generate, then we include in the price of doing energy business some measure of the degradation that energy production or use implies for our common property resource—the environment. But we shouldn't take the revenue and simply return it to the general fund. Instead we should target it for making energy more efficient and for developing renewable energy sources such as solar, or perhaps even passively safe nuclear. But in any case, let's direct the money—the way we direct our highway trust funds from gasoline taxes toward road construction—toward solving the problem fossil energy use creates."

One Australian got up to say that he didn't want any new overall taxes, so instead why couldn't we "add a tax on coal, but give a rebate to natural gas or solar so that there would be no *net* tax on society while we still provide the right incentives and disincentives for fuel uses?"

"That's a great idea," I said, "except that this doesn't produce new income, so we'd have to cut out some other activity in society in order to provide the funds needed to invest in renewable alternatives or increased efficiency."

"Take it out of welfare cheques," someone said.

"Cut the military," someone else said.

"Fire half the politicians," was the final suggestion.

Finally came the last report, on social impacts. The committee chairman reported that her members, including several high school students, had been asked to think up possible solutions, but also to state what they were going to do to "improve the greenhouse effect" in the next year. One student committed himself to write a report on it and discuss it with his class; another one promised to plant a tree and get all his friends to plant them too. One man said he would give a talk to his Rotary Club, and a college professor said she was going to learn much more about the issue so she could discuss it more intelligently with her students.

"But a greenhouse tax that increases the price of fuel or electricity would be unfair to the poor," the group leader said, "so the group has some concerns about that solution."

"I agree that equity is a critical issue and that we must not arbitrarily put undue burdens on those least able to pay," I replied, "but setting artificially low energy prices is a very bad way to subsidize poverty. There simply is no way to encourage rational economic behaviour if we don't include in the price of energy the costs of environmental disruption. However, what this equity question suggests is that increased prices of energy should also accompany decreased taxes for the poorest elements of society, so that while their net available resources are not reduced they too are encouraged to be more efficient. Perhaps some of the tax revenue from the coal use could be targeted for job retraining of coal miners." We need to be more creative in order to balance fairness against effectiveness.

In Jimmy Carter's presidency, his first energy plan called for taxes on energy imports as well as tax relief for the poor to prevent such inequitable impacts. This approach was ridiculed and ultimately defeated in Congress by those claiming it wasn't an energy policy but a tax policy. Indeed, if the greenhouse effect and protecting the global commons have any message at all, it is that in nature and in politics you can't do just one thing. The distribution of political responsibility into bureaucracies—of agriculture, energy, water, and so forth—simply cannot be allowed to continue to obstruct practical solutions to real-world problems that cut across these administrative fiefdoms. Similarly, the allocation of oversight responsibility in the U.S. Congress into agriculture, water, energy, or environment committees significantly complicates the ability of the Congress to legislate crosscutting solutions to an intercommittee problem like global warming. "It is not only that environmentalists need to accept political realities," I argued, "it is also that political institutions need to learn and accept environmental realities."

The whole trip was uplifting to me, despite the fact that the Australian coal supporters and water-resource planners insisted (as they do in the United States and other places) that there is too much uncertainty to justify action. "You can't expect people to respond to hypothetical situations," a typical press report quoted an Australian official as arguing. "Farmers really want something to hang onto, but we haven't got that yet. . . . It would be a big waste of time to have two or three people working on [the greenhouse effect]. Waiting is more constructive than building dams."[3]

In a sense, the experience brought into focus for me the notion of local versus national responsibility. It is unreasonable to expect local farmers concerned with a watershed, a dam project, or some other complex local resource-management issue, to plan radical strategies in response to the prospect of rapid climate change if we can't specify whether that change will make their district wetter or drier. The most we can hope is that local planners will attempt to build flexibility into their systems at a local level. But this in no way implies that we should wait for more detailed local forecasting skill before acting on a *national* scale. Change implies uncertainty and nearly everyone would support inhibiting the growth of uncertainty. But doing so would be a global-scale preventive measure, and fostering global prevention is a high-level, not a local, policy action. In this vein, national governments and international organizations could choose to set targets for lowered greenhouse gas emissions, and those emissions would then be lower in the future than if no targets had been set. Lower emissions would imply slower rates of climate change, which would mean less rapid and unpredictable changes with which the local planners would have to deal. Perhaps we need to turn around the popular bumper sticker "Think Globally, Act Locally" to read "Think Locally, Act Globally" if we are to slow down global warming.

The Australians I met, armed with a few hours' worth of working knowledge on the issues, kept coming up with ideas—and the same five or six tractable ones over and over again. If the policy professionals were the only ones offering them, I would be much less optimistic about the political tractability of these solutions. But the fact that so many people came up with the bare essentials of the very policies the professionals have tried to forward over many years of debate gives renewed hope that inherent good sense will dominate our political response to environmental policy requirements. Of course, first we have to get that few hours' worth of scientific information into all of our sometimes information-beleaguered citizenry.

POLICY PROCESS. Policy making with respect to complex problems is a multistep process that can be simplified into three overlapping stages. First comes *technical analysis,* in which scientific facts and other relevant information are assembled and analysed to estimate the likelihood of various potential consequences. Improving technical assessments of the probabilities and consequences of each problem can help to put policy making on a firmer scientific basis. However, this stage is ripe for exploitation by partisan groups practising the dubious art of "selective inattention"—that is, choosing from the wide range of available information those parts of the technical story that would seem to bolster their advocated solution. Even though the technical assessment stage of the policy process is often considered the most objective component, the selection and presentation of the "facts" still represent subjective value judgments. However, especially if the analyst is being balanced, the individual facts can at least be weighed on some objective basis.

The second stage, *policy analysis,* is more oriented toward decision making, though it is still technical and can be reasonably objective. In policy analysis, an effort is made to examine scientifically the varying consequences that might be associated with a range of alternative policies. Thus, the economic consequences of a 20% reduction in fossil fuel emissions might be evaluated with respect to strategies of fuel switching, development of renewable alternative energy systems, altered lifestyles, and so on. In this context, the best economic, engineering, or social scientific methods available would, or should, be used to form the differential analysis.

Choosing a mix of options is the final stage. This is the value-laden action of *policy choice.* No degree of scientific uncertainty in either of the first two steps can justify taking or not taking any specific action based on some objective criteria. Instead, we must turn to our own values in order to make the decision whether hedging against potential future change is worthy of the investments of present resources. Don't let anyone tell you that technical knowledge or uncertainty provides a "scientific" basis for policy choice, for science can contribute only to policy analysis, not to policy choice.

In the context of atmospheric problems such as acid rain,* ozone depletion, and global warming, we can break down policy making further into special subcategories. These are (1) engineering countermeasures, (2) adaptation, and (3) prevention. *Counter-*

*Normal rainfall is slightly acidic. That is because raindrops falling through our atmosphere, which contains carbon dioxide, makes a weak solution of carbonic acid. This acid is partly responsible for the weathering of rocks and the manufacture of carbonate sediments which, as we saw in Chapter 4, could be a long-term stabilizing mechanism for the earth's climate. However, other chemicals in the air, especially sulphuric and nitric acids, can easily be dissolved in cloud water, raindrops, or snowflakes and be deposited on the earth. During the last half century human industrial activity has rivalled and often outstripped natural sources such as phytoplankton and soil emissions as producers of acidic chemicals.[4] As a result, acidity in rainfall and snowfall in the northeastern part of North America, in Europe, and increasingly in the western United States and Japan has become a serious ecological problem. This acidity can damage lakes and has killed fish in Scandinavia, New England, and Canada; it also can leach minerals out of the soil and other structures. In soils, it can change the species composition; for example, promoting mosses that lead to bogs at the expense of trees. Acid rain has also created substantial political tension between normally friendly neighbours such as Canada and the United States or Scandinavia and Germany. If China proceeds with major coal use in old-style combustion devices, acidity could become a serious threat to downwind Japan.

measures include active steps to impede atmospheric change through a technical or engineering modification to the environment. An example would be the dumping of a chemical in the stratosphere to counteract the ozone-reducing effects of chlorofluorocarbons. *Adaptation,* often the favorite of economists, involves doing nothing active now to prevent the buildup of the pollutants but assuming that current resources will be used to maximize future wealth from which solutions will eventually flow more efficiently. *Prevention,* generally favoured by environmentalists, involves reducing the destructive factors—in this case, the insults to the atmosphere or environment that are believed to cause atmospheric change. Clearly, the latter is the most politically difficult response, since it requires that we redistribute our current resources to some extent thus altering accustomed levels of expenses or incomes. It will help to be specific by giving examples of these three responses in the context of the three major atmospheric problems: acid rain, ozone depletion, and global warming.

Countermeasures. The purpose of countermeasure strategies is not only to control pollutant emissions at the source, but to control the consequences of emissions by some compensatory antidotal technique. The complexity of the chemistry involved in estimating stratospheric-ozone depletion is so great that I consider it irresponsible to consider strategies at this time that attempt to counteract ozone depletion through the addition of one or more substances to the atmosphere. There is simply too much uncertainty.

The technique of liming areas that suffer from acid rain with acid-neutralizing liming chemicals reverses the local environmental impacts of acid deposition. The efficacy of liming depends on local conditions—for example, how long water in lakes and streams has been present and the feasibility of covering a large fraction of the runoff area with the liming agent. High-volume stream flow in springs and lakes with a high turnover rate are difficult to treat and large-scale liming is infeasible for both practical and environmental reasons. However, this countermeasure can be used to defend some sensitive areas until adaptive or preventive strategies can be brought into play.

Several proposals have been made for reducing the amount of CO_2 in the free atmosphere as a means of moderating the greenhouse effect. And there are ways for physically removing carbon from the energy-generation process before it reaches the atmosphere. One is prescrubbing—taking the carbon out of the fuels prior to combustion, leaving only the hydrogen to be burned. Of course, the residue would have to be dumped somewhere, presumably in the hole in the ground where the fossil fuel had been originally extracted. A second technology is postcombustion scrubbing—the removal of CO_2 from the emissions stream after burning but before release to the atmosphere.

Very little definitive research has been forthcoming with regard to either of these measures. To my knowledge, only one prescrubbing process has received any study at all—this is the "hydrocarb process," in which hydrogen is extracted from coal and the carbon is then stored for possible future use or buried. In this process, only approximately 15% of the energy in coal would be converted to hydrogen for use as fuel in existing coal power plants and there would be lots of residual solid coal material to store. Also, the hydrogen generated by this process would have to be transported, presumably through the existing pipeline system, which is currently limited to an annual flow of about 25 trillion cubic feet of gas in the United States, an amount insufficient to run the present utility business. (Similarly, if we switched to natural gas from coal, we would have to confront the question of pipeline capacity again.) In a report to the Department of Energy, Pacific Northwest Laboratory energy analyst James Edmonds and his colleagues concluded that our experience with this new process was not even sufficient to allow us to estimate its maximum possible cost. Nevertheless, these researchers estimated the range of capital costs to be somewhere between 0.5 and 2.5 trillion (1987) dollars (that is, $2,000 to $10,000 per capita in the United States) to provide 300,000 megawatts of electricity, the generating capacity needed to replace all the coal currently consumed in U.S. electrical-power generation. Postcombustion scrubbing, the Edmonds report concluded, is a "known but unapplied technology." At best, the capital costs for removing 90% of the CO_2 from the stack gases would be about half to perhaps one trillion dollars,[5] (some $2,000 to $4,000 per capita).

Pumping the carbon dioxide produced at industrial plants into the deep ocean was proposed in 1977 by Cesare Marchetti, a creative global problem solver of the International Institute of Applied Systems Analysis (IIASA) near Vienna, Austria.[6] This would reduce and delay the rise of carbon dioxide in the atmosphere, but would not prevent an eventual warming as some of it made its way back into the atmosphere.[7] And the work of removing the CO_2 at a plant would consume something like half the energy output of the power plant, making this strategy economically questionable, as the Edmonds estimates confirmed. Nevertheless, such ideas need careful study.

Using reforestation to make a kind of carbon bank would also capture carbon from the atmosphere, but decay or burning of the harvested trees decades later would have to be prevented—for example, by burial in mines—to prevent the re-release of the CO_2 into the atmosphere. Furthermore, vegetational carbon banks would have to compete with agriculture for land and nutrient resources. Forester Allen Solomon, at IIASA, and economist Roger Sedjo, at Resources for the Future in Washington, D.C., estimated that a land area approximately the size of Alaska would have to be planted with fast-growing trees over the next fifty years in order to take up about half the projected fossil-fuel-induced CO_2 at a cost of something on the order of $250 billion (only $50 per person for the global population). One criticism is that once the trees were fully grown they would no longer be taking up CO_2 very rapidly and thus would need to be cleared so that new trees could be planted to continue quick uptake. To sequester the carbon, old trees could be used for lumber, but not fuel, since the latter would release the CO_2. However, even if used for fuel a delay of fifty years or so, which represents a typical generational growth time of trees, would certainly stretch out the buildup rate of atmospheric CO_2, so delay should be considered a valuable component of CO_2 response strategy.

Proposals for actually counteracting the global warming from the enhanced greenhouse effect must rely on extensive geoengineering. Russian climatologist Mikhail Budyko suggested years ago that airplanes flying in the stratosphere could release dust or other aerosol particles that would reflect away part of the solar energy normally absorbed by earth, thus neutralizing a global warming trend. This could work, at least on a global average basis, but the actual mechanisms of warming and cooling would vary in space and time, implying that large regional climatic changes could still occur.[8]

In general, there are two major problems associated with environmental countermeasure strategies: (1) gaps in knowledge regarding environmental response may result in cures that are worse than the illnesses they are intended to combat and (2) anything that goes wrong could be blamed on the geoengineering action. For example, earlier I suggested that the current uncertainty in estimating the inadvertent consequences of human activities on global temperature is some factor of 2. Thus, it is logical that we might also make an error of equal size in estimating any deliberate global climate-modification attempt. For example, if GCMs have overestimated the CO_2 doubling impact by a factor of 2, global warming by 2050 might then only be 2° , not 4° C. Overestimation is certainly an even bet. And suppose we chose a 4° C global cooling as a climatic countermeasure but overestimated *that* by a factor of 2. In that case, we could produce an 8° C cooling—again, an even bet. If we combine the two possibilities, there would be a 25% chance of *over*estimating the global warming and *under*estimating the countermeasure—clearly, a cure worse than the disease! Considering this gamble plus the fact that it would be irresponsible to attempt countermeasures on a global scale without at least some form of "insurance" for potential losers, it is doubtful that geoengineering strategies could be politically viable.

In 1974, in an article Will Kellogg and I wrote on deliberate climate modification, we suggested, somewhat tongue-in-cheek, that in order to prevent political crises in the event that adverse climatic events occurred on the heels of a deliberate climate modification (any link would be impossible to actually disprove), the world would first have to set up some form of international "no-fault climate-disaster insurance."[9] Clearly, our political institutions have thought even less about how to deal with deliberate environmental modification than we have about learning how to share resources on the planet.

Adaptation. Adaptive strategies seek to adjust society to environmental changes without attempting to counteract or prevent those changes. Many different adaptation techniques address changing environmental conditions on all geographic scales. During the ozone debates of the 1970s, industry supporters argued that living with reduced ozone protection from extra ultraviolet radiation would simply be equivalent to moving a short distance toward the equator (since in the tropics there is less ozone overhead and thus more UV reaching the ground than in midlatitude). This, or course, ignores the fact that moving toward the tropics is a voluntarily assumed risk, whereas being forced to live with an ozone reduction is not. (In the 1980s, a secretary of the interior reportedly suggested that we all buy sunscreen and hats to adapt. Many less-than-amused letters and editorials in the media followed this report.) The concept of informed consent holds that individuals should be informed of and be allowed to decline risks to which they might be exposed for someone else's gain. Moving is clearly not a viable adaptation option for most people. Effective ultraviolet-blocking chemicals available over the counter can be applied to the skin, but even now they are not used heavily enough despite the well-known high rates of skin cancer among susceptible people at low latitudes (for example, white Australians). Moreover, adaptive strategies designed for animals and plants are unavailable.

Adapting ecosystems to acid deposition will be difficult. In agriculture, it might be possible to use or develop more tolerant species, but native species would have to adapt on their own. Furthermore, increasing the acidity of soils could help to speed the natural process of bog formation. Bogs, in which mosses grow at the expense of forest covers, tend to be preferentially located in acid soils. According to ecological geographer Lee Klinger, moss actually generates acid soils. Therefore, acidification could well lead to the rapid transformation of forests into boglands, where this new ecosystem would be well adapted to the acid pulse from human activities.[10]

Corrosion-resistant materials (for example, plastic pipes) could be used in new applications to prevent acid from destroying metal pipes, but it would take a long time to replace old materials (for example, marble used in buildings or statuary), even if replacement were desirable.

One way of adapting to acid deposition would be to financially compensate "losers" (of both environmental and economic resources, such as tourism) to help them adjust to a degraded situation. (This use of compensation would differ in intent from the compensation derived from penalty taxes. The latter, in addition to providing funds that could compensate people harmed by acid deposition, would also serve to reduce emissions and thus would promote prevention.) However, without detailed knowledge of the relationships between pollution sources and damage receptors, trying to make the individual polluting "winners" pay the "losers" would be divisive in itself. It is possible that a compensation fund supported by regional to national taxes or penalties on polluting fuels could be viable compensation strategies.

Those who think they are being injured by acid rain caused by someone else's pollution will not remain silent—witness the current debate both within and between the United States and Canada. Indeed, harsh words continue to pass between these countries regarding what should be done about transnational acid deposition. If prevention does not become a primary goal, then perhaps some form of compensation of the perceived losers might be used to moderate the political divisiveness of the problem. (But prevention is much simpler, I believe.)

In the case of global warming, we already have had considerable experience in adapting human institutions to climatic variability. The natural laboratory for this experience is composed simply of the year-to-year fluctuations in climate that have already created institutions, such as the Soil Conservation Service, crop insurance agencies, irrigation projects, and so forth. A number of such strategies, named and discussed in one of my earlier books, *Coevolution of Climate and Life,* are as follows:[11]

- Beware of generalizing from short-term records. Estimates of the likelihood or severity of climatic extremes should be based on as long a term as possible to maximize statistical reliability.

- Build diversity to provide stability. Maintaining diverse food, water, and energy sources is obviously a good hedge against large fluctuations in supplies. Developing diversity in crop strains can minimize overall vulnerability to extreme climatic or pest outbreaks locally.

- Improve genetic resources. Developing and testing a number of genetic varieties that can deal with a range of climatic, CO_2, nutritional, or other conditions is an important component of crop diversity, which is essential to sustainable food production.

- Match agricultural practices to crop-climate timetables. Careful matching of irrigation, fertilization, and pest controls to the growth stages of crops can increase production with minimal resource use and environmental pollution.

- Maintain adequate storage. Keeping sufficient reserves of food, water, or energy to guard against some degree of climate-induced risk is time-worn wisdom.
- Match agricultural credit and climatic time scales. The threat of climatic variability often causes under-investment in food production inputs such as fertilizer. This climate-defensive behaviour is simply a way for food producers faced with fixed debt repayments to hedge against foreclosure should yields or prices fluctuate wildly in anomalous climatic years. By matching debt repayment to production variations, the consequences of production fluctuations could be shifted from the producers alone to a larger part of society. In essence, this is a generalized crop insurance strategy.
- Maintain future productive potential. Conservation of today's soil and wild genetic resources will be necessary if long-term production is to be greatly increased.
- Maintain a diversity of international economic ties. Trade is a principal means of adapting to regional imbalances caused by climatic anomalies.

A strategy more active than adaptation but short of prevention—it can be called anticipatory adaptation, but has also been labelled build resilience—should be seriously considered even before major international efforts are implemented to control the CO_2 increase. Anticipatory adaptations are not the same as prevention; they are advance investments designed to facilitate adaptions or preventions in the future. Such investments would result from strategic decisions to accelerate activities that would otherwise proceed more slowly. Thus, anticipatory adaptation would tend to counteract the normal tendency to place less value on the future than on the present.

Accelerated research (physical, biological, and social) is an obvious anticipatory adaptation strategy. Of course, delaying stronger actions in favour of research itself entails risk—namely, committing society to adapting to more greenhouse gases and their effects (good and bad) than if actions had been taken earlier to reduce the impact. But accelerated research now could minimize the need for crash research programmes were adverse climatic effects to occur sooner or in more extreme form than expected. Moreover, it is typically easier to adapt to a predicted change than to a surprise, so accelerated research can make adaptation easier. Other adaptive strategies to predicted climatic change could involve building dikes to protect property against sea-level rise or improving the shelter of those who might be hurt by an increase in the number of extremely hot days. Other examples of anticipatory adaptation include the following:

- Building a diversity of political ties. Good rapport among nations will be absolutely necessary if an international consensus to limit the CO_2 increase is eventually sought.
- Accelerated development and testing of crop strains and agricultural practices for more efficient adaptation to higher CO_2 levels (recognizing, however, that pests and weeds may adapt in the same way).
- Starting work on possible compensation mechanisms to adjudicate disputes between perceived "winners" and "losers." Examples of what could arise as a result of climatic change are movement of grain-growing belts, coastal flooding from sea-level rise, and altered crop or forest productivity from CO_2 fertilization.

Such strategies for making future adaptation to environmental variability easier are sensible regardless of the eventual impacts of increases in CO_2 and other trace gases. Thus, the problem has links to other environmental and societal problems, such as the social impact of natural climatic variability[12] and national security as it relates to energy self-sufficiency.

Are There "Winners" and "Losers"? Before leaving the subject of adaptive strategies, it is important to look at the winner/loser question closely in a real-world context. Earlier, I told of the conference *Time* magazine arranged with scientists, politicians, and policy makers from all over the world to discuss critical environmental problems.[13] The event, which took place at NCAR, proved to be a fascinating few days of discussion of global warming, population growth, biological diversity, and waste disposal. Senator Albert Gore was a vocal participant in the working group on climate change. Senator Gore, the only presidential aspirant to strongly emphasize global warming and other environmental issues in the 1988 political campaign, strenuously objected to the whole concept of adaptation in general and "winners" and "losers" in particular. He had challenged NCAR political scientist Michael Glantz over this issue on several occasions. "I'd like to challenge this notion of winners and losers," Gore said. "Those who have been involved in this debate for a long while have come to see the winners and losers phrase as an impediment to increased awareness and a bar to meaningful action. And, in fact, I think one can make a very strong case that there are no winners at all. . . . The fact is that the change is so radical and so rapid that it disrupts the society and the political structure. I just think that it's a mistake to put it in terms of winners and losers. I think that the earth is a loser."

Gore's objections to the general category of adaptation was based on a simple concern: if we take comfort in the belief that adaptation is possible, then we will lose the political will to practice prevention. Sir Crispin Tickell, the British ambassador to the United Nations, elaborated on Gore's concern: "Before we move on, there's just one point I think needs to be pulled out of this discussion, which is if there are no winners or losers, or if the effect of short-term change is nearly always disruptive, which I think we can agree upon, then the world has got to be persuaded of that." He went on to emphasize the challenge of persuading a global population of which significant numbers worldwide are uninformed or have only lately begun to be informed of the phenomena at hand. Even more difficult, some believe that more CO_2 could make them better off, he reported.

Vladimir Sakharov, from Moscow, responded to the senator's plea by suggesting, "I would not want to be a winner and to be able to grow oranges at my villa not far from Moscow. I think we'd all be losers." However, he went on to say that if the West wanted the Soviet Union to reduce its pollution it would have to help by exporting modern, more efficient technology to the Soviets. The Reagan administration, in particular, strongly opposed the transfer of technology, especially computers and other advanced technologies, to the Soviet Union, fearing that it would be used in military applications. But, as Sakharov's remark suggested, if the Soviets lack efficient manufacturing- or energy-producing processes, many of which do require sophisticated computer equipment, they will have difficulty meeting low pollution standards.

Recognizing the dilemma, Senator Gore later suggested that perhaps a bargain could be struck whereby high-technology exports to the Soviet Union would be linked to substantial reductions of Soviet troop concentrations in Europe. To that concept Sakharov replied, "I would agree completely with what Senator Gore said concerning U.S./U.S.S.R. relations and reductions of spending on arms. Everyone needs money nowadays, both the developing and the wealthy countries. And we have to take it from somewhere, from the pocket. And the only pocket which I see is the military pocket. In the past, such statements were regarded as pure propaganda, but not anymore. Now it becomes reality."

It is important that the link articulated in this remarkable exchange—between economic development, environmental protection, and the conversion of military expenditures—proceed at once. After all, the resources needed for environmentally sustainable development are substantial—perhaps a hundred billion dollars a year—and large standing armies in Europe represent rampantly wasteful expenditures, to say nothing of the risks of fighting. It is in both alliances' interests to reduce these expenditures as much as possible. General Secretary Gorbachev's promises to unilaterally withdraw some Soviet forces from Eastern Europe are an encouraging start. We must verify these promises and follow them up with similar steps. The political transformation in Eastern Europe in 1989 is encouraging.

To return to Senator Gore's philosophical and political objections to adaptation in general and winners and losers concepts in particular, I chimed in somewhat less vehemently: "We're a very adaptable species, but we're better adapted at things when we know what's happening." The faster that climate changes, I went on, "the more likely it is that the sum of the effects will be negative. I don't agree completely, Al, that all potential effects are negative. It's really a problem of redistribution. And it's conceivable that you could open up lands in the north, for example, and plant [crops or trees]. True, the soils may not be as good as the ones they replaced. [But] I don't want to . . . say that absolutely nothing could be positive because then you're going to get somebody who's going to say it isn't true. What I want to ask is, What's the net effect? And I think the net effect is increasingly negative, the faster things change."

Our *Time* hosts brought the conversation back to adaptation later in the day, not fully agreeing with Gore's assessment that it would sap our will to practice prevention. I argued for anticipatory adaptation as part of our general strategy. Rather than waiting for changes to come, we could begin now to develop crop strains that could take advantage of CO_2 and to test those that are more widely adapted so that as we increase our knowledge base and forecast specific regional effects, we could facilitate future adaptations. In so doing, we become more flexible, more knowledgeable, and less vulnerable—while keeping alive our resolve to act when action became necessary. Finally, I concluded, an anticipatory adaptation effort makes sense, even if global warming "proves to be an infrared herring. You haven't wasted your money because it's still going to buy you a little bit less vulnerability to natural interannual invariability." In the discussion that followed, it was agreed that despite the environmental risks of discussing adaptation, some climate changes can't be prevented and it was indeed wise policy to recognize ways of minimizing the damages from an altered climate and of taking advantage of opportunities as they arose.

Prevention. Preventive strategies are those that actively limit emissions of substances perceived to be environmentally harmful. The primary preventive strategy designed to avoid damage to the ozone layer is the reduction or banning of all substitutable uses of CFCs. The Montreal Protocol of 1987 proposed a 50% cut in CFCs by the year 2000, but not all nations have signed the treaty. Furthermore, most scientific testimony suggests that at least a 90% ban is urgently needed if the ozone hole is to be permanently reversed. At the *Time* magazine meeting, Senator Gore proposed that the single best politically feasible goal for the environment would be a ban of up to 95% of all chlorofluorocarbons. (He introduced legislation in 1989 to do just that.) This would not only help protect the ozone layer but would cut emissions of a trace greenhouse gas that might in the long run be responsible for as much as 25% of the global warming. It is important to do what is imaginable first, Gore argued, and "it's unimaginable at the current time that we would cut global fossil fuel consumption by 50%. . . . So let's expand the limits of our imagination by building confidence that we can solve those aspects of the problem that are currently susceptible to imaginable solutions, such as stopping CFC production." Sakharov, you will recall from Chapter 7, noted that a CFC phaseout was not imaginable for the Soviet Union in the next five years, so even this basic solution will probably not be implemented fully in this century. But everyone at the conference readily agreed that more research into alternatives to CFCs, CFC recycling technologies, and their more efficient use was urgently needed.

With respect to the issue of ozone depletion, potential problems arising from a fleet of high-flying SSTs still exist, as do plans for a new generation of aircraft. It is therefore important to continue research on effects of nitrogen oxide emissions in the stratosphere. The use of artificial fertilizers in agriculture also generates atmospheric nitrogen compounds that can reach the stratosphere and possibly destroy ozone. It is imperative that we develop a better understanding of the global nitrogen cycle so we can better assess this potential problem and others we may yet uncover. As yet, actual prevention of these emissions seems remote.

Similarly, attempting to control acid rain by limiting emissions of pollutants requires that we know enough about acid deposition and its environmental effects to give emissions controls a reasonable chance of success. For beginning a prevention programme, it may be sufficient to know that a widespread decrease in acid deposition would follow widespread decrease in emissions, even though detailed relationships between sources and receptors are not available. Several groups (including the Canadian-United States Memorandum of Intent Working Group on Impact Assessment) have set acid- or sulphate-deposition limits below which little cumulative damage to aquatic ecosystems would be expected.

If present theories of the origin of acid rain are correct, then technologies to limit acid rain are available today. We can reduce sulphur dioxide emissions by switching to low-sulphur fuels. Unfortunately, only 20% of the world's petroleum reserves are low in sulphur, and these are currently being used at a disproportionately high rate. Switching U.S. midwestern power plants to low-sulphur coal could, in the absence of compensation mechanisms, cause economic displacement because much coal from the Midwest and Appalachia, the present fuel sources, has a high sulphur content. Nevertheless, not switching to low-sulphur coal or no-sulphur natural gas allows more acid deposition in regions such as the Northeast and Canada, which typically follow more stringent SO_2 emission practices.

It is possible to keep sulphur dioxide from reaching the atmosphere by washing the coal or by removing the SO_2 from the flue gas. Simple washing can remove about 50% of the sulphur. Further removal, up to 90%, requires high temperature and high pressures and can cost ten times more than washing. Flue-gas desulphurization (scrubbing) by reacting the effluent gas with lime or limestone in water can remove 80% to 90% of the sulphur but generates large quantities of solid waste.

Techniques for minimizing emission of SO_2 from coal-burning power plants have no effect on nitrogen oxide (NO_x) emissions. Oxides of nitrogen derive both from the burning of nitrogen impurities in fuel and from oxidizing the nitrogen normally found in combustion air. The percentage of NO_x generated by the burning of air, about 80% in the case of conventional coal-fired boilers, depends strongly on the temperature of combustion. Improving furnace design and combustion techniques can reduce NO_x emissions from stationary sources by 40% to 70%, although such methods are not in widespread commercial use now. Processes for removing NO_x from flue gases are only in the development stage. Emissions from automobiles can be decreased to some extent by improvements in the design of combustion chambers and the control of combustion mixtures. And exhaust-gas catalytic converters can be used to further limit emissions from mobile sources. Battery-powered cars could dramatically reduce air pollution in cities, but only if the electric power sources used to charge the batteries were themselves less polluting. Energy-efficient mass transit can also reduce mobile-source emissions of NO_x as well as CO_2.

There is no doubt that preventing acid deposition would be costly. Cost estimates for U.S. emission-control programmes range from $2 billion to $20 billion per year. Even the "small" $2-billion-per-year programme that could reduce acid deposition in New

York's Adirondack Mountains by, say, 25%, is considered far too expensive by some. Former Office of Management and Budget director David Stockman reportedly estimated that the programme would cost $6,000 to $10,000 per pound of fish saved. However, Stockman's cost-benefit argument ignores the question of what might be fair to those affected by acid rain. More important, he concentrated on only one form of damage from acid rain – a highly visible form, at that – apparently ignoring all other environmental, agricultural, health, and materials damages. The quantitative assessments of these additional damages, given the present levels of uncertainty, cover a broad range. It would be folly at present to make policy decisions based solely on a few detailed risk benefit studies for specific scenarios of control, mitigation, and compensation.

The divisiveness of the acid rain problem argues for some form of national or even international cost sharing to finance preventive measures. It can also be argued, however, that it is only fair that polluters pay a large fraction of the prevention costs – for example, by reducing the height of smokestacks so that more pollution would be deposited locally and less transported over long distances. Obviously, balancing the costs of prevention among the polluters, regions that are polluted, and those who are relatively unaffected will be a sensitive political task.

The electric power industry – particularly in the Midwest, where much of the high-sulphur coal is used that produces the acid rain that so upsets New England and Canada – is very unhappy about the prospect of spending tens of billions of dollars to scrub the sulphur out of coal. Moreover, the laws of thermodynamics tell us that you can't get something for nothing. Therefore, some energy penalty will be paid for the processes that remove the sulphur – to say nothing of the environmental problems that disposing of it will bring. Therefore, approximately 5% more coal would have to be burned to keep electricity production from these power plants at current levels if most of the sulphur is scrubbed out. More coal burning would only exacerbate the CO_2 problem in the name of lowering acid rain.

Despite the fact that scrubbing coal is the politically favoured acid rain solution at the moment, and despite the risk of alienating many of my environmental friends, I personally do not believe that scrubbing coal is the best solution to the acid rain problem in the U.S. Midwest power plants. I have already been told by industry representatives that the power companies have a "siege mentality," with their executives trying to hold onto their slim profit margins by stretching out the lifetimes of their power plants, delaying further capital investments, and "circling the wagons against all pressures to invest." If these already embattled executives are forced by government regulations into investing tens of billions of dollars to scrub sulphur from coal, then they will certainly not be inclined to replace their inefficient, old electricity-generating plants with the vastly more efficient new plants already available. Rather than scrubbing coal to take the sulphur out of power plants that already dump two-thirds of the heat energy as waste heat at the site, why not save the money to be used on scrubbers as an incentive—or, better, a requirement—to replace these plants as soon as possible with the much more efficient boilers, turbines, and the like, which dump only about half their heat into the local environment, transforming the rest into useful power. In that way, we would not be committed to a longer run of energy-inefficient, high-CO_2-, high-SO_2-producing power plants. (When I furthered this controversial view at the *Time* magazine conference, Senator Gore quipped, "I see you don't want to scrub sulphur from coal at the power plants, you want to scrub the power plants!" He agreed that stretching out the effective lifetime of current inefficient power-generating technology by decades was a worse environmental evil than pushing back acid removal by a few years.)

In the meantime, even before the plants were modernized, we could do some switching from coal to methane—natural gas—which would both dramatically reduce the acid rain problem and cut CO_2 emissions in half.[14] I realize that such a switch would be a staggering blow to the coal mining industry. But where is it written that anyone has an indefinite right to antisocial or anti-environmental employment? I know this is a very hard line to take, but I simply do not believe that we should prop up dinosaur industries for immediate political convenience when the health of the planet is at stake. To be sure, a fair-minded society would not expect one industry or group of workers to absorb such an economic shock without major financial assistance. I am happy to have my taxes on energy services increased for the purposes of job retraining, worker relocation, developing alternative industry in the region, and so on. Those new industries, by the way, would pose many fewer threats to the health of the miners than their present one does. In any case, a crucial aspect of our overall approach to atmospheric pollution is the recognition that partial solutions in one area could create greater problems in another. Quite simply, we must take an integrated look at all the consequences associated with our actions, and not continue in the piecemeal fashion that has dominated policy making in the past. Problems of the atmosphere represent a new dimension of long-lasting global change, which will require a new dimension of creative, integrated solutions.

One strong objection to energy taxes, or any other form of environmental regulation, is that they will interfere with the free market. This has been the dominant U.S. ideology of the 1980s. However, those who hold this view do not recognize that the current free market is in fact not free, but rigged. How can the present market, which determines the price of electricity, coal, and so on, possibly represent the true costs of these fuels or the energy produced from them if there is no charge for the environmental damage they do? True, it is difficult to evaluate quantitatively the economic consequences of acid rain or global warming, but it is outrageous, given the wide range of damage estimates from billions to trillions of dollars, that our present economic system has chosen *zero* as the so-called "external cost" of pollution. Clearly, the risk of disruption of the atmosphere and climate has a price that is greater than zero, and that price should be reflected in the cost of fuels. Therefore, we need user fees—taxes, to be honest—on fuels in proportion to how much they pollute. Coal, for example, should be taxed most, since it usually is the worst CO_2, acid rain, and local air pollution offender. Oil is intermediate, depending upon the degree of impurities contained, and natural gas is the least offensive. Our failure to charge these fuel producers and consumers for the damage they do to the global atmosphere accounts for the distortion in the current "free market." It is completely wrong to argue that such taxes themselves would distort the market, as we can expect those whose interests would be damaged by correcting past distortions to claim. The obvious fair solution is to internalize the so-called externalities and restore a truly free market.

A number of years ago the EPA conducted a study in which

it assumed an *imposed* tax on fuels[15]. The agency concluded pessimistically that such a tax would do no more than delay the date at which CO_2 would double and dismissed this result as unsatisfactory. I disagree with that conclusion on two grounds. First, as I remarked earlier, delay would be no small accomplishment since it is not absolute change so much as the rate of change we should fear. The point is, the more rapid the change, the harder it is to predict and the more likely it is to damage ecosystems and society.

Secondly, if such a tax were really designed to account for the "externality" of environmental damage from certain fuels, rather than taking the revenue and returning it to the general fund where it would mingle with all the other diffuse dollars in that massive pot, we should target the money specifically for actions designed to correct the damages done by fuel burning. Let's invest it, for example, in helping the poor to insulate their houses, developing new industries in coal mining areas, developing alternative crop strains, making power plants more efficient, making our manufactured goods less energy-intensive, and speeding up the deployment of renewable energy systems. Targeting tax money in this way is not a new or radical idea—petrol taxes, for example, are earmarked to maintain and construct highways.

Nevertheless, uttering the "T-word" has been political suicide for many politicians in recent years, so it remains to be seen whether this very important policy measure will receive the serious consideration it urgently deserves. At the *Time* magazine conference, *Time* science correspondent Richard Thompson asked Senator Gore, "Since we're trying to draw up a realistic agenda, I wanted to ask Senator Gore how likely it is you'll get a CO_2 fee or a petrol tax in the Bush administration?" "Read my lips," Gore responded, echoing Bush's famous campaign tag line promising no new taxes, "not likely. But there are possibilities to consider: rebates that adjust for income levels, shifting revenue sources, and—just to throw one idea out off the top of my head—a petrol tax coupled with significant adjustments to the earned-income tax-credit period so that most Americans will not have an increased tax burden but the society as a whole will get incentives to stop the irresponsible environmental activity."

To many, the single most important and effective means of preventing a rapid and large buildup of CO_2 in the atmosphere is to make existing efficiency standards for energy use a strategic goal of nations, particularly those nations with high per capita fossil fuels consumption.[16] Few national strategic options are more important than energy conservation, since it will help reduce the impact of many problems. Specifically, increasing energy efficiency can

- Reduce atmospheric pollution on nearly all scales.
- Enhance national security through increased energy independence.
- Reduce the environmental effects of carbon dioxide and acid rain, thereby reducing the effort required to adapt to such conditions.
- Reduce risks of uncertain but plausible climatic catastrophes, such as coastal flooding and increases in the probability or severity of extreme heat-stress periods or tropical cyclones.
- Buy time to study climatic change and assess climatic-impact problems and develop non-fossil alternative energy sources, upon which the ultimate prevention of CO_2, acid rain, and local urban air pollution depend.
- Improve the competitiveness of manufactured products by reducing the energy it takes to make them. At present, the United States uses over two times more energy in manufacturing than Japan, West Germany, or Italy.[17] Energy expenditure helps keep the cost of production of U.S. products higher than their competitors, especially as the price of energy increases.

Developing non-fossil energy sources and improving efficiency in all energy sectors should be viewed in part as high-priority strategic (rather than purely market) investments. Whether the mechanisms to accomplish this should be subsidized research and development (for example, on solar photovoltaic cells or passively safe nuclear reaction designs that cannot melt down), tax incentives to reduce fossil fuel emissions, or elimination of existing hidden subsidies to existing nuclear or fossil fuel technologies is an important tactical question that must be debated further by appropriate private, academic, and legislative institutions.

Even though at national levels non-market investments are possible and already desirable, greenhouse gas buildup is nonetheless fundamentally a global problem in both cause and effect.

Moreover, it is inextricably interwoven with the overall problem of global economic development and cannot be left out of debate on population, resources, environment, and economic justice. Rich nations cannot ask poor nations to abandon their development plans because of potential global warming without making comparable sacrifices to help provide international economic equity. Since developed countries are by far the major producers of the CO_2 related to fossil fuel use, their disproportionate per capita use of energy must be part of the bargaining process surrounding the control of global emissions. Therefore, any global strategies for preventing a CO_2 buildup will require international cooperation between rich and poor nations on the transfer of knowledge, technology, and capital.

The mounting burden of debt is a chief impediment to global development in the Third World. It is very difficult for countries to invest in more initially expensive energy-efficient equipment when they can hardly afford to pay back the interest on loans from other countries. It is becoming increasingly apparent among those who study development problems that somehow environmental problems in developing countries will not be solved until the debt problem is also solved. One creative solution, originally proposed by Thomas Lovejoy, now of the Smithsonian Institution in Washington, D.C., has been labelled "debt-for-nature swaps," in which underdeveloped countries would agree to turn over tracts of tropical rain forest to developed countries in exchange for forgiving some portion of their debt.

Somehow, the world needs to reverse the horrendous current situation, in which the primary flow of capital is from the poor nations to the richer nations rather than the reverse. Despite the strong support at the *Time* magazine meeting for ways to connect environmentally benign development and the relief of Third World debt, Sir Crispin Tickell added an appropriate note of caution. He pointed out "that the debt problem is not really *a* debt problem. It is about 132 debt problems. And each of these problems is really quite different. . . . One of the principle reasons for the Mexican debt is that the Mexicans themselves have big investments in the United States, above all real estate in the United States. If you look at the debt problems of Botswana or the debt problems of the poorest African countries, you have a completely different situation." For the poorest countries, he suggested lowering interest rates, rescheduling debts, forgiving debts of the very poorest, and essentially converting loans into grants. But "when you come to the Latin American debtors and the Philippine debtors and the other debtors, and the Yugoslavs, and of course the United States to some extent—because the United States, as Senator Gore reminded us this morning, is now the world's biggest debtor—you have a completely different set of issues in every case." Sir Crispin's advice was to keep debt in mind as an environmentally linked problem but not to prescribe general solutions applicable everywhere. The better approach, he suggested, was to press the World Bank to yield more resources for environmentally sound development by placing environmental conditions on its loans. He also noted that a number of the current loans had been made by private banks, not governments, and that these private banks should be allowed, in fact required, to deal with the bad debts without being bailed out by governments, except perhaps at a very large discount.

Population growth rates will be another point of contention in the dialogue between developed and developing countries. This problem is relevant to the CO_2 issue simply because total emission is the per capita emission rate times total population size (See Chapter 5). If there is a movement toward parity between rich and poor in per capita use of fossil fuels in the next century, then population growth (which is occurring predominantly in the Third World) will become as important a factor in the CO_2 climate problem as high per capita fossil fuel use is today in the developed nations.[18]

In *The Genesis Strategy,* I built on the notion of a "planetary bargain" offered by U.S. diplomat Harlan Cleveland by proposing a "global survival compromise." The idea was that developed countries would control their disproportionate use of natural resources and provide capital and technology to help developing countries improve their economic base. At the same time, the less developed countries would be expected to reduce population growth rates markedly and raise their standards of living.[19] In *The Coevolution of Climate and Life,* Randi Londer and I phrased the compromise as follows:

> [It] must be conceded that for the poor of the world, at least in the short run, more energy—and certainly more energy services—can mean a better quality of life. If energy use—even fossil fuel based—can be focused on providing services that improve health

> care, education, and nutrition for the bulk of the people in less-developed nations, then we would personally endorse its increase in LDCs [less developed countries] as a worthwhile gamble despite the environmental risks of more CO_2. But it is essential, we believe, that all such risks taken for the sake of economic development in LDCs help bring about some rapid form of demographic transition—that is, a sharp decrease in population growth rates—for, as population or affluence grows, so usually does pollution. In other words, we would be willing to accept a little additional pollution now if it meant that ultimately the world population would stabilize sooner and thus future pollution levels would be lower for any per capita standard of consumption. A stable population is, of course, an essential component of a sustainable future. . . .
>
> [Any] plan with that end would have to be implemented quickly (over decades). Then reasonable, sustainable levels of per capita energy consumption might be obtained in these countries before world population becomes so large that the atmospheric concentration of CO_2 (and other pollutants) would become high enough possibly to cause major disruptions to health, food production, water supplies, or an irreversible sea level rise. The urgency of this problem underscores just how intimately connected the CO_2 climate problem is with global economic development strategies. Of course, accepting this trade-off of a little more CO_2 growth now (and for several decades more) for stable concentrations later on suggests that we must have in mind some long-range target figures for both a stable population size and per capita energy use. Policy makers should try to agree on such long-range targets before we gamble that it is worth the risk to increase pollution now in order to achieve the improved quality of life necessary to a population size that is sustainable (and achieved without famines or Draconian coercion).[20]

Of course, such negotiations to create a global survival compromise would not be simple or short-term. But if buildup of CO_2 and other trace gases is not viewed as part of the negotiation process for global development, it is extremely unlikely that substantially greater buildup will be prevented, except by the calamity of war, fortuitous technological advances in alternative energy systems, and major programmes to increase energy efficiency.

TOWARD POLICY ACTION. Policy instruments to restrict CO_2 emission are, of course, at an early stage of discussion and study. Their ultimate viability at the national or international level is unknown. At the national level, several studies have suggested that taxing fossil fuel use might be effective, although, as noted earlier, some think only marginally so. Others propose massive reforestation programmes or debt-for-nature swaps. At a grander, global level, Will Kellogg and Margaret Mead proposed in 1977 the ultimate need for a "law of the air," whereby all nations would agree to limit CO_2 emissions to some negotiated level.[21]

The concept of a law of the atmosphere got a major boost recently in Toronto, Canada, at the conference called Changing Atmosphere: Implications for Global Security, which took place in late June 1988.[22] More than three hundred technical experts, scientists, lawyers, government emissaries, economists, industrialists, policy analysts, and planners from forty-six countries came together at the invitation of the Canadian government to discuss what might be done about atmospheric change. The meeting was truly a watershed in atmospheric policy issues, since it combined the environmental, scientific, industrial, bureaucratic, ministerial, and political functionaries with a bevy of media. This meeting was a happening because of the remarkable unanimity across a wide spectrum of people (who rarely meet and when they do often disagree) that it was no longer premature to consider actions to curb the rapid rates of atmospheric change. Indeed, in closing remarks to the assembly, I cited the unity and spirit of the meeting, referring to it as "the Woodstock of CO_2." This consensus for action included the normally conservative director general of the United Kingdom Meteorological Office, the minister of development of Indonesia, the prime ministers of Canada and Norway, as well as the more activist representatives of environmental organizations. At the crux of the published conference summary statement was a section entitled "A Call for Action":

> *An action plan for the protection of the atmosphere* needs to be developed, which includes an international framework convention, encourages other standard-setting agreements, and national legislation to provide for the protection of the global atmosphere. This must be complemented by implementation of national action plans that address the problems posed by atmospheric change (climate warming, ozone layer depletion, acidification, and the long-range transport of toxic chemicals) at their roots.

In addition to calling for a rapid ratification of the Montreal Protocol on substances that deplete the ozone layer, the meeting did something that was a first in the context of these events—it went beyond rhetoric and gave specific, quantitative targets for emissions reductions—in essence, the guts of a law of the atmosphere. The conference summary called for reduction of *"CO_2 emissions by approximately 20% of the 1988 levels by the year 2005 as an initial global goal."* It was suggested that the developed countries, at present the greatest polluters, should lead the way. It also pointed to energy efficiency as well as ways to reduce non-CO_2 greenhouse gases as requiring priority. The report also acknowledged the special difficulties the Third World countries will have, noting the need to "extend technology transfer with particular emphasis on the needs of the developing countries." To help fund this effort, it suggested the establishment of a "World Atmosphere Fund," financed by taxes on fossil fuel consumption in industrialized countries with the subsequent resources used to help implement the action plan. Encouragingly, it was a representative of industry who first suggested that fund, not an environmentalist or a delegate from the Third World. The 20% reduction in emissions below the 1988 level will probably prove to be the most enduring component of this meeting, primarily because it is numerically specific. Already, that recommendation has been incorporated in nearly identical form into several bills submitted in 1988 to the U.S. Congress to control CO_2 emissions. But it did not come easily at the meeting.

When the draft statement of the conference was first distributed to the dozen working groups that were simultaneously thrashing out their own reports, it created considerable consternation. As the chairman of the "Futures and Forecasting" group, I read this recommendation to my very diverse working-group members. These included the director general of the U.K. Meteorological Office, the head of Australia's Bureau of Meteorology, the secretary-general of the World Meteorological Organization, African and Middle Eastern industrialists, intellectuals, environmentalists, and U.S. EPA officials. The more conservative elements of the group immediately objected to the idea of a specific numerical target, complaining that scientific uncertainties did not permit us to specify numbers. It is true, they all agreed, that the uncertainty is not so vast that it is premature to call for action, but they were uneasy with any numerical targets.

Environmentalists in the group were ecstatic at the specific targets, feeling that only with numbers did the meeting have a chance of any enduring contribution. A normally environmentally oriented representative from the EPA said he agreed with the thrust of the recommendation but felt that the report would "lose its credibility" with most policy planners and bureaucrats, since any specific quantitative regulatory recommendation had to be based on detailed studies showing how this rule would impact on the economic and social sectors of various countries. I agreed that there was some risk of damaging the credibility of the report but felt that naming numbers was critical to drawing substantial attention to it. I then proposed a "snivelling compromise," whereby the final report would include the target numbers, but rather than couching them in a "we should" statement, it would frame a proposition whereby in a few years (say, 1992), at an international convention, all nations would have to show why—if they had not—they could not reduce their emissions by at least 20%. That way, we could preserve the political impact of naming a number without losing credibility because we could cite no comprehensive backup studies as support for the recommendation.

My working group accepted the compromise and I presented this view to the plenary. That presentation evoked the predictable reaction—industrial representatives still thought naming numbers was too specific to be justified and environmentalists chided me for trying to weaken the most important part of the conference report. Ultimately, the recommendation was carried, but the words "as an initial goal" appeared in the final report to convey the notion that the numbers should not be viewed as the rigid opinion of the participants. I am satisfied with the compromise and feel this report will help to stimulate various national and international organizations to press for a law of the atmosphere. (U.S., Japanese and British opposition kept the 20% commitment out of the final text of a November 1989 ministerial conference on climate change held in Noordwijk, Holland.) Of course, before we can *reduce* global CO_2 emissions, we first need to stop them from growing—a more realistic goal.

The Toronto conference, with its number naming, has already influenced a serious attempt to quantify the economic implications of specific reduced greenhouse targets. In December 1987, Fred Koomanoff, of the U.S. Department of Energy, commissioned economist Jae Edmonds and a number of his colleagues to examine the question of emission reductions. Their report for the department highlights the divisive nature of the suggestion that CO_2 emissions can easily be decreased by some 20% in twenty years, but it also helps to articulate the disagreements between those who insist that energy must increase for prosperity to grow and those who insist with equal vehemence that in a "least-cost" solution to our energy and environmental problems per capita energy use would decrease.[23]

Edmonds and his colleagues begin the report by noting that in the United States we currently produce 1.25 billion metric tons of carbon each year in the form of carbon dioxide. To meet the Toronto suggestion, they explain, we have to reduce that target to approximately 1 billion metric tons per year. To reach the very stringent suggestion of Vermont Senator Stafford's draft legislation in 1988, which calls for a 50% reduction in CO_2 from the United States early in the next century, we would need to reduce emission to about 630 million metric tons per year. To begin analysing how the United States might do this, the Edmonds group broke down U.S. energy use in 1985 by type of fuel in the economic sector. For example, residential and commercial energy use, half of which involves natural gas, is about 16%. Industrial energy use is 21%, roughly evenly divided between oil and gas, with a substantial minority depending on coal. Transportation is also about 20%, most of which is derived from oil. Electric utilities account for 34%, the bulk of which is produced by coal, with half that amount again from nuclear, and about half that amount from natural gas. These categories account for most energy use in the United States. Since coal is the most inefficient fuel, it produces the greatest amount of CO_2 per unit energy. Therefore, any predicted growth in electric utilities or industrial applications that depend on coal would substantially increase projected U.S. CO_2 levels. Any switches to natural gas or shifts in residential and commercial sectors to a dependence on gas or, to some extent, nuclear or solar power would decrease the future projected CO_2 amounts.

These energy economists then cited another Department of Energy study, which made two alternative projections of future U.S. energy use, a business-as-usual plan and an energy-efficiency plan. Both these projections suggested substantial *increases* in U.S. production of CO_2 and consumption of fossil fuels over the next 20 years. This study, the DOE's National Energy Policy Plan (NEPP) projection to 2010, was prepared in 1985 and is not a document in which environmental effects are the primary motivation for the projections. The NEPP reference case predicts an increase in energy between 1985 and 2010 of about 30%. While projected oil and gas consumption remain relatively constant over this period, coal consumption increases dramatically—by more than 100%. CO_2 emissions increase from 1.25 billion metric tons per year in 1985 to about 1.73 billion metric tons in 2010. This is a 38% increase in CO_2. Compare it with the 1975 to 1985 experience, in which dramatic gains in energy efficiency in the United States actually lowered our fossil fuel emissions while at the same time our gross national product increased—this was forced by the OPEC price rise. The DOE's defence of these dramatic growth projections by the year 2010 are based on the fact that 1975–1985 was the only period in which economic growth and energy growth remained relatively unlinked; historic periods showed the reverse trend. The DOE authors viewed this period as an aberration. The main reason for the projected 38% increase in CO_2 generation to 2010 is a more than doubling of the coal use in electric utilities and a near doubling of coal use in the industrial sector. The Edmonds group also noted that coal was the fifth largest U.S. export in 1982 and that there would be substantial pressure to maintain this source of foreign revenue, given our large trade deficit.

The NEPP's second case, its "high-efficiency" case, still increased CO_2 production to 1.48 billion metric tons per year, some 230 million tons more than the 1985 emissions. This is hardly the 20% reduction called for by the Toronto delegates or the 50% called for by Senator Stafford! The major difference in the two NEPP cases is that the high-efficiency case uses less coal. As Edmonds and colleagues noted, "this notwithstanding, the high-efficiency case provides little cause for optimism about the potential for substantial reductions in CO_2 emissions, as compared with 1985 levels, without market intervention in the period 2010."

Edmonds and his colleagues then contrast this high-growth Department of Energy forecast with those of four international energy analysts who share a very different vision of the world's energy future. They are José Goldemberg, Thomas Johansson, Amula Reddy, and Robert Williams. Their book, *Energy for a Sustainable World* (referred to here as ESW),[24] outlines their view of a future in which population grows roughly at the same rate as in the DOE study, gross national product grows similarly (to a somewhat lower ceiling), but, in sharp contrast to the NEPP projections, the ESW model shows a decline in energy-growth rates and thus a decline rather than an increase in CO_2 emissions. The

Edmonds team diagnoses the difference and lays it at the doorstep of radically different assumptions on the part of the ESW and NEPP people as to the prospect for energy efficiency and the conversion of future economies to activities that are less energy-intensive. Whereas the DOE analysts forecast U.S. energy production under various assumptions about the economic viability and technological availability of energy systems, the ESW team, in the words of Edmonds *et al.*, "specifies a feasible energy future consistent with broad societal goals of economic efficiency, equity, environmental soundness, self-reliance, peace, and long-term viability. It includes policy prescriptions to achieve these objectives. . . . Energy service demands are developed based on an assumed future level of per capita income. The analysis then seeks to identify technologies consistent with that income level and the above mentioned societal goals."

Robert Williams, a Princeton energy analyst and one of the authors of the study, wrote a letter to Jae Edmonds critical of the characterization of the ESW study as uniquely value laden. Wrote Williams,

> Even those who try to make business-as-usual forecasts do [value-laden] analysis in the sense that a persistence of existing energy policies provides the basis for their forecasts. . . . We believe that energy policies should be shaped in ways that are supportive of, or at least consistent with, broad societal goals of economic efficiency, equity, environmental soundness, self-reliance, peace, and long-term viability. To the extent that evolving energy strategies are inconsistent with the achievement of these goals, we recommend market interventions—e.g., eliminating energy supply subsidies and moving towards marginal cost pricing of energy to promote economic efficiency; encouraging international assistance agencies to promote energy efficiency in their aid programmes to help remove the spectre of unaffordable energy as a constraint on development; levying a petroleum product tax or oil import fee to help promote self-reliance and reduce the chances of war arising from competition for Middle-East oil; levying a carbon tax to help cope with the greenhouse problems; increasing support for research and development on end-use efficiency to help sustain a stream of energy-saving innovations into the market place.

To the credit of Edmonds and his colleagues, they printed as an appendix to their report the entire criticism from Robert Williams, which went into substantially more detail than the above quote. But Edmonds and his team explained well the essence of the difference between the high-efficiency world of the Goldemberg report and the growth and pollution world of the Department of Energy report, showing that the NEPP "assumes that historical trend in the relationship between the demand for energy services and manufacture will continue. That is, even if energy prices remain constant increased production will result in less than proportional increases in the demand for energy services as changes in the industrial composition are projected to continue." In other words, a shift in the U.S. economy from energy- and materials-intensive activities such as steel manufacturing to information-intensive activities such as designing computers will continue to improve our gross national product while freeing us from our dependence on such environmentally disruptive activities as mining or coal burning. The Department of Energy report does not deny this trend, but it does project it at historical rates, whereas the ESW group projects substantially faster rates based upon what they believe to be plausible (and even cost-efficient) assumptions for the improved petrol mileage of cars and efficiency in the production of energy in power plants, in industrial applications and in home heating, lighting, and other sectors as well. For example, the NEPP study assumes that, despite their successful 1988 campaign to reduce the stringency of petrol-mile standards, U.S. manufacturers will continue to improve the average energy efficiency of cars. The study suggests that as America's fleet of ageing vehicles is replaced with newer cars with better petrol mileage, CO_2 emissions from automobiles between 1985 and 2010 will change very little, since the DOE authors assumed additional miles would be driven. In fact, the NEPP efficiency case assumed that by the year 2010 new cars would average 51 miles per gallon and would penetrate 50% of the U.S. market. The ESW team, on the other hand, believe that new car efficiencies could be even greater than that, with a fuel economy for the average vehicle of 75 miles per gallon. Even more important is the assumption of NEPP authors that the U.S. economy will reduce its dependence on energy at the historical rate of about 1.7% per year, whereas the ESW authors believe that active efforts to make our economy less dependent on energy inputs could be accomplished at a rate of some 4% per year. They also assume very high-efficiency lighting and the rapid deployment of electric-power-generating stations that are 50% or more efficient than the present capital stock.

Jae Edmonds and his colleagues concluded that both these opposing studies are plausible, and the likelihood of their occurrence will depend more upon whether the prevailing philosophy in the next seven decades is rooted in business-as-usual, with gradual improvements in efficiency continuing at historic rates, as assumed by the DOE office, or in major policy efforts to redirect energy production and end use toward environmental quality, source reliability, and the other values listed by José Goldemberg and his colleagues. Edmonds's group also chided both sets of study authors for incomplete studies of the costs of their various assumptions, although they do attempt some cost calculations.

Finally, the Edmonds group examined what would happen if the United States unilaterally chose to reduce its emissions and the rest of the world maintained on present trends. In 1950 the U.S. CO_2 emissions were approximately 40% of the global values. By 1975 this share fell to about 25%, and by the late 1980s it was at about 22%. If the U.S. held emissions constant at 1985 levels, a reduction of 15% from the NEPP forecast emissions in 1995 and a 28% reduction from the forecast emission in 2010, then global emissions would be reduced by only 3% in 1995 and 6% in 2010. Even for the case where U.S. emissions are reduced by 50% below the 1985 levels, global emissions would continue to grow and would be cut by less than 15% in the year 2010. (This, of course, makes the assumption that world emissions will continue to grow.) To be sure, savings of 5% or 10% should never be dismissed as minor, but four or five such reductions would amount to a 25% or more improvement, and ten such reductions could cause substantial delay in the rate of greenhouse gas buildup. However, it should not be assumed that the United States would be solely acting in a globally altruistic manner without company—if it were acting alone, it could levy tariffs on foreign products to compensate for the unilateral U.S. assumption of environmental costs.

Significant indirect effects are also possible, since it is likely that if the United States introduced technological improvements to reduce CO_2 emissions, then the ultimate long-term cost reductions and the competitive advantage they would give the U.S. would certainly spur imitation by our competitors. This response would then reinforce U.S. reduction of emissions, leading toward a worldwide effect.

But one could also argue in reverse, suggesting an indirect effect in which lowered U.S. energy demand would lower energy prices, thereby spurring other parts of the world to use more. This example shows the complexity of detailed forecasts, but also makes a more important point: no nation acting alone can substantially reduce the rate of buildup of greenhouse gases. Yet if each nation waits for another to act, all will continue to wait and the rate of buildup will be uncomfortably rapid. Such a "tragedy of the commons" can only be avoided by international agreements, in which some nations take the lead to set the moral (and economic) tone for the rest of the world with the full expectation that others will—by agreement, it is hoped—follow suit.

Thirteen years ago, I briefed an industrial group on the potential problems of greenhouse gas emissions. One international oil company executive responded this way: "I'm not a bad citizen of the world, and I don't want my company to contribute to significant global pollution. But if I were to act alone on behalf of my company in a way that reduced emissions and raised our prices, my board of directors would fire me. Furthermore, if no other company or country was required to do that, then people would simply buy around us and we'd be ruined, and no help would go to the world. Therefore, we mustn't do something that destroys the competitive balance across the world through unilateral action, but instead must agree internationally that everybody pays the same penalty for their pollution so the market is not unfairly distorted."

I was delighted with his response, but asked why, if oil executives were so enlightened, they continued to have lobbyists work in halls of parliaments and congresses around the world to block additional taxes on energy use and regulations to improve the efficiency of production equipment. The executive conceded that that was a fair point and said that "as world perceptions on the seriousness of this issue increase, then energy companies should, for their own good, join in international efforts to distribute the burden of pollution reduction fairly around the world." What he was getting at, about ten years ahead of the fashion then current, was that without a law of the atmosphere it would be very difficult to get any one nation to reduce fossil fuel emissions substantially, since such reductions could cost selected groups unduly in comparison to other competitors.

We should not think it will be easy to achieve a law of the atmosphere. For example, in the plenary session at the Toronto conference, a representative from the Maldive Islands, a small group of about 1,200 tiny islands south of India, complained that even a 1-metre rise in sea level would doom most of his 200,000 people to migration or death, and that anything we did would probably already be too late. An African delegate said that he didn't wish to be preached to by the West about what he could do with his trees or his development plans, that indeed it was the turn of his country and others like it to develop, and that the environment would just have to be secondary—as it was in our Industrial Revolution. A representative from India agreed that changes in the climate could be very serious for his country, but angrily snapped at one Westerner: "We may have three times your population, but you use twelve times the energy per capita we do, so you still have four times more reduction to do before we worry about our population." At that stage a Chinese scientist got up to chide the Indian and note that China had done something about population and therefore was entitled to develop coal, whereas India was working too little on population or technology efficiency improvements. Finally, when Minister Emil Salim from Indonesia was giving a lucid account of the developmental difficulties faced by that country and admitting forthrightly why deforestation was not in its long-term interest, he fielded a question from climatologist L. Danny Harvey, from the University of Toronto: "If deforestation is not in your interests, then why don't you just stop?" Danny asked. Salim retorted, "Those trees bring us $2.5 billion of foreign exchange a year that we absolutely need for our development. So if you could provide us with some alternative source, we would be very interested."

This brief sampling of remarks illustrates the delicacy that will be needed to bring about a major law of the atmosphere. Nevertheless, the Toronto meeting showed that despite the lurking dangers, the vast majority of delegates agreed that action is needed. And in a remarkable and welcome change of direction, British Prime Minister Margaret Thatcher recently endorsed the notion of reducing greenhouse gas emissions, "even though this kind of action may cost a lot, I believe it to be money well and necessarily spent because the health of the economy and the health of the environment are totally dependent upon each other."[25]

Perhaps the U.S. experience with tradable polluting rights can yield some encouragement on the law-of-the-atmosphere debate. A tradable polluting right means that each company that produces a product that has a certain pollutant as a by-product is restricted to emit only a limited amount of that pollutant. However, the law does not tell the company *how* to limit its emissions—that is, whether to produce less product or invest in more control technology. In fact, the permits are tradable. In other words, if a company is clever, it can invent a process that keeps its total emissions below its limitation and then actually sell the remaining "license to pollute" to another company that may not be as advanced. Such tradable permits offer companies strong incentive to get their own pollution down, for then they can market their remaining rights to somebody else for extra capital. An analogy with the law of the air seems clear: if one country were able to dramatically improve its energy efficiency or develop renewable alternatives ahead of schedule, it might produce less total CO_2 or fewer CFCs than permitted by a law of the atmosphere. It then could sell its remaining CO_2 or CFC quota to other nations not as advanced, thereby rewarding itself for its own ingenuity or efficiency. Such an incentive system built into a law of the atmosphere obviously needs further exploration,[26] but first, of course, we must decide what the total global emissions should be and how that total should be distributed among nations. I have no illusions that this will be anything but a drawn-out process, probably taking at least as long as the decade it took to get the Montreal Protocol signed. But it is hard to imagine a global agreement that could be more far reaching or of greater fundamental importance to the greenhouse century.

This discussion clearly suggests that global atmospheric problems can be prevented—or at least slowed—through a variety of mechanisms. Major immediate efforts to implement global prevention strategies are still not assured, however, given current political alignments, economic investment strategies, and prevailing philosophies of nations. But even if such atmospheric problems cannot be substantially prevented, a hierarchy of policy actions over the next few decades could help to minimize societal vulnerability to atmospheric changes and perhaps even help some sectors to take advantage of altered conditions. The atmospheric problems examined here represent in large part a fundamental challenge to the philosophy of individual, corporate, and state interests. These interests normally react to problems on time scales that are much shorter than those associated with the effects of ozone depletion, acid rain, and a buildup of CO_2 in the atmosphere. What may be needed before major policy initiatives are forthcoming is a re-evaluation of the present balance between narrow, short-term economic interests and long-term national and global concerns. The setting of strategic goals such as increased energy efficiency and the development of non-fossil energy supplies seems of the utmost importance to help reduce the major atmospheric pollution problems. If short-term return on investment is the principal goal of individuals, organizations, or nations, then we are likely to be forced to adapt to the range of consequences, good and bad, that will inevitably be thrust upon us if present trends continue. It is doubtful that current market-driven forces alone, that is, no charge for pollution, will encourage sufficient investment to prevent, or even substantially reduce, the potentially serious long-term consequences of the increase in atmospheric pollution. Furthermore, many economists or planners argue that the uncertainties are so large it is difficult to assess the impacts of climate change. Harvard economist Tom Schelling offers an example of the viewpoint that current uncertainties, the rapid adaptability of society, and the possibility of benefits from atmospheric change do not justify action:

> As emphasized earlier, any single nation that imposes on its consumers the cost of further fuel restrictions shares the benefits globally and bears the costs internally. For only the very largest fuel-consuming nations, probably for only the Soviet Union and the United States, might it be in the national interest unilaterally to suppress further the use of fossil fuels in the interest of mitigating climate change. And even that trade-off is certain to look unpersuasive to consumers paying current fuel prices. Some global rationing scheme that enjoyed the participation of the major producers and consumers of fossil fuels would be required if there were to be severe action at the national level.
>
> In the current state of affairs the likelihood is negligible that the three great possessors of the world's known coal reserves—the Soviet Union, the People's Republic of China, and the United States of America—will consort on an equitable and durable program for restricting the use of fossil fuels through the coming century and successfully negotiate it with the world's producers of

> petroleum and with the fuel-importing countries, developed and developing. It makes sense therefore to anticipate changing climates. In any event, no regime for further restricting fossil fuels would hold emissions constant, so climate change is what we should expect.[27]

I have argued with Schelling and other economists that uncertainty *per se,* or even the prospect of scattered benefits, does not justify, in my value system, gambling on unprecedented global change. At a personal level, none of us knows whether he or she will break our leg, get pneumonia, or remain healthy; yet most of us invest in health insurance. Of course, we could wait until we got sick before we tried to buy health insurance, but we would find that no insurance company would sell it to us at that stage. On a national scale, we all, from every spectrum of the political range, believe that national security deserves some public expenditure. It isn't some economic calculus or cost-benefit analysis that tells us that we need to make national security investments; rather, it is the strategic logic of hedging against perceived future threats to our liberty or security.

Again, in my value system, the prospect of climatic change occurring on a global scale ten to fifty times faster than typical natural average rates of change is not one we should relish. The possibility of major environmental surprises increases with the rate at which climate changes. Moreover, if there are things we can do to slow down this rate of change that simultaneously will provide multiple benefits, then it would seem logically compelling to take them seriously. Only foolish or desperately poor people would buy no health insurance and only the foolish would spend most of their income on premiums. Likewise, only foolish countries would over- or under-invest in security. To me, the question is not whether to adopt a strategic policy of protecting the atmosphere, but rather how much to invest. That is where the tie-in strategy becomes so important. Rather than spending vast resources for the sole purpose of averting atmospheric change, I have argued throughout these pages that the prudent course to follow would be making high-leverage investments, a common business practice that seeks to earn multiple benefits on the same investment. Using fuel efficiently not only reduces CO_2 injection, but also cuts acid rain, reduces the health effects of air pollution in cities, reduces the dependence of our energy security on unreliable resources, and improves our long-term competitiveness by cutting the energy cost of manufactured products. It also is a form of insurance against the possibility of catastrophic climatic changes, should some nasty surprises actually materialize—and their probability of occurrence is not negligible. Like an insurance policy, however, this protection does not come free. It requires a premium. It is my opinion that a few tens of billions, or perhaps hundreds of billions of dollars, spent annually around the world on such planetary insurance, with its multiple benefits, is an investment that is long overdue. That is only a small fraction of world expenditures on defence. Viewed in this light, the current trend toward East–West reductions in tensions and troops is also an environmental benefit.

Perhaps, with the East-West tensions being reduced dramatically, the time is now right to convene a Presidential (and Congressional as well) Commission on Conversion and Transition. The Commission should be charged with developing a timetable for implementing practical steps to convert, say, 5% of the defence budget each year until 2000 AD into investments in environmentally sustainable development.

Other strategic hedges include investments to improve the potential safety of nuclear reactors and the reliability and cost of solar electric machines. Substituting natural or biologically produced gas for coal in regions sensitive to air pollution would also have the diplomatic benefit of repairing a fracture in our relations with Canada. But to find more gas, we need to provide incentives for people to look for it. In short, no one gets something for nothing. But if an investment produces multiple benefits and provides insurance against the prospect of catastrophic loss, then certainly it is the kind of policy option that prudent individuals, families, companies, states, nations, and international organizations must consider carefully and long.

Finally, there is an ethical question associated with the atmospheric problems we have discussed here: Do we have the right to commit future generations to unprecedented atmospheric perturbations without actively attempting to prevent, impede, or at least anticipate them? We are insulting the atmospheric environment faster than we are comprehending the effects of those insults. Some of the uncertain consequences of our insensitivity could be serious and very long lasting.

To be sure, further institutional actions are needed to build a scientific consensus on physical, biological, and social aspects of atmospheric pollution issues. Perhaps even more important is the need for public awareness and understanding, which could lead to demands for political action. The prospects for alleviating most foreseeable atmospheric problems are good, but it isn't certain that anything much more aggressive than research funding will be instituted on a large enough scale before the atmosphere has itself performed its own experiments, now under way, with all of life on earth inside this unique laboratory—unless, of course, enough people demand otherwise.

"Are we entering the Greenhouse Century?" I asked in the subtitle of this book. It should be clear by now that I believe we've been in it for a while already, but admit that it will take a decade or so more of record heat, forest fires, intense hurricanes or droughts to convince the substantial number of sceptics that still abound. Unfortunately, while the antagonists debate, the greenhouse gases keep building up in the atmosphere. I wonder what we will say to our children when they eventually ask what we did—or didn't do— to create the Greenhouse Century they will inherit.

Epilogue

The Global Warming Debate Heats Up

Since the first appearance of *Global Warming* in September 1989 there has been vociferous debate among scientists, economists, industrialists, and environmentalists about how serious the global warming issue is. This sometimes acrimonious exchange has spread confusion among the public and politicians alike about the credibility of greenhouse effect warnings and has, at least temporarily, delayed action on the problem. Therefore, I think it is worthwhile to summarize briefly what has happened since the initial publication of *Global Warming,* and why the ensuing debate hasn't changed the fundamental conclusions of the original edition. Indeed, while many scientific questions remain open to dispute, some of the bitter criticisms denying the immediacy of the problem are becoming recognized for what they are: attempts by special interest groups or ideologues to delay global actions that might adversely affect them.

Eighteen years ago, when I first began to address the public policy implications of still uncertain climatic forecasts, a controversial and well-known scientist took me aside: "You'll be able to judge the magnitude of your impact on society," he warned, "by the position and vehemence of your critics." Getting people's attention is one thing, however; getting action on public policy is another.

Critics maintain four principal objections to the likelihood of global warming. First, the scientific basis for projecting future climate change is so uncertain that no responsible scientists would dare propose immediate policy responses. Second, those who suggest that a hundred years or more of unprecedented climate change (what I have called the Greenhouse Century) is being built into the future are just "environmental activists" whose ideological agenda aims to destroy the free market system. On the other hand, those who argue that there are unlikely to be any significant effects are, by contrast, thoughtful senior scientists protecting the public's interests. Third, decade-to-decade temperature changes over the past 100 years aren't consistent with climate model predictions of the effects of increasing greenhouse gases; thus the model projections are probably exaggerated. And fourth, it's too expensive to do anything about global warming anyway. The only thing to do now is wait and see what happens—after all, the changes probably won't be great and may be beneficial even if they occur. So say the most strident critics.

The debate became so intense that in 1990 the U.S. National Academy of Sciences invited the principal scientific opponents of activism on global warming to debate the scientific "establishment" that has consistently reaffirmed its confidence in its own 1.5°C to 4.5°C warming estimates for the middle of the next century. President Bush's Chief of Staff, John Sununu, cited these oppositionist critics to justify a go-slow approach to joining international efforts to regulate emissions of atmospheric pollutants that have the potential to cause unprecedented climate change in the twenty-first century. As long as the U.S. refuses to limit its emissions, Great Britain, France, Canada, Japan, and other nations are also unlikely to act. Thus, the U.S. refusal to agree to specific curbs on some of its emissions is having a major impact on the shape of the Greenhouse Century.

The debate has strayed far from reason and civility on occasion. *Forbes* magazine, for example, placed an ad on the back page of the *New York Times* (February 7, 1990), praising itself for courageous journalism in debunking global warming as "Hype Not Heat" and belittling the issue with the headline, "No Guts, No Story." "Global warming effects," *Forbes* said, "would be at worst minimal." Editorial cartoons in support of warming controls advocates escalated the media circus. One showed Sununu as a devil whispering in the President's right ear, "To hell with the future, let's go for short-term profit." Bush is flicking an angel, William Reilly, the Environmental Protection Agency administrator, off his left ear as Reilly reminds him of his campaign promise to be the "Environmental President." When debate degenerates to such inanity, it's no wonder the public and most politicians are confused about what the real problem is, let alone what to do about it.

The irony is that very little has changed scientifically during the past several years. Nevertheless, when new technical data surfaces that apparently reduces the magnitude of the problem, it is welcomed by global warming opponents as a major finding. New data that reinforces global warming concerns is ignored by the critics, but promoted by environmentalists. Sides are taken without putting new information in perspective, with the result that confusion grows. Unfortunately, while such a polarized debate makes entertaining op-ed page reading and grabs ratings on TV, it clarifies little of the real scientific controversy or of the broad consensus on basic issues within the scientific community.

Typical projections of global warming possibilities (see Figure 7, page 105) into the twenty-first century have been drawn by a group of scientists convened by the International Council of Scientific Unions. They show warming from a moderate half degree Celsius (.9°F) up to a catastrophic 5°C (9°F) or greater before the end of the next century. I do not hesitate to call the latter figure catastrophic because it is the magnitude of warming that occurred between about 15,000 and 5,000 years ago: from the end of the last ice age to our present interglacial epoch. It took nature some 5,000 to 10,000 years to accomplish that warming, which was accompanied by an approximately 100-metre (330-foot) rise in sea level, long distance migration of forest species, radically altered habitats, extinction and evolution of species, and other major environmental changes.

Critics of immediate policy responses to global warming are quick to point out the uncertainties that could reduce the average projections of climate models (such as the middle line on Figure 7). Indeed, most climate modellers include similar caveats in their papers. Many critics, including the authors of a report for the Marshall Institute[1] (a Washington-based think tank best known for its advocacy of President Reagan's "Star Wars" Strategic Defense Initiative), hardly mention that the sword of uncertainty has two edges: that is, the same inexactitude in physical or biological processes that makes it possible for the present generation of models to overestimate future warming effects is just as likely to cause the models to underestimate change. [I wrote a letter putting this Marshall Institute report in perspective at the request of Alan Hecht, a deputy administrator in the U.S. Environmental Protection Agency. Since that letter has already been widely circulated (after someone obtained it under the U.S. Freedom of Information Act), I reproduce it here in the notes.][2]

The public policy dilemma is how to respond in the absence of conclusive evidence of the effects of global warming. It is my opinion that the scientific community will not be able to provide definitive information over the next decade or so about the precise timing and magnitude of century-long climate changes, especially if research efforts remain at current levels. Policy makers must decide how much information is "enough" to act on and what measures to take to deal with the plausible range of environmental changes. Unfortunately, the probability of such changes cannot be precisely estimated by analytical methods. Rather, we must rely on the intuition of experts, which is why obfuscating media debate impedes policy development.

Fortunately, making intuitive scientific judgments is the purpose of such deliberative bodies as the National Research Council of the U.S. National Academy of Sciences (NAS) and the International Council of Scientific Unions. NAS, for example, regularly convenes a range of experts to estimate the probabilities of various scenarios of change. The deliberations of these panels are removed from the cacophony of media debates that typically highlight only the extreme opposite positions. Half a dozen such assessments[3] over the past ten years have all reaffirmed the plausibility of unprecedented climate change building into the next fifty to a hundred years. In 1990 a United Nations-sponsored group of several hundred international scientists, the Intergovernmental Panel on Climate Change (IPCC), also reaffirmed that plausibility.[4]

I've mentioned that the National Academy of Sciences has regularly produced studies from a broad cross-section of the scientific community of possible climate changes from greenhouse gas build-up. Since these assessments have been used by both politicians and conservationists to justify serious consideration of policy actions, some critics have tried to discredit the National Research Council (NRC), which compiles such assessments for NAS.

One such critic, Richard Lindzen, a meteorologist at the Massachusetts Institute of Technology, is himself a member of the National Academy. Nearly 20 years ago Lindzen criticized the Department of Transportation's Climatic Impact Assessment Program to evaluate the environmental impact of high-flying supersonic transports (SSTs). In 1989 he went so far as to try to discredit the Academy consensus on global warming by impugning the credentials of its committees:

> The National Academy consists of about 1,500 members who are elected on the basis of their scientific accomplishments. Election to the Academy is considered a high honour. When, however, the Academy responds to inquiry, it does so through committees of the National Research Council. Oddly enough, there's only a minuscule representation of Academy members on these committees (approximately 6%). Thus, the connection of so-called National Academy Reports to the Academy is itself tenuous and uncertain. In principle, these reports are reviewed by academicians, but the selection of these reviewers is fairly arbitrarily handled by the NRC staff and the presidents of the Academies of Science and Engineering and the Institute of Medicine.[5]

In other words, to Lindzen the Academy committees are suspect because many of their members have not met the standards for election to the Academy—high academic standards to be sure. But Academy membership is largely based on disciplinary specialization and outstanding contributions in a narrow field, rather than on an individual's capacity for multidisciplinary synthesis. But the latter quality is also essential for integrative assessments of complex subjects such as climate change. That is one reason why the Academy casts its net widely in selecting committee members.

Lindzen went on to suggest that his intuitive understanding of how the atmosphere worked led him to believe that its "response to doubling of carbon dioxide may readily be ⅛ to ¼ — or even less—of what is suggested" by the National Research

Council consensus of 1.5°C to 4.5°C warming if CO_2 doubled. Lindzen based this "⅛ to ¼" statement upon his intuitive scientific judgment and offered no calculations in the peer-reviewed literature to back it up. Nevertheless, opinion page articles proliferated following Lindzen's statements, and my telephone was busy with reporters and others seeking my response to his assertions that global warming was a vastly overblown environmental scare.

Finally, the December 25, 1989 issue of *Forbes* magazine featured an article whose scientific objectivity was emblazoned colourfully on the cover: "Global Warming Panic: A Classical Case of Overreaction."[6] Excerpts appeared in newspaper opinion pages for weeks following. The piece was by Warren Brookes, an economic journalist who combines sharply worded anti-global warming scientific and policy opinions with *ad hominem* attacks. It's not surprising that such stuff is written, but it is surprising to me that it appears on the opinion pages of respectable newspapers without any attempt to put this nonscientist's scientific arguments in the perspective of the broad consensus. What newspapers tend to do for "balance" is print the contrary views of comparably extreme advocates of radical environmental policies or *laissez-faire* economics. The result is increased confusion and further loss of objectivity and perspective, as I observed in Chapter 7 on "Mediarology."

What Brookes does is to cite the many uncertainties surrounding global warming projections – usually attributing the caveats to the critics, whose impeccable credentials he flaunts. Then he selectively quotes from "environmental activists" (like Jim Hansen or me). But more important than its *ad hominem* attacks, the *Forbes* article does not indicate that most of the scientific uncertainties mentioned are as likely to make our estimates for the future worse as they are to make it better. Unknown phenomena are as likely to cause increases as decreases in current estimates of future changes. Furthermore, Brookes does not refer to the substantial amount of information that validates climate models, including their capacity to reproduce the large seasonal cycle of surface temperature and to reproduce many significantly different climates from glaciological history. He overlooks the fact that ice ages and interglacial warm periods have seen 5°C (9°F) temperature changes marked by 25% increases in CO_2 at the warm times and 25% decreases in the cold times (see Figure 2, page 40). These strong circumstantial pieces of evidence are what motivate me, and most other scientists involved in National Research Council or IPCC studies, to believe there is a substantial probability of unprecedented warming building into the next century.

Brookes, along with other critics, also points out the lack of significant warming in the lower forty-eight United States during the past century. "That news alone," Brookes asserts, "should have cooled off the global warming movement."[7] What he and others who raise this issue ignore is that while the U.S. warmed in the west and cooled in the east, if Alaska were included in their estimates (which it was not), the U.S. as a whole would have warmed up more than 0.3°C, close to the global average. What the data actually shows is that natural climate variability is significant on a regional basis. For example, had one chosen to observe north central Asia, one would have noticed a 40-year warming trend vastly in excess of the rest of the globe. It is as deceptive to suggest that the Asian trend proves the world is warming up faster than the models project, as it is to allege that the absence of an appreciable temperature trend in the lower forty-eight United States proves that the greenhouse effect does not apply to the world—or even to the U.S. This kind of scientific silliness couldn't survive in serious debate, yet it is reprinted in news stories and opinion pieces.

As mentioned before, because of public doubt aroused by the *Forbes* story and other commentaries, the National Academy of Sciences and National Academy of Engineering organized debates to inform their most recently convened panel of the critics' views. Lindzen and others were present at the first debate, along with the "establishment scientists": James Hansen, Director of NASA's Goddard Institute for Space Studies; V. Ramanathan, a climate expert from the University of Chicago; and Jerry Mahlman, Director of the National Oceanic and Atmospheric Administration's Geophysical Fluid Dynamics Laboratory in Princeton. Also attending was my NCAR colleague, Kevin Trenberth, frequently cited as a critic of global temperature trend data.

Trenberth started the discussion, showing how difficult it is to estimate the world's average surface temperature. He noted that ocean temperatures inferred from ships' records in the pre-1900s period were problematic, since they were obtained from thermometers placed in buckets dropped over the side of the boat. Some buckets were made of leather, some of wood; some measurements were taken on the windward side, some to the leeward side of the ship, all of which would affect the readings. He also showed that the fraction of the oceans covered by ship tracks before 1900 was 10% to 20% at best. One of the Academy panel members asked him if he felt there was any utility at all in the ocean temperature data before 1900 for the purpose of global trend analysis. "Not much," he said. In fact, he commented, the corrections due to faulty measurement techniques are typically larger than the inferred climatic trends, which is why no one pays serious attention to them.

Nevertheless, Brookes, Lindzen and other critics have prominently cited M.I.T. meteorologist Reginald Newell's study suggesting that ocean temperatures were as warm in the 1850s as in the 1950s; thus they argue that no global warming has taken place over the past 150 years. But as noted by Trenberth, pre-1890 thermometer data is not usually regarded as credible in scientific assessments of global trends; coverage is not global and the measurements themselves are unreliable. These reservations did not stand out in the critics' citation of Newell's analysis.

I asked all the assembled scientists at this Academy debate my favourite polling question: "What is your estimate of the probability that the next century will see a global warming of 2°C (3.6°F) or more?" All the atmospheric scientists present, including Lindzen, agreed that there would be warming. He, however, felt the most likely extent of warming was between 0.5°C to 1°C rather than 2°C to 4°C as in typical Academy assessments—and the IPCC estimate. Atmospheric scientists Hansen, Mahlman, Trenberth and Ramanathan all agreed that 2°C was certainly a reasonable number for the twenty-first century. They assigned the occurrence of a 2°C warming a probability between 60% and 90%-95%. Lindzen was the only exception; his probability for 2°C warming was 25%—the lowest estimate I've yet heard from any knowledgeable atmospheric scientist. Indeed, several people in the audience—aware of his assertions that global warming is likely to be "⅛ to ¼" of what the Academy presently estimates—expressed surprise to hear that his estimate of the probability of unprecedented warming in the next century differs from mine by less than a factor of 2.5 (25% for him versus 60% for me). One science writer observing the Academy debate for a major newspaper said to me later, "This 'great debate' is a phoney; you guys really disagree scientifically much less fundamentally than most people think." I was glad he discovered it.

Perhaps the strangest aspect of the National Academy debate occurred the morning before the debate itself. Warren Brookes, writing an editorial in the *Washington Times* entitled "Greenhouse Showdown or Show Trial?"[8] blasted the upcoming session by challenging the roster of debaters and the credentials of the Academy committee members who called for the debate. "Only one of the three panelists asked to prepare papers is from the dissenting side," said Brookes. That, of course, was not true; in addition to Lindzen there was Trenberth, and retired Yale University forester William Reifsnyder, all of whom had been mentioned as critics. Moreover, Robert Jastrow of Dartmouth and William Nierenberg of Scripps Institute of Oceanography, co-authors of the Marshall Institute critique of global warming, refused to attend the meeting, though invited. Nierenberg, however, apologized for his absence by letter and said he was certain that his views would be adequately represented by the three critics present.

Nevertheless Brookes went on: "The most serious 'offenders' to be tried at today's meeting are three of the nation's most prominent senior scientists who won't even be there, Robert Jastrow of Dartmouth, founder of the Goddard Institute for Space Studies at NASA, Frederick Seitz, past president of NAS and William Nierenberg, former director of Scripps Institution (*sic*) of Oceanography." Brookes accused the Academy of being "eager to trash" Jastrow, Seitz and Nierenberg, authors of the Marshall Institute report that many believe was used by White House Chief of Staff John Sununu to prevent E.P.A. Administrator William Reilly from convincing President Bush to join in an international commitment to reduce greenhouse gas pollution. Brookes concluded by accusing some in the Academy of coming "dangerously close to using Lysenkoist tactics to anyone who dares to dissent from the prevailing political/scientific wisdom." [Lysenko was the Soviet biologist who faked genetic research results to match Stalinist ideology.]

Interestingly these personal attacks were not reciprocated by those accused of Lysenkoism. Perhaps they did not feel the ideological antagonism expressed by Brookes in *Forbes:* "As Marxism is giving way to markets, the political 'Greens' seem determined to put the world economy back into the red, using the greenhouse effect to stop unfettered market-based economic expansion." To me, this says that *Forbes* wasn't publishing a courageous exposé about science, as its newspaper ads boast ("No Guts, No Story"), but rather a defence of its ideology in the guise of science journalism. Everyone is entitled to an ideology, and, to his credit, Brookes admits his in this one sentence, even if he buries it in the middle of a six-page article. What does bother me is his cavalier way of taking selected bits of science out of context to cut and paste them into a slick-sounding article in support of a blatantly ideological viewpoint—a criticism that applies to the writings of some environmental activists as well.

Scientists share responsibility with the media for often failing to communicate complex issues clearly to the public. What the general public, as well as politicians and bureaucrats, do not recognize is that most scientists spend their time arguing about what they don't know. Scientists generally consider discussions of accepted ideas boring and a waste of time. This is because the scientific method operates by constant questioning, particularly of issues not yet well substantiated. But if the public and its representatives do not understand the process of scientific inquiry, then they will have difficulty interpreting the "duelling scientists" debates, let alone deciding whether debaters are honest or ideologically driven.

We scientists simply have to spend more time differentiating accepted information from what is reliably believed to be true and, most important, from what is highly speculative. The public version of the global warming debate rarely separates those components clearly, thereby leaving the false impression that the scientific community overall is in intellectual disarray, when, in fact, the IPCC and the National Research Council's consensus of 1.5° to 4.5°C warming in the next century still reflects the best estimate of a wide range of knowledgeable scientists. This estimate includes recent United Kingdom studies that halved the "best guess" from over 5°C to around 2.5°C.[9] Perhaps some new discovery will push the best guess higher again next week, but meanwhile the 1.5° to 4.5°C consensus warming range endures. Unless we communicate what we do know along with what we don't know the public policy process is subverted in confusing debate that inadequately represents the true nature of informed opinion.

It is difficult for the media to do what I sometimes wish they would: back off the concept of "balance" in favour of the concept of "perspective." If an issue is complicated, it simply is not enough to play off "all sides," particularly if the opinions of the majority of the experts—the people who create the consensus—are left out. Moreover, that consensus information must be expressed in terms of its probability. Very few scientists would say they believe the future climate *will* warm up from 1.5°C to 4.5°C; rather, most believe that to be reasonably probable. Conveying issues in probabilistic terms with the range of views in perspective is necessary if scientific opinion is to be communicated accurately, rather than as a misleading debate among feuding scientists, or occasionally as a travesty perpetrated by polemicists and ideologues.

Let's return to the details of the debate. Another criticism of global warming projections has been the imperfect match between the warming of the earth and the smooth increase in greenhouse gases over the past hundred years (see Figure 4, page 85). It has been alleged that since most warming in the twentieth century took place up to the 1940s, followed by a cooling at the time the global greenhouse gases were increasing at their fastest, the decade-by-decade temperature trends in the twentieth century cannot therefore be attributed to greenhouse gas build-up.

At first reading that sounds like a valid criticism; but there are several flaws in the argument. First, nature always fluctuates. Several tenths of a degree Celsius warmings and coolings over decades are part of the natural record and, indeed, are normal. Scientists call these fluctuations "climatic noise." These are not predictable as far as anyone can tell, since they appear to be caused largely by the internal redistribution of energy among the principal reservoirs: the atmosphere, oceans, ice, and land surfaces. Therefore, natural fluctuations could partially explain the sharp warming up to the 1940s, the Northern Hemisphere's cooling to 1975, and possibly even the spectacularly rapid re-warming of the 1980s—the warmest decade in the 100-year instrumental record. Secondly, we do not know precisely what other potential climatic forcings (that is, processes that could force the climate to change) have been doing over the past 100 years. These include energy output from the sun, atmospheric particles from volcanic eruptions, or particles generated by human activities—or as University of Wisconsin climatologist Reid Bryson likes to put it, the "human volcano."

This "forcings" problem is akin to a criminal investigation in which the whereabouts of only one suspect is known and the activities of the other possible suspects were not carefully observed. In this case, the "crime" is the 100-year 0.5°C (0.9°F) warming trend and the leading "suspect" is the known greenhouse gas increase. Unfortunately, since we do not have quantitatively accurate ways of knowing precisely what the other potential forcings may have been (that is, the unwatched "other suspects"), we can't rule out some possible role for them. Some scientists, such as James Hansen and myself, have led efforts to estimate what volcanic or solar forcings may have contributed to twentieth-century temperature trends.[10] Indeed, such estimates improve the match of our computer model simulations to observed twentieth-century temperature trends. However, as all of us have admitted, without more quantitatively reliable information these exercises can do nothing more than sketch out plausible rather than definitive results.

Incidentally, my own simulation result, which used temperature trend data up to the early 1980s, suggested the best fit to the data was such that a CO_2 doubling would cause only about 1.5°C to 2°C global warming—the lower end of the 1.5° C to 4.5°C range cited in most U.S. National Academy of Science reports or by the ICPP. However, it was noted at the time that this could not be taken as strong evidence that CO_2 doubling would result in such a moderate climate warming because we know that the twentieth century record is marked by natural noise, unmeasured alternative forcings and uncertainties in how much temperature change actually occurred. Most scientists still agree that without ten to twenty more years of thermometer, solar, air pollution, and volcanic observations it's difficult to pin down anything quantitatively to very high reliability, say 99% confidence.

Arguing that climate models have been unable to predict a detailed sequence of decade-by-decade temperature fluctuations has been a favourite tactic of some global warming critics.[11] But this is akin to arguing that because we can't predict the individual rolls of a pair of dice, we also can't predict the odds of getting any two faces on any roll. All gamblers know better. Though we know the statistics for rolling a particular number, we certainly wouldn't be expected to know the sequence of numbers on successive rolls, even for slightly loaded dice. In short, those who argue that an absence of an exact match on a decade-by-decade basis between observed temperature fluctuations and greenhouse gas build-up demonstrates that greenhouse effect sensitivity of models is wrong are simply off in their logic. Such agreement should not be expected in detail as long as a large degree of climatic noise continues to make up much of the decade-by-decade temperature record and as long as we lack precise data on other non-greenhouse gas forcings.

Fortunately, we are now measuring the sun, volcanos and pollution-generated particles, and can thus account better for their effects. In other words, we finally are checking up on the "other suspects." Thus, as greenhouse gases continue to build up in the future, if greenhouse warming does not take place at roughly the predicted rate during the 1990s and into the twenty-first century, then indeed it will be possible to argue on the basis of some direct evidence that the effect predicted by today's models is off line. Personally, I'll be surprised if there is a major error that overrides the 1.5°C to 4.5°C warming projections.

Let me next address the final, and perhaps the most important, criticism made against action to slow global warming; that immediate policy steps to cut CO_2 emissions are too expensive. The *Forbes* newspaper ad suggests, for example, that if we cut CO_2 emissions the U.S. will be bankrupt and the Third World impoverished. Indeed, as I discuss frequently in *Global Warming,* there is substantial Third World opposition to the prospect that they may be deprived of their own industrial revolutions, and of the economic growth experienced in the Victorian period by the then-developing countries using cheap and dirty coal. Since developing Third World countries such as India and China have abundant coal supplies, they would like to use them as low-cost means to industrialization. In 1990, however, these countries have between them some two billion people, whereas in the nineteenth century the entire world didn't have two billion people. The global impact of the developing countries' use of coal to produce even a quarter of our current industrial standard of living would be greater than was ours. Needless to say, such arguments are not met sympathetically in China or India.

It is sensible, I believe, to argue that developing countries need not repeat the western experience of industrialization with smog-choked cities, acid rain, and inefficient power production, given that modern technology has many better solutions. Unfortunately, developing countries typically respond that high-tech, efficient machinery is more expensive than the traditional options available to them. This dilemma makes obvious the need for a bargain. Countries with technology and capital must provide resources to developing countries, which in return must keep population growth under control and work toward industrial development with the lowest polluting, most efficient technologies, even if they cost more initially.

There have been international efforts afoot to draft an agreement requiring the developed nations to decrease their carbon dioxide emissions by, say, 20% by the year 2000, and to cut the projected emissions growth rates in developing countries. This has been strongly opposed by the United States, echoed by Japan, the U.K., the U.S.S.R., and some other countries. The Japanese are unhappy since they're already twice as energy efficient as the U.S. They claim it would cost them much more to cut carbon dioxide by 20% than the U.S., since our very inefficiency gives us more opportunity to cut cheaply.

Still less efficient developing countries could produce far less growth in carbon dioxide pollution by using efficient, modern technologies instead of older, cheaper ones. This sets up a possibility for creative international management that might not only eliminate Third World opposition to global emissions limitations, but also could encourage competition among these nations to be the venue for future emissions cuts funded by developed countries; cuts that in turn could buy the developed nation out of its reduction requirement by funding even larger CO_2 reductions in now-energy inefficient developing countries.

I recently discussed this idea with a Japanese economist, Yoichi Kaya, an energy analyst from the University of Tokyo. He tried to assess what a significant cut in CO_2 emissions might imply for world and Japanese economic growth. Kaya concluded that for energy-efficient countries (like Japan) substantial cuts could severely lower economic growth—unless ways were found to increase the rate at which the economy becomes less dependent on energy growth. (This is what environmentally-oriented energy analysts already argue is essential for environmentally sustainable development; see discussion in Chapter 8.) He also showed that developing countries' economic growth would require efficient energy or else they would pollute severely.

I met with Professor Kaya in Tokyo in November 1989, while the Noordwijk Conference, an international environmental meeting in the Netherlands, was producing daily headlines about how the U.S., Japan, and the United Kingdom were balking at specific emissions cuts. Professor Kaya and I agreed that equal percentage emissions limitations for each nation may make international politics simpler, but that from a global point of view, reducing emissions by a fixed fraction for all nations may be neither the most cost effective plan, nor the fairest. Unequal fractional cuts may sound unfair to some nations, but what determines fairness, we felt, is not how much CO_2 emission each cuts at home, but *who pays* for all the cuts. To be sure, the rich nations should pay a disproportionate share of the costs since they have been responsible for over half of the CO_2 pollution to date. But energy-efficient countries like Japan, Germany, and Italy may not be the logical first places to look for big CO_2 emissions savings. However, we agreed it would not be politically easy to negotiate a system where some nations pay and other nations cut their emissions (or limit emissions growth rates in the case of Third World nations) by greater amounts. But from the point of view of the minimum global investment for the maximum amount of global pollution reduction, such a strategy probably is most efficient and consideration of the issue should be part of the international negotiating process.

In other words, supposing each developed country had to reduce its CO_2 impact on the world by something equivalent to 20% of its present production. Let's say, continuing the Japanese example, that their quota is to cut 100 million tons of CO_2 annually. Why not structure an international agreement so that the Japanese need to be responsible either for reducing 100 million tons of CO_2 from their own industries, or else paying to cut 150 million tons in another country (or some combination of both)? The obvious candidate is China, since Japanese investment in China for efficient energy production would both reduce acid rain over Japan as well as global CO_2 build-up. It is likely to be much cheaper per unit of CO_2 saved for the Japanese to improve

Chinese energy efficiency, since China is starting out so inefficient, than for Japan to improve its own efficiency. At the same time the Chinese would receive extra development assistance, the Japanese would get less acid rain and could buy out of their emissions reduction quota without having to cut emissions at home. Moreover, the Japanese would be creating friendship and markets for their products in the future, whereas the Chinese would be getting more efficient machines with lower long-term operating costs, thereby improving their economy and competitive posture into the twenty-first century. In other words, everybody wins. But first, we need a world emissions agreement that provides incentives for such bargaining and trading to take place. That agreement is what the U.S., the U.K., the U.S.S.R., and Japan were balking at in 1989.

Other similar ideas include issuing "tradeable (or leaseable) permits," giving everyone in the world the right to emit a certain amount of CO_2 or some other greenhouse gas.[12] These permits could be traded for cash, food, energy efficient products, etc. As of the signing of the global agreement, all nations would have equal per capita CO_2 emissions rights. That implies a certain amount of total national emissions. In the future a country could sell or trade these rights or exercise them for development, or increase their population and thereby limit their future emissions per capita. In other words, a fixed per capita emissions right that goes into effect at the signing date would dramatically reduce Third World suspicions that they were being singled out to bear the immediate burden of emission controls.

Critics of emissions reductions cite the supposed annual costs of global warming reduction at tens of billions of dollars—too much to be worth the benefits of climate change abatement. But they often neglect the additional non-greenhouse-effect benefits of emissions reductions: reduced acid rain; reduced air pollution; reduced balance of payments deficits; and lower long-term operating costs of efficient equipment, which reduces the energy costs of manufactured products and enhances competitiveness. Critics who simply cite the potential capital costs of CO_2 reduction write newspaper stories about how many billions or trillions it will cost and scare people away from action. But they often present a very unbalanced view of the distribution of benefits that come with greenhouse gas abatement. Unfortunately, it is very difficult to communicate these benefits in window sticker length headlines or in sound bites on the evening news, which is often all the time this complicated story gets.

For example, John Sununu, defending his role in persuading President Bush not to agree to specific CO_2 emissions reductions at Noordwijk, said on TV in 1990: "There's a little tendency by some of the faceless bureaucrats on the environmental side to try and create a policy in this country that cuts off our use of coal, oil and natural gas. I don't think America wants not to be able to use their automobiles."[13] I agree we don't want to cut "off our use of coal, oil and natural gas"; nor do we want to abandon our cars. But, Mr. Sununu, who does? Not one proposal from any bureaucrat, National Academy of Sciences committee or even environmental group I know of ever proposed such an absurd policy. Rather, most talk about giving up (or heavily taxing) petrol guzzling cars and switching to less polluting, more energy-efficient equipment, regardless of the fuel used. But some studies have suggested that switching to less polluting energy systems could cost "$800 billion, under optimistic scenarios of available fuel substitutes and increasing energy efficiency, to $3.6 trillion under pessimistic scenarios . . . between now and 2100." This quote from the February 1990 "Economic Report of the President" to Congress was based on the initial results of the first wave of economic model simulations.

Because of the controversy over the reliability of such models, the National Academy of Sciences ran a debate (following the climate debate mentioned earlier) among some economic forecasters and their critics. What emerged was very revealing. First, over 110 years (1990 to 2100) even a trillion dollars in CO_2 reduction costs, which sounds very expensive, is less than $10 billion each year—only a few per-cent of the current annual defence budget. Moreover, Robert Williams, an energy technology specialist from Princeton University, pointed out that the so-called "optimistic scenario" of $800 billion in costs to cut CO_2 emissions was based on very *pessimistic* assumptions about the rapidly decreasing costs of renewable energy systems such as solar, wind, or biomass power. Furthermore, with the exception of one effort by Yale University economist William Nordhaus, none of the other modelling simulations attempted to estimate the direct environmental benefits of our supposed trillion dollar investment in CO_2 emission controls. It is unconscionable that some critics of global warming action would cite these dubious cost estimates without so much as mentioning the potential benefits of slowing CO_2 emissions. Nordhaus, though, by balancing costs and benefits in his model runs, argued that cutting annual CO_2 emissions by as little as 10% or as much as 47% would actually produce benefits greater than the costs.[14] His model, however, was admittedly crude, laden with unprovable assumptions, and unable by itself to provide quantitatively reliable information for making policy choices.

Cross-examination of the economists by Academy committee members also revealed that their economic models had not been tested to see how they performed in predicting the economic consequences of historical events such as the 1973 OPEC oil price rise. I was shocked that such tests had not been done, and dismayed that some administration officials were actually citing these premature, unvalidated economic model results for costs of emissions controls as an alleged rational basis for making national policy.[15]

In summary, then, the greenhouse effect, the heat-trapping properties of the atmosphere and its gases and particles, is well understood and well validated as a scientific principle. Indeed, it is as good a theory as there is in the atmospheric sciences. Moreover, in late 1989, A. Raval and V. Ramanathan at the University of Chicago, used satellite observations to study the important water vapour greenhouse feedback mechanism, a process that is central to most models' estimates of some 3°C plus or minus 1.5°C equilibrium warming from a doubling of CO_2. They conclude, "The greenhouse effect is found to increase significantly with sea surface temperature. The rate of increase gives compelling evidence for the positive feedback between surface temperature, water vapour and the greenhouse effect; the magnitude of the feedback is consistent with that predicted by climate models."[16] *In other words, the heat-trapping capacity of the atmosphere is well understood and well measured on earth, and much of the sometimes polemical debate in the media over the greenhouse effect has little basis in reality.* This empirical confirmation of the natural greenhouse effect, which is consistent with the greenhouse effect of climate models, stands in stark contrast to the theoretical arguments of some critics. They believe that their untested conceptions of temperature-water vapour processes in parts of the tropics will reduce present model estimates of global warming by a factor of four or so.

It is well known that the 25% increase in CO_2 documented since the industrial revolution, the 100% increase in methane since the industrial revolution, and the introduction of man-made chemicals such as chlorofluorocarbons (also responsible for stratospheric ozone depletion) since the 1950s should have trapped about two extra watts of radiant energy over every square metre of earth. That much is accepted by most climatological specialists. Less well accepted, however, is how to translate those two watts of heat into "x" degrees of surface temperature change, since this involves assumptions about how that heat will be distributed among surface temperature rises, evaporation increases, cloudiness changes, ice changes, and so forth. The factor of two to three uncertainty in global temperature rise projections cited in the National Research Council's reports reflects a legitimate estimate of uncertainty held by most in the scientific community. Indeed, recent attempts by a British group to mimic the effects of cloud droplets halved their model's sensitivity to doubled CO_2, but the results remained well within the often-cited 1.5°C to 4.5°C range. However, the authors of the study wisely pointed out that "although the revised cloud scheme is more detailed, it is not necessarily more accurate than the less sophisticated scheme."[17] I have never seen this forthright and important reservation quoted by global warming critics who cite the British work as a reason to lower our concern by 50%. Nor, in the spring of 1990, was a NASA group's satellite estimate of global temperature change for the 1980s properly cited in the media as confirming, rather than questioning, instrumental records of global warming.[18] Finally, as explained in the original edition of *Global Warming*, predicting detailed regional distribution of climatic anomalies—that is, where and when it will be wetter and drier, whether floods will occur in the spring in California or forest fires in Siberia in August—is highly speculative, although plausible scenarios can be given.

While climatic models are far from fully verified for future simulations, the present seasonal and ancient climatic simulations, along with satellite observations of atmospheric heat trapping, are strong evidence that state-of-the-art climatic models already have considerable predictive skills. An awareness of what models are and what they can and can't do is probably the best we can ask of the public and its representatives. Then the tough policy problem is how to apply society's values about risk taking in choosing to face the future, given the possible outcomes that climatic models foretell.

The global warming debate takes in both science and politics. But it is essential for the public to understand that disagreements over what to do about the prospect of global warming (a political value issue) are far greater than over the approximate probability that unprecedented climate change is being built into a Greenhouse Century (a scientific debate). Nothing that has happened since the first publication of *Global Warming* has changed the strong consensus among scientists that climatic changes unprecedented in the 10,000-year era of human civilization are a good bet to happen. The more we debate and the longer we delay slowing down the greenhouse gas emissions, the greater the magnitude of climatic change that we and the rest of life on Earth will have to cope with. We are still marching relentlessly into the Greenhouse Century.

Notes

Chapter 1

1. K. Emanuel, "The Dependence of Hurricane Intensity on Climate," *Nature* 326 (1987): 483–85.
2. These quotes can be found in H. Schwarz and L. Dillard, "Urban Water," Chap. 15 in *Climate Change and U.S. Water Resources*, ed. P. E. Waggoner (New York: Wiley & Sons, 1990): 341–66.
3. Fires in Alberta in 1950 did indeed generate so much smoke that lights had to be turned on in the middle of afternoon baseball games in the U.S. Midwest. The fact that smoke from such fires was later observed high in the atmosphere over England by Royal Air Force pilots made this incident part of the controversy over "nuclear winter," the notion advanced by P. J. Crutzen and J. W. Birks, "The atmosphere after a Nuclear War: Twilight at Noon," *Ambio* 11 (1982): 115–25. This idea postulates that smoke generated from fires in the aftermath of a nuclear war could blanket much of the globe, blocking out sun and causing winterlike conditions in any season (as suggested by R. P. Turco, O. B. Toon, T. P. Ackerman, J. B. Pollack, and C. Sagan, "Nuclear Winter: Global Consequences of Multiple Nuclear Explosions," *Science* 222 [1983]: 1283–92). For a recent review of the nuclear winter controversy, see S. H. Schneider and S. L. Thompson, "Simulating the Climatic Effects of Nuclear War," *Nature* 333 (1988): 221–27. A description of the 1950 Alberta fires can be found in H. Wexler, "The Great Smoke Pall," *Weatherwise* 3 (September 24–30, 1950): 6–11.
4. The Colorado River Compact allotments are no fiction. Indeed, the framers of the compact actually allocated a fixed amount of water rather than a percentage of water, without knowing that the compact was drawn up after a historic high-flow period, which made those fixed water allocations highly questionable. An interesting history of the compact and this issue is given by B. G. Brown, "Climate Variability and the Colorado River Compact: Implications for Responding to Climate Change," in *Societal Responses to Regional Climatic Change: Forecasting by Analogy*, ed. M. H. Glantz (Boulder, Colo.: Westview Press, 1988), 279–305.
5. L. O. Mearns, R. W. Katz, and S. H. Schneider calculated that in July in Dallas, Texas, a 3° F rise in mean temperature could increase the probability of five or more consecutive days above 100° F (38° C) from today's value of about one year in three to a future value of two years in three ("Changes in the Probabilities of Extreme High Temperature Events with Changes in Global Mean Temperature," *Journal of Climate and Applied Meteorology* 23 [1984]: 1601–13).

Chapter 2

1. S. H. Schneider, "Climate Modeling," *Scientific American* 256 (1987): 5, 72–80.
2. For a fuller discussion of the physics of the greenhouse effect and a list of other articles or books on the subject, please see S. H. Schneider and R. Londer, *The Coevolution of Climate and Life* (San Francisco: Sierra Club Books, 1984), Part 2.
3. I first heard this term used by microbiologist Lynn Margulis (*The Climate Crisis,* "Nova" television programme transcript, Boston, WGBH, 1983).
4. C. Sagan and G. Mullen, ("Earth and Mars: Evolution of Atmospheres and Temperatures," *Science* 177 [1972]: 52–56) first proposed a supergreenhouse effect to explain how the earth was not a frozen ball even though the sun may have been some 30% less luminous 4 billion years ago—what some scientists call the "faint early sun paradox." More recently, this has been reviewed by J. F. Kasting, O. B. Toon, and J. B. Pollack, "How Climate Evolved on the Terrestrial Planets," *Scientific American* 258 (February 1988): 90–97.
5. M. I. Budyko and A. B. Ronov, "Chemical Evolution of the Atmosphere in the Phanerozoic," *Geochemistry International 1979,* 16 (1980): 1–9. The data also appear in M. I. Budyko, *Climatic Changes* (in Russian) (Leningrad: Hydrometeorological Publishers, 1974), 128.
6. J. M. Barnola, D. Raynaud, Y. S. Korotkevich, and C. Lorius, "Vostok Ice Core Provides 160,000-Year Record of Atmospheric CO_2," *Nature* 329 (1987): 408–14.
7. This discussion was derived from Schneider and Londer, *Coevolution,* 363–64.
8. J. F. Kasting and J. C .G. Walker, "Limits on Oxygen Concentration in the Prebiological Atmosphere and the Rate of Abiotic Fixation of Nitrogen," *Journal of Geophysical Research* 86 (1981): 1147–58; H. D. Holland, *The Chemical Evolution of the Atmosphere and Oceans* (Princeton, N.J.: Princeton University Press, 1984).
9. J. E. Lovelock, *The Ages of Gaia: A Biography of Our Living Earth* (New York: W. W. Norton & Company, 1988), 126–51.
10. M. B. McElroy and R. J. Salawitch, "Changing Composition of the Global Stratosphere," *Science* 243 (1989): 763–70.
11. V. Ramanathan, R. J. Cicerone, H. B. Singh, and J. T. Kiehl, "Trace Gas Trends and Their Potential Role in Climate Change," *Journal of Geophysical Research* 90 (1985): 5547–66.
12. The hearing transcript quoted in this chapter comes from the following: S. H. Schneider, Prepared Statement, "The Greenhouse Effect: Do We Need Major Federal Action Now?". In *Federal Agency Response to Global Climate Change: The National Energy Policy Act of 1988,* Hearings before the U.S. Senate Full Committee on Energy and Natural Resources, August 11, 1988, (Washington, D.C.: U.S. Government Printing Office, 1989), 39–72. Press coverage of the event was nationwide because of several wire service stories.
13. P. Shabecoff, "U.S. Data Since 1895 Failed to Show Warming Trend," *New York Times* (January 26, 1989): 1. This article summarizes a controversy generated when government scientists published a record showing virtually no temperature or precipitation trends in the United States. The article went on to point out that the United States is only a few per-cent of the world's area, and therefore one could conclude nothing positively or negatively about the greenhouse effect from the United States alone. Nevertheless, this record was widely cited by those sceptical about the greenhouse effect as evidence that it posed no serious threat.
14. P. E. Waggoner, *Climate Change.*
15. R. Kerr quotes a number of scientists who point out that despite the urban heating bias, global warming of at least 0.4° C seems likely over the past century ("Global Warming Is Real," *Science* 243 [1989]: 603).
16. Hansen's written testimony, "Modeling Greenhouse Climate Effects," before the U.S. Senate Committee on Commerce, Science, and Transportation; Subcommittee on Science, Technology, and Space, presented on May 8, 1989, had the following paragraph added by the Office of Management and Budget: "Again, I must stress that the rate and magnitude of drought, storm, and temperature change are very sensitive to the many physical processes mentioned above, some of which are poorly represented in the GCMs. Thus, these changes should be viewed as estimates from evolving computer models and not as reliable predictions." (See page 4.) Quite frankly, I don't

substantially disagree with the sentiments in the sentence, but Jim Hansen did. Since this was a scientific, not a political opinion, I therefore believe it was inappropriate for the government to have inserted the sentence since it distorted a scientist's scientific views. The Office of Management and Budget also added (see page 7) the following sentence: "One point that remains scientifically unknown is the relative contribution of natural processes and human activities to [CO_2 and methane increases since 1850]." The latter is a severe misrepresentation of current knowledge in my opinion, equivalent to the tobacco industry's hollow defence of smoking because the detailed physical links between smoke and cancer are still debated, despite overwhelming statistical evidence of the linkage. There is virtually no doubt that the 25% increase in CO_2 and the doubling of methane since the Industrial Revolution are largely caused by human activities. What is "scientifically unknown" is the precise amount of change due to fossil fuel combustion, cattle, rice paddy agriculture, deforestation, mining, or organic matter changes in soils.

17. Hansen's frustration was expressed in P. Shabecoff, "Scientist says U.S. Agency Altered His Testimony on Global Warming." *New York Times,* May 8, 1989: p. 1. This article instigated voluminous press coverage of the hearing, transforming it from a scientific discussion to a political forum.

Chapter 3

1. S. H. Schneider with L. E. Mesirow, *The Genesis Strategy: Climate and Global Survival* (New York: Plenum, 1976).
2. L. A. Frakes, *Climates Throughout Geologic Time* (Amsterdam: Elsevier, 1979), 261.
3. L. W. Alvarez, W. Alvarez, F. Asaro, and H. V. Michel, "Extraterrestrial Cause for the Cretaceous-Tertiary Extinction," *Science* 208 (1980): 1095–1108.
4. The first temperature calculation from a hypothesized earth-asteroid collision was made using a simple one-dimensional climate model: J. B. Pollack, O. B. Toon, T. P. Ackerman, C. P. McKay, and R. P. Turco, "Environmental Effects of an Impact-Generated Dust Cloud: Implications for the Cretaceous-Tertiary Extinctions," *Science* 219 (1983): 287–89. Later on, a three-dimensional model was applied by Starley Thompson and Curt Covey that showed that the deep subfreezing temperatures predicted in the one-dimensional model of Pollack et al. were likely to be moderated because of the heat retained in the oceans, even if a massive dust cloud from an asteroid impact obscured the sun for many months (S. L. Thompson and C. Covey, "Global Climatic Effects of Atmospheric Dust from Large Bolide Impacts," in *Global Catastrophes in Earth History: An Interdisciplinary Conference on Impacts, Volcanism, and Mass Mortality: Geological Society of America Special Paper,* eds. V. L. Sharpton and P. Ward [submitted]). However, as the dust particles banged into each other and grew in size, they may very well have created a supergreenhouse effect causing 10° or 20° C warming shortly after a 20° C cooling. As if such a cooling-heating shock weren't enough to kill the dinosaurs, Thompson and Paul Crutzen added a chemical calculation to include the effects of oxides of nitrogen produced by the very high temperatures generated by shock waves when an asteroid exploded on impact at very high speeds. These oxides of nitrogen could destroy most of the ozone on the planet for several years. Therefore, survivors on the surface would also have to deal with very high doses of ultraviolet radiation (S. L. Thompson and P. J. Crutzen, "Acute Climatic and Chemical Effects of Large Bolide Impacts," in *Global Catastrophes,* Sharpton and Ward, submitted).
5. P. Hut, W. Alvarez, W. P. Elder, T. Hansen, E. G. Kauffman, G. Keller, E. M. Shoemaker, and P. R. Weissman, "Comet Showers as a Cause of Mass Extinctions," *Nature* 329 (1987): 118–26.
6. S. H. Schneider, "Climate Modeling."
7. J. Imbrie and K. P. Imbrie, *Ice Ages: Solving the Mystery* (Short Hills, N.J.: Enslow, 1979).
8. Barnola et al., "Vostok Ice Core."
9. For example, see Schneider and Londer, *Coevolution,* 33–91, for a discussion of the chemical composition of snow, its relationship to temperature, and the difficulties associated with these reconstructions.
10. Briefly, the temperature at which snowfall was deposited can be inferred by study-

ing the relative amount of the heavy isotope of oxygen (O^{18}) in proportion to the primary isotope of oxygen (O^{16}). When times are cold, the heavier oxygen, O^{18}, remains in the ocean or is rained out before the water vapour reaches the very cold poles. Therefore, snow deposited during colder glacial times contains less O^{18} than snow deposited during warmer times, which is what samples taken from deep polar ice cores show us. However, there are a number of complications in the interpretation of O^{18} as a palaeothermometer; we have to consider the temperature at which the water evaporated, the temperature at which it precipitated, and its transport path, as explained in chap. 2 of Schneider and Londer, *Coevolution,* 33–91.

11. There is some dispute as to the actual dates of the so-called Eemian interglacial. Ice core dates are based upon the indirect method of constructing mathematical models of ice flow in order to estimate where and when on the surface of Antarctica the snow fell that was found in an ice core 100,000 years later. Such models cannot be exact and therefore the dating becomes increasingly less accurate the further back we go. Other methods, such as looking in ocean planktonic sediments and relating them to mathematical reconstructions of orbital elements, are believed by some to be more precise: J. Imbrie, A. McIntyre, and A. C. Mix, "Oceanic Response to Orbital Forcing in the Late Quaternary: Observational and Experimental Strategies," in *Climate and Geo-Sciences. A Challenge for Science and Society in the 21st Century,* eds. A. L. Berger, S. H. Schneider, and J. C. Duplessy (Dordrecht, the Netherlands: Kluwer Academic Publishers, 1989):121–164. Nevertheless, some geologists have suggested that the last interglacial may have begun 10,000 years earlier than the ice-core researchers suggest and 20,000 years earlier than the ocean-core researchers suggest based upon the examination of fossil water in caverns in the U.S. Great Basin (I. J. Winograd, B. J. Szabo, T. B. Coplen, and A. C. Riggs, "A 250,000-Year Climatic Record from Great Basin Vein Calcite: Implications for Milankovitch Theory," *Science* 242 [1988]: 1275–80). In any case, even if the absolute dating in the ice cores were incorrect, the close correlation between the temperature reconstruction and CO_2 does suggest a coupling of these variables. Another element of debate is how rapid the onset of a glacial age is. A number of years ago the idea of "instant glacierization"—possibly brought about by a rapid disintegration of part of the Antarctic ice sheet which then spread icebergs into the ocean—generated controversy. Both the data and the dating are disputed, but the issue of rapid climate change remains interesting (for a summary see Schneider and Londer, *Coevolution,* 84–91).
12. COHMAP Members (P. M. Anderson et al.), "Climatic Changes of the Last 18,000 Years: Observations and Model Simulations," *Science* 241 (1988): 1043–52.
13. R. Peters, ed., *Proceedings of Conference on the Consequences of the Greenhouse Effect for Biological Diversity* (New Haven, Conn.: Yale University Press, in press).
14. This average rate of temperature change is calculated simply by assuming that in the northeastern United States the temperature changed by 10°–20° C over a 10,000-year period. Alternatively, it could be argued that there were rapid periods of deglaciation that took place within the 10,000-year transition time from the glacial to the interglacial. For example, on a global average basis we could assume that the 5° C global temperature warming from the end of the ice age to the present interglacial took place in only 5,000 years, which would still yield a rate of 1° C per 1,000 years. If one insisted that the fastest rates of change on a global scale took place over a time of only 2,000 years or so, then one still comes up with natural, global average rates of change on the order of a few degrees Celsius per 1,000 years.
15. W. R. Emanuel, H. H. Shugart, and M. P. Stevenson, "Climatic Change and the Broad-Scale Distribution of Terrestrial Ecosystem Complexes," *Climatic Change* 7 (1985): 30–43; J. Pastor and W. M. Post, "Response of Northern Forests to CO_2-Induced Climate Change," *Nature* 334 (1988): 55–58; D. B. Botkin and R. A. Nisbet, "Projecting the Effects of Climate Change on Biological Diversity in Forests," in *Consequences of the Greenhouse Effect,* Peters; A. M. Solomon, "Transient Response of Forests to CO_2 Induced Climate Change: Simulation Modeling Experiments in Eastern North America," *Oecologia* 68 (1986): 567–579.
16. Imbrie and Imbrie, *Ice Ages* (see especially chaps. 5 and 6).
17. A. Berger, "Accuracy and Frequency Stability of the Earth's Orbital Elements During the Quaternary, Part 1 in *Milankovitch and Climate,* eds. A. Berger, J. Imbrie, J. Hays,

G. Kukla, and B. Saltzman (Dordrecht, the Netherlands: Reidel Publishing, 1984) 3–40.

18. J. R. Bray, "Volcanic Triggering of Glaciation," *Nature* 260 (1976): 414–15; R. Bryson and T. Murray, *Climates of Hunger: Mankind and the World's Changing Weather* (Madison, Wis.: University of Wisconsin Press, 1977).
19. The possibility that processes internal to the climatic system could be responsible for large changes such as glacial/interglacial cycles was suggested by MIT theoretical meteorologist Edward Lorenz, based on his classic paper that delineated his discovery of the field of mathematical chaos. What Lorenz later suggested was that climate change may not be determined by external forces, but could be an irregular or chaotic occurrence of the redistribution of energy among the major reservoirs: the atmosphere, oceans, and ice sheets. In 1968 he applied this idea to the climate in a short paper ("Climatic Determinism," *Meteorological Monographs* 5 [1968]: 1).
20. G. J. MacDonald, "Variations in Atmospheric Carbon Dioxide and Ice Age Climate," in *Preparing for Climate Change, Proceedings of the First North American Conference on Preparing for Climatic Change: A Cooperative Approach,* 27 to 29 October, Washington, D.C. (Rockville, Md.: Government Institutes, 1988) 108–17; L. Klinger, "Successional Change in Vegetation and Soils of Southeast Alaska," thesis, Department of Geography and Institute of Arctic and Alpine Research, University of Colorado, April 1988; see also G. J. MacDonald, "Role of Methane Clathrates in Past and Future Climates," *Climatic Change* (in press).
21. C. Emiliani, "Ice Sheets and Ice Melts," *Natural History* 89 (1980): 82–91.
22. V. Ya. Sergin, "Numerical Modeling of the Glaciers-Ocean-Atmosphere Global System," *Journal of Geophysical Research* 84 (1979): 3191–3204; W. F. Ruddiman and A. McIntyre, "Oceanic Mechanisms for Amplification of the 23,000-Year Ice-Volume Cycle," *Science* 212 (1981): 617–27.
23. D. Pollard, "Ice-Age Simulations with a Calving Ice-Sheet Model," *Quaternary Research* 20 (1983): 30–48.
24. M. Ghil, "Internal Climatic Mechanisms Participating in Glaciation Cycles," in *Climatic Variations and Variability: Fact and Theory,* A. Berger, ed., (Dordrecht: Reidel, 1981) 539–557. Ghil's article summarizes a long series of studies conducted by Ghil and colleagues (for instance, those by E. Kallen, C. Crafoord, H. LeTreut, and K. Bhattacharya; L. D. D. Harvey and S. H. Schneider, "Sensitivity of internally-generated climate oscillations to ocean model formulation," in *Milankovitch and Climate,* Part 2, A. Berger, 1984; see also H. LeTreut and M. Ghil, "Orbital forcing, climate interactions, and glaciation cycles," *J. Geophys. Res. 88* (1983):5167–5190; Saltzman, B. and Sutera, A., 1987, The mid-Quaternary climatic transition as the free response of a three-variable dynamical model, *J. Atmospheric Science 44*:236–241; Birchfield, G. E. and Grumbine, R. W., 1985, "Slow" physics of large continental ice sheets, the underlying bedrock and the Pleistocene ice ages, *J. Geophys. Res. 90*:11294–11302; Oerlemans, J., 1982, Glacial cycles and ice sheet modelling, *Clim. Change 4*:353–374; Ghil, M., 1988, Deceptively-simple models of climate change, to appear in Berger, A. L., et al., *Climate and Geo-Sciences*; Hyde, W. T. and Peltier, W. R., 1987, Sensitivity experiments with a model of the ice age cycle: the response to Milankovitch forcing, *J. Atmos. Sci. 44*:1351–1374.
25. J. D. Hays, J. Imbrie, and N. J. Shackleton, "Variations in the Earth's Orbit: Pacemaker of the Ice Ages," *Science* 194 (1976): 1121–32.
26. J. E. Kutzbach and B. L. Otto-Bliesner, "The Sensitivity of the African-Asian Monsoonal Climate to Orbital Parameter Changes for 9,000 Years B.P. in a Low-Resolution General Circulation Model," *Journal of the Atmospheric Sciences* 39 (1982): 1177–88.
27. Benson, L. V., Currey, D. R., Dorn, R. I., Lajoie, K. R., Oviatt, C. G., Robinson, S. W., Smith, G. I., Stine, S., "Chronology of Expansion and Contraction of Four Great Basin Lake Systems During the Past 35,000 Years," in Benson, L. V., and Meyers, P. A. (eds.), special issue of *Palaeogeography, Palaeoclimatology, and Palaeoecology,* in press.
28. G. R. Coope, "Fossil Coleopteran Assemblages as Sensitive Indicators of Climatic Changes During the Devensian (Last) Cold Stage," *Proceedings of the Philosophical Transactions of the Royal Society of London B* 280 (1977): 313–40.
29. An entire conference on abrupt climate change was convened primarily under the leadership of Wolfgang Berger of Scripps Institute of Oceanography. The Proceedings contains many interesting papers on this issue (W. H. Berger and L. Labeyrie, eds., *Abrupt Climatic Change,* NATO/NSF Workshop, Biviers [Grenoble], [Dordrecht, the

Netherlands: Reidel Publishing, 1987]): In particular, for modelling studies of the Younger Dryas, see S. H. Schneider, D. M. Peteet, and G. R. North, "A Climate Model Intercomparison for the Younger Dryas and its Implications for Paleoclimatic Data Collection," 399–417. Also see D. Rind, D. Peteet, W. S. Broecker, A. McIntyre, and W. F. Ruddiman, "Impact of Cold North Atlantic Sea Surface Temperatures on Climate: Implications for the Younger Dryas Cooling (11–10k)," *Climate Dynamics* 1 (1986): 3–33; see also W. F. Ruddiman and A. McIntyre, "The North Atlantic Ocean During the Last Deglaciation," *Palaeogeography, Palaeoclimatology, and Palaeoecology* 35 (1981): 145–214.

30. W. F. Ruddiman and A. McIntyre, "Warmth of the Subpolar North Atlantic Ocean During Northern Hemisphere Ice-Sheet Growth," *Science* 204 (1979): 173–75.
31. E. A. Boyle and L. D. Keigwin, "North Atlantic Circulation During the Last 20,000 Years Linked to High-Latitude Surface Temperatures," *Nature* 330 (1987): 35–40.
32. W. S. Broecker, M. Andree, W. Wolfli, H. Oeschger, G. Bonani, J. Kennett, and D. Peteet, "The Chronology of the Last Deglaciation: Implications to the Cause of the Younger Dryas Event," *Paleoceanography* 3 (1988): 1–19; W. S. Broecker, "Unpleasant Surprises in the Greenhouse," *Nature* 328 (1987): 123–26.
33. J. D. Hamaker and D. A. Weaver, *The Survival of Civilization* (Burlingame, Calif.: Hamaker-Weaver, 1982). More recently Larry Ephron, of the Institute for the Future, summarized the views of this group in his book, *The End: The Imminent Ice Age and How We Can Stop It* (Berkeley, Calif.: Celestial Arts, 1988).
34. Sir G. Simpson, "Ice Ages," *Nature* 141 (1938): 591–98.
35. Therefore, I certainly would not deny that the Hamaker-Weaver hypothesis has some mathematical chance of being valid. However, my scientific judgment (intuition, really) is that the probability of this scenario is small. Nevertheless, I do agree with Hamaker and Weaver that climatologists should not parochially ignore the potentially important interactions between the atmosphere, oceans, ice, biosphere, and soils. I also applaud their contribution of the soil demineralization issue to the already long list of possible internal influences on the glacial/interglacial cycles. A number of others have dissented from the conventional wisdom about the magnitude of warming caused by the greenhouse effect. The most prominent is Sherwood Idso, an agricultural scientist who works for the U.S. government in Phoenix, Arizona. I discussed a number of issues dealing with Idso's antiestablishment criticisms in Schneider and Londer, *Coevolution*, 328–30. For more recent references to this continuing one-man dissension, please see S. H. Schneider, "Can Modeling of the Ancient Past Verify Prediction of Future Climates," *Climatic Change* 8 (1986): 117–20; S. B. Idso, "A Clarification of My Position on the CO_2-Climate Connection," *Climatic Change* 10 (1987): 81–86; G. L. Potter, J. T. Kiehl, and R. D. Cess, "A Clarification of Certain Issues Related to the CO_2-Climate Problem," *Climatic Change* 10 (1987): 87–96; S. B. Idso, "Me and the Modelers: Perhaps Not So Different At All," *Climatic Change* 12 (1988): 93–94. Finally, for a discussion of a very recent challenge to the idea that coal burning produces a net global warming, see endnote 17 of chapter 4.
36. C. B. Schultz and J. C. Frye, eds., *Loess and Related Eolian Deposits of the World* (Lincoln, Neb.: University of Nebraska Press, 1968); in particular see chap. 8 by C. B. Schultz and chap. 9 by A. L. Lugn.
37. W. W. Kellogg and R. Schware, *Climate Change and Society: Consequences of Increasing Atmospheric Carbon Dioxide* (Boulder, Colo.: Westview Press, 1981). Kellogg's map plots the areas of agreement between himself and K. W. Butzer (see Butzer's article, "Adaptation to Global Environmental Change," *Professional Geographer* 32, no. 3 [1981]: 269–78).
38. J. E. Kutzbach and P. J. Guetter, "The Influence of Changing Orbital Parameters and Surface Boundary Conditions on Climate Simulations for the Past 18,000 Years," *Journal of the Atmospheric Sciences* 43 (1986): 1726–59.
39. J. E. Kutzbach and F. A. Street-Perrott, "Milankovitch Forcing of Fluctuations in the Level of Tropical Lakes from 18 to 0 kyr BP," *Nature* 317 (1985): 130–34.
40. W. L. Prell and J. E. Kutzbach, "Monsoon Variability over the Past 150,000 Years," *Journal of Geophysical Research* 92 (1987): 8411–25.
41. COHMAP Members, "Climatic Changes."
42. H. H. Lamb, *Climate: Present, Past and Future*, vol. 2 (London: Methuen); J. M. Grove, *The Little Ice Age* (Hampshire, U.K.: Associated Book Publishers, 1988).
43. T. M. L. Wigley, M. J. Ingram, and G. Farmer, eds., *Climate and History: Studies in Past*

Climates and Their Impacts on Man (Cambridge, U.K.: Cambridge University Press) 1981; T. K. Rabb, "Climate Society in History: A Research Agenda," in *Social Science Research and Climate Change: An Interdisciplinary Appraisal,* eds. R. S. Chen, E. M. Boulding, and S. H. Schneider (Dordrecht, the Netherlands: Reidel Publishing, 1983).

44. H. C. Fritts, *Tree Rings and Climate* (New York: Academic Press, 1970), 405–407.
45. For an expanded treatment of the embedding of humans in the environment please see S. H. Schneider and L. Morton, *The Primordial Bond: Exploring Connections Between Man and Nature Through the Humanities and Sciences* (New York: Plenum, 1981).

Chapter 4

1. Lovelock, *Ages of Gaia* (New York: W. W. Norton & Company).
2. L. Margulis, *Symbiosis in Cell Evolution* (San Francisco: Freeman Press, 1981).
3. J. Lovelock, "The Independent Practice of Science," *Coevolution Quarterly* (Spring 1980): 26.
4. L. Margulis and J. E. Lovelock, "Is Mars a Spaceship, Too?" *Natural History* (June/July 1976): 86–90.
5. Lovelock, "Independent Practice," 28.
6. Classical accounts of the evolution of geological or geochemical conditions on earth can be found in the following textbooks: F. Press and R. Siever, *Earth* (San Francisco: Freeman Press, 1982) and Holland, *Chemical Evolution.*
7. J. E. Lovelock and L. Margulis, "Atmospheric Homeostasis by and for the Biosphere: The Gaia Hypothesis," *Tellus* 26 (1973): 2.
8. Lovelock, *Ages of Gaia,* 134.
9. Sagan and Mullen, "Earth and Mars."
10. T. Owen, R. D. Cess, and V. Ramanathan, "Enhanced CO_2 Greenhouse to Compensate for Reduced Solar Luminosity on Early Earth," *Nature* 277 (1979): 640–41. See also M. H. Hart, "The Evolution of the Atmosphere of the Earth," *Icarus* 33 (1978): 23–29. These issues are discussed at greater length in Schneider and Londer, *Coevolution,* 222–93.
11. J. C. G. Walker, P. B. Hays, and J. F. Kasting, "A Negative Feedback Mechanism for the Long-Term Stabilization of Earth's Surface Temperature," *Journal of Geophysical Research* 86 (1981): 9776–82.
12. Kasting, Toon, and Pollack, "How Climate Evolved." However, David Schwartzman at Howard University and Tyler Volk at New York University ("Biotic Enhancement of Weathering and the Habitability of Earth," *Nature,* Vol. 340, 1989: 457–459) argued that biological activity in the soil substantially speeds up the seemingly inorganic rate of weathering. This implies that the world would have had much more CO_2 and could have been much warmer today if it weren't for the presence of soil biota. However, Heinrich Holland (in S. H. Schneider and P. Boston, eds., *The Science of Gaia,* M.I.T. Press, Cambridge, MA; in press) argues that these effects were overestimated by Schwartzman and Volk. Thus the debate continues.
13. E. J. Barron, S. L. Thompson, and S. H. Schneider, "An Ice-Free Cretaceous? Results from Climate Model Simulations," *Science* 212 (1981): 501–508.
14. *Goddess of the Earth,* "Nova" television programme transcript, Boston, WGBH, 1986. P. R. Ehrlich, "Coevolution and Its Applicability to the Gaia Hypothesis," in *Proceedings of the Chapman Conference on the Gaia Hypothesis,* S. H. Schneider, and P. Boston, eds., accepted.
15. J. W. Kirchner, a joint philosophy-physics major at the University of California at Berkeley, delivered a dramatic and controversial dissection of the Gaia hypothesis in an evening session on epistemology in San Diego. Kirchner's article (J. W. Kirchner and J. Harte, "The Gaia Hypotheses: Are They Testable? Are They Useful?") will appear in *The Science of Gaia,* eds. S. H. Schneider, and P. Boston, (M.I.T. Press, in press); and also J. W. Kirchner, "The Gaia Hypothesis: Can It Be Tested?" *Reviews of Geophysics* Vol. 27 (1989): 223–235. Lovelock reacted in print to Kirchner a year and a half after the conference: J. E. Lovelock, "Hands Up for the Gaia Hypothesis," *Nature,* Vol. 344, 1990: 100–102.

16. S. Twomey in *Inadvertent Climate Modification,* Report of the Study of Man's Impact on Climate (SMIC) (Cambridge, Mass.: MIT Press, 1971), 229.
17. R. J. Charlson, J. E. Lovelock, M. O. Andreae, and S. G. Warren, "Oceanic Phytoplankton, Atmospheric Sulphur, Cloud Albedo and Climate," *Nature* 326 (1987): 655–61. For an interesting debate, see the criticism of Charlson et al. by S. E. Schwartz ("Are Global Cloud Albedo and Climate Controlled by Marine Phytoplankton?" *Nature* 336 [1988]: 441–45) as well as the reply by Charlson et al. submitted to *Nature* ("Pollution Sulphates and Cloud Albedo," *Nature,* Vol. 340, 1989: 436–438). One possible implication of the changes in cloud albedo from particles has been argued by Robert Cess of the State University of Stony Brook, New York. He has suggested that clouds over water are brighter nearer to the U.S. East coast than in the central Atlantic Ocean, possibly from pollution generated in the U.S. and Canada. Patrick Michaels of the University of Virginia has speculated that this cooling could explain why the Northern hemisphere has warmed up less than the Southern hemisphere this century. It will take another decade or so of such data to move this idea beyond speculation.

    As this book was going to press in late 1989, I received a prepublication copy of a draft ms. by Yoram Kaufman of the NASA Goddard Space Flight Center in Greenbelt, Maryland, and two colleagues, Robert S. Fraser and Robert L. Mahoney. "Fossil Fuel and Biomass Burning Effect on Climate/Heating or Cooling," Y. J. Kaufman, R. S. Fraser, R. L. Mahoney, *Journal of Climate,* submitted. It was an impressive attempt to quantify an issue that has been speculated on by Sean Twomey for over two decades: that extra particles from air pollution in the atmosphere could overseed clouds, thereby increasing their albedo and in turn cooling the earth. The NASA authors combined assumptions about coal-burning—which emits both CO_2 and SO_2—into a complex chain of events that leads to both increased cloud albedo and greenhouse warming. Based on many links and admittedly speculative assumptions they cautiously concluded that cooling could dominate heating from coal burning. However, this conclusion was based on their best guesses for CO_2 and SO_2 removal and injection rates, the effects of SO_2 on cloud albedo, how much of the hemisphere could be reached by such SO_2 pollution (CO_2 is globally distributed whereas SO_2 is concentrated near industrial regions), and how many particles were in the atmosphere in the first place before coal was burned (which is an important assumption since the more particles that are initially present, the less effect newly additive ones have in making extra cloud drops).

    Although their analysis was careful to show how critically dependent the conclusion quoted above was to even small changes in these hard-to-validate parameters, they still offered a very controversial conclusion: "although we should acknowledge that the large emission of particulates and trace gases into the atmosphere is a dangerous one, and may result in climate changes, it may be too early to predict the direction and the possible twists of the effect." Kaufman and colleagues went on to argue that biomass burning, because it produces virtually no SO_2, contributes only to climate heating. I applaud their caution, welcome the quantification of a long-debated idea, but disagree strongly with the notion that coal burning is *as likely* to produce cooling as warming.

    First of all, they assumed a fairly short life time for CO_2 added to the atmosphere (less than 100 years) whereas I believe it is possible to be as much as several hundred years (which would significantly reduce the cooling effect. Furthermore, they assumed that one-third of the planet could be significantly affected by cloud albedo increases from coal burning (which I suspect is too high—possibly by a factor of 2). Whether the cloud albedo is sensitive to particle pollution depends on the relationship between cloud drop concentrations and particle concentrations. The authors rightly point out that this relationship is marked by a great degree of uncertainty. They neglected entirely the effect of soot particles which are produced along with SO_2 in many industrial applications in which coal is burned, especially if no pollution control devices are used. Such particles can substantially darken clouds adding to global warming, but like the cooling counterpart quantifying the change for any effect of pollution on clouds in the real world is highly speculative. As acid rain from coal burning forces increased controls on SO_2 emissions, any SO_2-induced cooling to date will cease, and further unrealized warming could soon follow. My objection is not to the overall work, but to the comparable probability they gave the warming/cooling outcome given the very solid physical basis for global scale heat-trapping from increased CO_2 versus a highly speculative difficult-to-verify chain of assumptions connected to a limited regional effect. This is why I switched from thinking in the early 1970s that particulate pollution was likely to cool the planet more than greenhouse gases would warm it to my present view that greenhouse gas increases are likely to dominate in the long run all countervailing regional effects on a global basis. In 1989, Tom Wigley (T.M.L. Wigley, "Possible Climate Change Due to SO_2-Derived Cloud Condensation Nuclei," *Nature,* Vol. 339, 1989:365–367.) addressed this same question by asking if the Southern Hemisphere had warmed up more than the Northern Hemisphere. This is important because if greenhouse gases were the only climate mechanism at work, then the Northern Hemisphere should have warmed up more since it has less ocean and thus less heat capacity to retard the rate of warming. However, if SO_2 pollution or some other largely Northern Hemispheric pollutant were responsible for cooling in this hemisphere, then the Southern Hemisphere may have warmed up more. Wigley's attempt to analyse differences between Southern and Northern Hemisphere temperature trend data was somewhat inconclusive because the data are too noisy and the signals too small for certainty. But it did appear likely that at least the Northern Hemisphere was not warming up more rapidly than the Southern Hemisphere. Moreover, a slight cooling trend in the Northern Hemisphere took place between the 1950s and the 1970s. Since SO_2 emissions, largely a Northern Hemisphere phenomenon, went up sharply from 1950 to 1975 (when Northern Hemisphere surface temperatures declined) but declined in light of environmental control requirements thereafter (with the 1980s being the warmest decade on record), one can make a circumstantial argument that the Northern Hemisphere was indeed held back from model predicted amounts of global warming during the 1950–1980 period because of SO_2-induced changes in cloud brightness. This is still speculation, but it is plausible speculation. In a decade or so, we shall all know for sure.

18. Steve Warren, in reading through a draft manuscript of this book, commented that during the ice age the sea ice extent was greater and temperatures were colder. He also noted that when sea ice grows, pockets of salty water form and salt is rejected back to the ocean. During the ice age, this condition could have favoured a variety of DMS-producing plankton whose habitat is the sea ice region. If this species increased, Warren pointed out, it would not necessarily be a global sulphate signal that was found in the ice cores, but rather a local connection between a particular DMS-producing species and the expansion of sea ice that would accompany the ice age. The issue remains to be resolved.
19. MacDonald, "Role of Methane Clathrates." See also R. Revelle, "Methane Hydrates in Continental Slope Sediments and Increasing Atmospheric Carbon Dioxide," in *Changing Climate, Report of the Carbon Dioxide Assessment Committee,* National Research Council (Washington, D.C.: National Academy Press, 1983), 252–61.
20. Klinger, "Successful Change."
21. As the principal convener of this American Geophysical Union's (AGU) Chapman Conference, I can report that the AGU council did not approve the idea without considerable debate—and I received quite a few complaints from scientists for scheduling the conference even after the AGU finally approved it. However, the AGU deserves high praise for ignoring the gripes and helping to organize what turned out to be a spectacularly successful scientific get together.
22. See S. H. Schneider for some of the history of the debate ("A Goddess of the Earth: The Debate over the Gaia Hypothesis," *1988 Yearbook of Science and the Future* [Chicago: Encyclopaedia Britannica, 1987], 30–43). See also my cover story Gaia piece in the May 1990 issue of *Environment* magazine.
23. D. A. Lashof, "The Dynamic Greenhouse: Feedback Processes That May Influence Future Concentrations of Atmospheric Trace Gases and Climatic Change," *Climatic Change 14* (1989): 213–242.
24. P. R. Ehrlich and P. H. Raven, "Butterflies and Plants: A Study in Co-evolution," *Evolution* 18 (1965): 586–608.

CHAPTER 5

1. R. A. Berner, A. C. Lasaga, and R. M. Garrels, "The Carbonate-Silicate Geochemical Cycle and Its Effect on Atmospheric Carbon Dioxide over the Past 100 Million Years," *American Journal of Science* 283 (1988): 641–83; M. I. Budyko, A. B. Ronov, and A. L. Yanshin, *History of the Earth's Atmosphere* New York: Springer-Verlag, 1987).
2. SMIC, *Inadvertent Climate Modification.*
3. One day in 1974, an American (C. E. Leith) who had just visited Leningrad handed me a three-page review by Budyko of an early draft of an article I was writing with my colleague Robert Dickinson. In that article, we attempted to summarize and guide the newly emerging field of climate modelling (S. H. Schneider and R. E. Dickinson, "Climate Modeling," *Reviews of Geophysics and Space Physics* 12 [1974]; 447–93). I was glad to eventually get the comments from Budyko, no matter how unconventionally they were conveyed.
4. Budyko, Ronov, and Yanshin, *History*, 92.
5. J. Hansen and S. Lebedeff, "Global Trends of Measured Surface Air Temperature," *Journal of Geophysical Research* 92 (1987): 13, 345–13, 372.
6. J. Hansen and S. Lebedeff, "Global Surface Air Temperatures: Update Through 1987," *Geophysical Research Letters* 15 (1988): 323–26; P. D. Jones and T. M. L. Wigley, personal communication, 1988. For a discussion of the urban heat island corrections, see T. R. Karl and P. D. Jones, "Urban Bias in Area-averaged Surface Air Temperature Trends," *Bulletin of the American Meteorological Society* 70 (1989): 265–70.
7. Kerr, "Global Warming Is Real."
8. We have already discussed how ancient palaeoclimatic information may be used in this regard, more as a guide than as a quantifiable predictor. In particular, such ancient climates can be helpful to future forecasts, but only to the extent that they help validate the tools (that is, models) that must be used to predict the response of the future climate to realistic scenarios based on increasing greenhouse gases. In Chapter 3, we saw Will Kellogg's attempt to define a warm-world analogy by looking at the climatic optimum as a metaphor for a warm earth. Although this proved valuable to understanding how climate can change with large-scale changes in radiative heating of the earth, and thus provided an important validation technique for climate models, this empirical forecasting method probably has little to say about the distribution of regional details of climate change over the next century. Other scientists have attempted to infer what a warm-year or warm-decade earth might look like by identifying the warmest years or warmest decade in this century and then seeking consistent patterns of altered aridity or wetness that might correlate with those warm years. European climatologist Jill Jager (Williams) looked at regional temperature and precipitation anomalies for different seasons in the Northern Hemisphere when the Arctic region north of 65° latitude was warm compared to the long-term mean from 1900 to 1969. (J. Williams, "Anomalies in Temperature and Rainfall During Warm Arctic Seasons as a Guide to the Formulation of Climate Scenarios," *Climatic Change* 2 [1980]: 249–66). She found large and fairly coherent anomalies in temperature in other parts of the hemisphere when the Arctic was warm; she also detected anomalies in precipitation distribution. Tom Wigley and colleagues and Barrie Pittock and Jim Salinger (who studied the Southern Hemisphere) also used similar techniques (J. M. Lough, T. M. L. Wigley, and J. Palutikof, "Climate and Climate Impact Scenarios for Europe in a Warmer World," *Journal of Climate and Applied Meteorology* 22 [1983]: 1673–84; A. B. Pittock and M. J. Salinger, "Towards Regional Scenarios for a CO_2-Warmed Earth," *Climatic Change* 4 [1982]: 23–40). For example, Jager and Kellogg compared the distributions of climate anomalies based on five consecutive years whose average was warmer than the long-term mean (J. Jager and W. W. Kellogg, "Anomalies in Temperature and Rainfall During Warm Arctic Seasons," *Climatic Change* 5 [1983]: 39–60). They compared distributions to the soil-moisture pattern generated from one of Syukuro Manabe's computer models, finding some gross similarities in dry and wet anomaly regions. This study provided at least some circumstantial evidence that consecutive-year analysis may prove to be a potentially useful empirical forecast tool. Unfortunately, all these techniques suffer from the same obvious limitation: they are based on patterns of climatic anomalies obtained from real data under past short-

term warming fluctuations. These warmings could well have been caused by internal redistributions of energy among atmosphere, oceans, land, and ice. Thus, there is no reason to believe that such patterns should be identical to those that would occur if the climate changes were forced by increases in greenhouse gases.

9. See Mearns and colleagues for an expanded discussion of alternative prediction methods (L. O. Mearns, P. Gleick, and S. H. Schneider, "Climate Forecasting," in *Climate Change,* Waggoner).
10. R. E. Dickinson, "How Will Climate Change? The Climate System and Modelling of Future Climate," in *The Greenhouse Effect, Climatic Change, and Ecosystems,* eds. R. Bolin, B. R. Doos, J. Jager, and R. A. Warrick (New York: Wiley & Sons, 1986), 207–70. Recent results reported by John Mitchell and colleagues of the U.K. Meteorological office (Mitchell, J.F.B., C. A. Senior and W. J. Ingram, "CO_2 and Climate: A Missing Feedback?" *Nature, 341,* 1989:132–134) using a new cloud prediction scheme, give a CO_2 doubling value of about 1.9°–2.7°C, about half of their previous results of a little over 5°C. Of course, as the authors themselves pointed out, new is not necessarily better, so the climate sensitivity debate still remains unshakeably in the 1°–5°C range for an equilibrium doubling of CO_2.
11. Although a weather forecaster need not consider, for example, the small day-to-day changes in ice, ocean temperature, forest cover, or circulation of the sea, such changes affecting the lower atmosphere must be considered by the predictor of atmospheric changes from one season or one decade to another. On the other hand, changes in the earth's orbit occur over thousands of years and are negligible when considering climate changes during less than a few millennia. Similarly, when computing an ice age cycle, one need not take into account continental drift, which becomes significant only on even longer time scales of tens of millions of years.
12. J. M. Mitchell, Jr., C. W. Stockton, and D. M. Meko, "Evidence of a 22-Year Rhythm of Drought in the Western U.S. Related to the Hale Solar Cycle Since the 17th Century," in *Solar-Terrestrial Influences on Weather and Climate,* eds. B. M. McCormack and T. A. Seliga (Dordrecht, the Netherlands: Reidel Publishing, 1979), 125–43. H. van Loon and K. Labitzke, "Association Between the 11-year solar cycle, the QBO, and the Atmosphere. Part II: Surface and 700 mb in the Northern Hemisphere in Winter," *Journal of Climate* 1 (1988): 905–920.
13. There is a rich history of speculation on the important role of volcanic dust in the stratosphere on cooling the earth's climate. For example, Benjamin Franklin referred to a "dry fog" from an Icelandic volcano that he observed from his diplomatic post in Paris in the eighteenth century (see the discussion in Chapter 7 of Schneider and Londer, *Coevolution*). For a more recent survey of possible volcanic dust influences on climate, see R. S. Bradley, "The Explosive Volcanic Eruption Signal in Northern Hemisphere Continental Temperature Records," *Climatic Change* 12 (1988): 221–44.
14. J. Shukla and Y. Mintz, "Influence of Land-Surface Evapotranspiration on the Earth's Climate," *Science* 215 (1982): 1498–1501; A. Henderson-Sellers and V. Gornitz, "Possible Climatic Impacts of Land Cover Transformations, with Particular Emphasis on Tropical Deforestation," *Climatic Change* 6 (1984): 231–57; A. Henderson-Sellers, R. E. Dickinson, and M. F. Wilson, "Tropical Deforestation: Important Processes for Climate Models," *Climatic Change* 13 (1988): 43–68.
15. E. Lorenz, "Climate Predictability," in *The Physical Basis of Climate and Climate Modeling,* Report of the GARP Study Conference, Stockholm, 29 July–10 August, GARP Publication Series No. 16 (1975): 132; R. C. J. Somerville, "The Predictability of Weather and Climate," *Climatic Change* 11 (1987): 239–46.
16. It has amazed professional meteorologists that some customers are willing to pay to convincing entrepreneurs what these meteorologists consider outrageous sums of money for daily forecasts well beyond the theoretical limit that most weather scientists believe prediction is possible. I once asked a staff meteorologist of a large agribusiness corporation if his company paid tens of thousands of dollars for private, long-range daily weather forecasts. I'm afraid so, he shrugged, even though I strongly object on the grounds that it is theoretically impossible to have any local weather forecasting skill further than a few weeks. But corporate executives are supposed to be hard-nosed business people, I replied, so why would they want to throw $10,000 out on a set of well-hyped but nonetheless random numbers? I thought the same thing,

he said, until I found out that the vice president has to make weather-dependent decisions in advance. For example, should he deploy his fleet of tractors with tyres appropriate for muddy fields or dry fields in the spring? Should stored grain be scheduled for shipment by barge or by railroad, a weather-dependent question, influenced by the ease of navigation down the Mississippi? These decisions have to be made even though we can't reliably forecast the daily weather months in advance. But, he continued, the weather isn't the only factor in these decisions: economic questions, demand projections, cropping patterns, and so forth also figure in. Therefore, the vice president probably has an idea of what decision he'd like to make based on those nonweather factors and he usually goes ahead and makes it. A few thousand dollars for a forecast serves as an insurance policy, my friend explained. If the vice president's decision turns out badly, he can take the weather forecast out of the drawer and show it to the board of directors: if the forecast was wrong, he has a perfect excuse for making the wrong decision with the blame falling on someone else—the meteorologist. As incredible as it may seem, my friend concluded, it really isn't irrational from the perspective of the executive to make a small investment (by corporate standards) in meaningless long-range forecasts that could come in handy later on. In any case, even though most atmospheric scientists do not believe the details of daily weather can be forecast with more than random skill much beyond a few weeks, this does not also mean that the longer term prediction of average weather—that is, climate—is a hopeless task.

17. The temperature difference between winter and summer for the continental United States is about 25° C (45° F), a magnitude many times larger than the difference (i.e., the anomaly) of any one summer and its long-term average. Furthermore, the causes of short-term climatic anomalies can be very different from the causes of changes in the long-term means over many years. Likewise, it is much more difficult to forecast how next summer will be different from average than it is to forecast how the mean summertime temperature will change in thirty years if CO_2 increased 50%. Much of this depends upon how the climate system is forced to change and how predictable the forcing is.
18. E. M. Rasmusson and J. M. Hall, "El Niño, Pacific Ocean Event of 1982–1983," *Weatherwise* 36 (August 1983): 167–75.
19. K. E. Trenberth, G. W. Branstator, and P. A. Arkin, "Origins of the 1988 North American Drought," *Science* 242 (1988): 1640–45.
20. Lorenz, "Climatic Determinism"; J. Gleick, *Chaos* (New York: Viking, 1987).
21. K. C. Land and S. H. Schneider, "Forecasting in the Social and Natural Sciences: An Overview and Analysis of Isomorphisms," *Climatic Change* 11 (1987): 7–31.
22. For example, a global grid 4.5° latitude by 7.5° longitude with nine vertical levels takes some ten hours of time on a current supercomputer to compute a year of weather in thirty-minute time steps. In this example, divide the surface of the earth into a grid of 1,920 squares, each 4.5° latitude by 7.5° longitude. At 40° latitude each square is about 500 by 640 km (about 310 by 400 miles). Then divide the atmosphere above each square into nine strata. The calculation of ten years of climate statistics in these 17,280 "boxes" updated by thirty-minute increments takes some one hundred hours (at $1,000 per hour) on a Cray XMP computer (which can add nearly a billion numbers in one second!).
23. The issue of so-called "cloud feedback" (a climate feedback mechanism that affects the sensitivity of the climate to outside forcings like greenhouse gas increases) is undoubtedly the most contentious. For a short history of this issue, please see Schneider and Londer, *Coevolution*, 214–17. For more recent results of the effect of clouds on the climate as observed from satellites, see V. Ramanathan, R. D. Cess, E. F. Harrison, P. Minnis, B. R. Barkstrom, E. Ahmad, and D. Hartmann, "Cloud-Radiative Forcing and Climate: Results from the Earth Radiation Budget Experiment," *Science* 243 (1989): 57–63.
24. Two kinds of empirical validation can be mentioned. First of all, general circulation models produce daily maps of weather, and indeed these models, which associate clouds with storms, do have considerable predictive skill for several days. Secondly, clouds form and dissipate in hours to days; therefore, if there were a very large cloud feedback signal, it certainly would affect climate sensitivity on the time scale of a

season. Yet, models do very well in reproducing the seasonal cycle of surface temperature, which provides an empirical verification that the sum of all fast feedback processes in models is probably simulated to a rough factor of 2 (see the discussion in S. H. Schneider, "The Greenhouse Effect: Science and Policy," *Science* 243 [1989]: 771–81). A very important empirical confirmation of model performance in the so-called "water vapour-greenhouse effect" (positive) feedback process was recently published: A. Raval and V. Ramanathan, "Observational Determination of the Greenhouse Effect, *Nature* 342 (1989): 758. This paper was a major blow to critics of model-estimated global warming projections.

25. In 1989, a Washington D.C.-area group wrote a controversial report arguing that so much research progress could be made in three to five years' time that it was scientifically unjustifiable to implement any policy response to the buildup of greenhouse gases other than research (*Scientific Perspectives on the Greenhouse Problem,* Washington D.C.: George C. Marshall Institute, 1989). I strongly objected to this three to five year estimate, and think ten to twenty years is a more realistic range in which substantial progress can be expected, especially in reducing uncertainties surrounding prediction of time-evolving regional climatic changes. Presently, the scientific community is groping with the organization of a multidisciplinary research effort known as the International Geosphere Biosphere Program (IGBP) or, more generally, Global Change (T. F. Malone and J. G. Roederer, eds., *Global Change,* [Cambridge, Mass.: Cambridge University Press, 1984]). One aim of this effort is to get scientists from different fields to collaborate on the joint research necessary to develop a reliable predictive capacity of the regional changes that will occur over the next several decades. In addition to collaboration, to be useful this effort must proceed at an accelerated pace. However, such an effort may be difficult to achieve; academic institutions are usually organized by department, and rewards such as promotion, salary, and grants traditionally go to those who stay close to and publish original work in their discipline. This system discourages the brightest young people from entering interdisciplinary, problem-driven work. And although there are hundreds of qualified scientists around the world willing to work together at an accelerated pace on multidisciplinary problems, they need to be assured that long-term support and rewards would attend their efforts. In my opinion, if the academic community is left to its own traditions and existing resources, it will not mobilize quickly enough; to achieve a truly interdisciplinary, problem-driven effort, Congress and various science agencies here and abroad will need to fashion appropriate incentives. While I can't promise that such an effort would achieve worldwide consensus on the reliability of regional forecasts in the next decade, I have no doubt that it could substantially shorten the time it takes to build consensus on regional climate change prediction or, at worst, declare that such prediction is impossible. Either conclusion would put policy making—and certainly adaptive response planning—on a firmer factual basis and thus be worthwhile (S. H. Schneider, "The Whole Earth Dialogue," *Issues in Science and Technology* 4, no. 3 [1988]: 93–99).

26. P. R. Ehrlich and J. P. Holdren, "Impact of Population Growth," *Science* 171 (1971): 1212–17.

27. B. Bolin, "How Much CO_2 Will Remain in the Atmosphere?" in *Greenhouse Effect,* Bolin et al., 93–155.

28. R. Cicerone, "Methane Linked to Warming," *Nature* 334 (1988): 198. There is confusion about the lifetimes of CH_4 versus CO_2 in the atmosphere. It is true that a typical molecule of either gas is removed from the air in about ten years, but their subsequent fates are very different. CH_4 is oxidized by the chemical radical OH into (eventually) CO_2, but when some CO_2 molecules are removed by photosynthesis in a green plant or by uptake into the oceans, they typically reappear or cause other molecules of CO_2 to reappear shortly thereafter. Thus, the typical lifetime of an increase in atmospheric CO_2 *concentration* is some ten times longer than the lifetime of an individual CO_2 molecule. Therefore, it is inappropriate to compare the lifetime of CO_2 molecules to CH_4 molecules. Instead we must compare the lifetimes of the concentration anomalies, which differ by roughly a factor of ten. However, if CH_4 removes enough OH to permit the buildup of another greenhouse gas, tropospheric ozone, then CH_4 could have a worse greenhouse effect than these figures suggest.

29. Ramanathan et al., "Trace Gas Trends."

30. Ramanathan and colleagues have noted that uncertainty in projecting the trace greenhouse gases methane, chlorofluorocarbons, tropospheric ozone, nitrous oxide, and a number of other species is sufficient enough that they could increase the estimate for CO_2 alone by as much as a factor of 3 or perhaps as little as a factor of 1.5 ("Trace Gas Trends," Ramanathan et al.). The interactions among these gases make the use of "equivalent CO_2" a useful but still highly simplified concept that requires us to keep in mind that the potential interactions among these gases could well have other feedback effects on ozone, air pollution, and climate.
31. An encouraging suggestion by Firor implies that reducing CO_2 emissions could initially provide a bonus by reducing the airborne fraction—however, increasing CO_2 emissions could impose a penalty by increasing the airborne fraction, thus exacerbating the greenhouse effect (J. Firor, "Public Policy and the Airborne Fraction—Guest Editorial," *Climatic Change* 12 [1988]: 103–105). Although Firor's arguments are based on complicated processes, they can be summarized by noting that the oceans take up something like 2.5 of the approximately 5.5 billion tons of carbon injected into the atmosphere as CO_2 each year. Thus, if humans were somehow able to agree to cut their emissions from the present 5.5 to about 2.5 billion tons per year, then—for a few years at least—it is reasonable to expect the oceans to take up about the same 2.5 billion tons. However, others have argued that rapid biochemical adjustments in the ocean would soon limit this advantage as surface ocean waters mixed with deep waters. Nevertheless, I agree with Firor that qualitatively, at least, a quick drop in human emissions of CO_2 probably would get a bonus from ocean uptake just as a quick increase in CO_2 would get a penalty in the form of increased airborne fraction (neither effect having been considered on Figure 6). Despite all this uncertainty, most projections of total CO_2 concentration continue to suggest a doubling sometime between the years 2030 and 2080, although many other scenarios are possible, as Fig. 6 shows. Figure 6 is limited by an assumption of fixed airborne fraction of about 50%. This has been reinforced by a calculation by L. Danny Harvey at the University of Toronto (L. D. D. Harvey, "Managing Atmospheric CO_2," *Climatic Change,* Vol. 15 (1989):343–381). However, both Harvey's and Firor's conclusions are still quite controversial. In addition, debate over the carbon cycle intensified in 1989 when Taro Takahashi at Columbia University, Inez Fung at Goddard Institute for Space Studies and Peter Tans at National Oceanographic and Atmospheric Administration in Boulder argued that the oceans are taking up only about 1.5 billion tons of carbon as CO_2 annually and that mid-latitude forests are absorbing several billion tons. Undoubtedly this will spur major renewed debate and one hopes further research into the important question of carbon cycle behavior.
32. Woodwell has made these arguments in the informal literature, see for example, G. M. Woodwell, "Biotic Implications of Climate Change: Look to Forests and Soils," in *Preparing for Climate Change, Proceedings of the First North American Conference on Preparing for Climate Change: A Cooperative Approach,* October 27–29, 1987, Washington, D.C., ed. Climate Institute (Rockville, Md.: Government Institutes, 1988), 53–54.
33. Lashof, "Dynamic Greenhouse." See also A. Andreae and D. Schimel, eds., *Dahlem Workshop on Exchange of Trace Gases Between Terrestrial Ecosystems and the Atmosphere,* Berlin, February 19–24, 1989, 347.
34. The highest (5.2°C) and the lowest (1.9°C) values of recent GCM computed surface temperature change from a fixed CO_2 doubling are both attained with the same model: The United Kingdom Meteorological Office model used by John Mitchell and his colleagues. The lowest value was obtained by modifying the model's cloud parameterization scheme in several ways. This result created quite a media stir when published. For example, MIT meteorologist and greenhouse effect critic, Richard Lindzen cited the lowered sensitivity of Mitchell's model as having resulted from "an improvement." However, the British authors of the changes wisely recognized that "changes" are not necessarily "improvements" until they are validated: "Note that although the revised cloud scheme is more detailed it is not necessarily more accurate than the less sophisticated scheme. Our work highlights how different parameterization schemes can affect the sensitivity of the climate to CO_2." [Mitchell, J.F.B., C. A. Senior and W. J. Ingram, "CO_2 and Climate: A Missing Feedback?" *Nature,* Vol. 341, 1989): 132–134.] In any case, both of Mitchell's group's newest results are within the often cited 1.5° to 4.5°C range

for climate sensitivity to CO_2 doubling. Dickinson, "How Will Climate Change?"; Board on Atmospheric Sciences and Climate, Commission on Physical Sciences, Mathematics, and Resources, National Research Council, *Current Issues in Atmospheric Change, Summary and Conclusions of a Workshop October 30–31, 1986* (Washington, D.C.: National Academy Press, 1987).

35. J. Hansen, G. Russell, A. Lacis, I. Fung, and D. Rind, "Climate Response Times: Dependence on Climate Sensitivity and Ocean Mixing," *Science* 229 (1985): 857–59 (see especially Fig. 2). For a discussion of the delay time problem, see also L. D. D. Harvey and S. H. Schneider, "Transient Climate Response to External Forcing on 10^0–10^3 Year Time Scales. Part I: Experiments with Globally Averaged, Coupled, Atmosphere and Ocean Energy Balance Models," *Journal of Geophysical Research* 90 (1985): 2191–2205.
36. S. Manabe and R. Wetherald, "Reduction in Summer Soil Wetness Induced by an Increase in Atmospheric Carbon Dioxide," *Science* 232 (1986): 626–28.
37. W. Kellogg and Zong-Ci Zhao, "Sensitivity of Soil Moisture to Doubling of Carbon Dioxide in Climate Model Experiments. Part I: North America," *Journal of Climate* 1 (April 1988): 4, 348–66.
38. National Research Council, *Current Issues*, 9.
39. K. Bryan, S. Manabe, and M. J. Spelman, "Interhemispheric Asymmetry in the Transient Response of a Coupled Ocean-Atmosphere Model to a CO_2 Forcing," *Journal of Physical Oceanography* 18 (1988): 851–67.
40. S. H. Schneider and S. L. Thompson, "Atmospheric CO_2 and Climate: Importance of the Transient Response," *Journal of Geophysical Research* 86 (1981): 3135–47.
41. W. Washington and G. Meehl, "Climate Sensitivity due to Increased CO_2: Experiments with a Coupled Atmosphere-Ocean GCM," *Climate Dynamics* Vol. 4 (1989):1–38.
42. S. L. Thompson and S. H. Schneider, "CO_2 and Climate: The Importance of Realistic Geography in Estimating the Transient Response," *Science* 217 (1982): 1031–33.
43. D. Rind, R. Goldberg, and R. Ruedy, "Change in Climate Variability in the 21st Century," *Climatic Change* 14 (1989): 5–37.
44. L. O. Mearns, S. H. Schneider, S. L. Thompson, and L. McDaniel, "Analysis of Climate Variability in GCMs: Comparisons with Observations and Changes in Variability in $2 \times CO_2$ Experiments," *Journal Geophysical Research* (accepted).
45. For examples of the seasonal cycle performance of climate models, see M. Schlesinger and J. F. B. Mitchell, "Climate Model Simulations of the Equilibrium Climatic Response to Increased Carbon Dioxide," *Reviews of Geophysics* 25 (1987): 760–98; or Mearns et al., "Analysis of Climate Variability."
46. Schneider, "Climate Modeling."
47. It is possible that both the models and the observed temperature trends are correct and not inconsistent in the following sense: nonlinear processes in the climate system (for example, cloud feedback or sea ice feedback) could produce different sensitivity of the climate for different amounts of greenhouse forcing. That is, so far only a 25% increase in CO_2 and slightly larger increases of other trace gases are coincident with only ½° C of warming, which suggests a CO_2-doubling equivalent temperature increase of only about 1.5–2° C (R. L. Gilliland and S. H. Schneider, "Volcanic, CO_2, and Solar Forcing of Northern and Southern Hemisphere Surface Air Temperature," *Nature* 310 [1984]: 38–41; see also, T.M.L. Wigley and S.C.B. Raper, "Natural Variability of the Climate System and Detection of the Greenhouse Effect," *Nature* 344 [1990]: 324–327). On the other hand, if we wait long enough for an equivalent doubling of CO_2 in the atmosphere, it might very well prove a climate sensitivity in the 2° to 4° C warming range that present GCMs suggest, even though there is apparently lesser sensitivity to smaller increases in greenhouse gases. That at least is a possibility. Other possibilities are that the natural climate variability due to random or even chaotic internal dynamic processes within the climate system has produced a confounding noise level that has made the sensitivity detection difficult (for example, see the discussion in Schneider, "Science and Policy"), or that sulphate particles produced along with coal burning could create a cooling effect that would, at least on a regional basis, offset some of the global warming from CO_2 emission when coal is burned (see the discussion in endnote 17 of Chapter 4). An increase of up

to ½ of 1 per-cent in the solar energy output was speculated on as the cause of the twentieth century warming trend by the Marshall Institute authors (see note 25). However, those authors neglected to emphasize that this could not have been reliably observed by available instrumentation up to the 1970s, and that a solar energy output decrease, which could have opposed any greenhouse warming, is just as likely to have occurred. However, temperature fluctuations (whether or not they are associated with very large changes in sunspot number over the past 10,000 years or so) have not resulted in global temperature changes any larger than about 1°–2°C, since no changes larger than this have occurred since the ice age ended. Thus, if CO_2 and other trace gas projections are even remotely correct, whatever solar effect was or was not present in the twentieth century, or might be present in the twenty-first, would likely be swamped by global warming in the Greenhouse Century.

Chapter 6

1. E. Boulding, S. H. Schneider, in *Carbon Dioxide Effects, Research and Assessment Program: Workshop on Environmental and Societal Consequences of a Possible CO_2-Induced Climate Change*, Report 009, CONF-7904143, U.S. Department of Energy (Washington, D.C.: U.S. Government Printing Office, October 1980), 79–103; R. S. Chen, E. M. Boulding, and S. H. Schneider, eds., *Sociel Science Research and Climate Change: An Interdisciplinary Appraisal* (Dordrecht, the Netherlands: Reidel Publishing, 1983).
2. This distinction between multi- and interdisciplinary was made by R. S. Chen ("Interdisciplinary Research and Integration: The Case of CO_2 and Climate," in *Social Science Research,* Chen, Boulding, and Schneider). See also S. H. Schneider, "An International Program on 'Global Change': Can It Endure?—An Editorial," *Climatic Change* 10 (1987): 211–18; Schneider, "Whole Earth Dialogue."
3. All the quotes in this chapter from the congressional hearing on the DOE's impact assessment plans come from the following: *Carbon Dioxide and Climate: The Greenhouse Effect,* Hearing before the Subcommittee on Natural Resources, Agriculture Research and Environment and the Subcommittee on Investigations and Oversight of the Committee on Science and Technology, U.S. House of Representatives, 97th Congress, First Session, No. 45, July 31, 1981 (Washington, D.C., U.S. Government Printing Office, 1981).
4. L. B. Lave, "Mitigating Strategies for CO_2 Problems," in *IIASA Report CP-81-14,* May 1981, IIASA (Laxenburg, Austria: International Institute for Applied Systems Analysis, 1981), vi. Lave argued that uncertainties in projecting regional climate changes are so great that he favours adaptation as the primary response to the prospect of greenhouse warming.
5. The question-and-answer period—in which the greatest fireworks occurred—is found on pages 69–98 in the hearing transcript, "Carbon Dioxide and Climate," *U.S. House of Representatives.*
6. J. B. Smith and D. Tirpak, eds., *The Potential Effects of Global Climate Change on the United States: Draft Report to Congress,* vols. 1 and 2, U.S. Environmental Protection Agency, Office of Policy, Planning, and Evaluation, Office of Research and Development (Washington, D.C.: U.S. Government Printing Office, October 1988).
7. Waggoner, *Climate Change.*
8. G. I. Pearman, *Greenhouse: Planning for Climate Change* (Melbourne, Victoria, Australia: CSIRO Publications, 1988).
9. See Smith and Tirpak, *Potential Effects,* chap. 8, "Water Resources."
10. J. R. Hanchey, K. E. Schilling, and E. Z. Stakhiv, "Water Resources Planning Under Climate Uncertainty," in *Cooperative Approach,* First North American Conference on Preparing for Climate Change, 399.
11. P. Gleick, "Vulnerabilities of Water Systems," chap. 10 in *Climate Change,* Waggoner.
12. C. W. Stockton and W. R. Boggess, *Geohydrological Implications of Climate Change on Water Resources Development* (Fort Belvoir, Va.: U.S. Army Coastal Engineering Research Center, 1979).

13. M. H. Glantz and J. H. Ausubel, "Impact Assessment by Analogy: Comparing the Impacts of the Ogallala Aquifer Depletion and CO_2-Induced Climate Change," in *Societal Responses,* Glantz, 113–42.
14. P. H. Gleick, "Regional Hydrologic Consequences of Increases in Atmospheric CO_2 and Other Trace Gases," *Climatic Change* 10 (1987): 137–60.
15. See Smith and Tirpak, *Potential Effects,* vol. 1 , chapter 4, pp. 24 and 41.
16. D. Marchand, M. Sanderson, D. Howe, and C. Alpauch, "Climatic Change and Great Lakes Levels," *Climatic Change* 12 (1988): 107–33. See also, Smith and Tirpak, *Potential Effects,* vol. 1, chap. 5, "Great Lakes."
17. Smith and Tirpak, *Potential Effects,* vol. 1, chap. 6, "Southeast."
18. A. B. Pittock, "Actual and Anticipated Changes in Australia's Climate," in *Greenhouse,* Pearman, 35–51.
19. B. P. Sadler, G. W. Mauger, and R. A. Stokes, "The Water Resource Implications of a Drying Climate in South-Western Australia," in *Greenhouse,* Pearman, 296–311. The section quoted is on p. 308.
20. Schwarz and Dillard, "Urban Water."
21. Sadler, Mauger, and Stokes, "Water Resource Implications," 309.
22. L. van der Leij, "Holland's New Horizons," *Holland Herald* 23, no. 5 (May 1988): 18.
23. G. P. Hekstra, "Regional Consequences and Responses to Sea-Level Rise," in *Greenhouse Warming: Abatement and Adaptation,* eds. N. J. Rosenberg, W. E. Easterling, III, P. Crosson, and J. Darmstadter (Washington, D.C.: Resources for the Future, 1989).
24. Sir C. Tickell, *Climatic Change and World Affairs* (Cambridge, Mass.: Center for International Affairs, Harvard University, 1986).
25. Hekstra, "Regional Consequences."
26. K. D. Cocks, A. J. Gilmour, and N. H. Wood, "Regional Impacts of Rising Sea Levels in Coastal Australia," in *Greenhouse,* Pearman, 105–20; quoted passages are found on p. 108.
27. Emanuel, "Dependence of Hurricane Intensity on Climate."
28. K. P. Stark, "Designing for Coastal Structures in a Greenhouse Age," in *Greenhouse,* Pearman, 161–76; quoted passage is on 172.
29. Smith and Tirpak, *Potential Effects,* vol. 2, chap. 9, "Sea Level Rise." The rest of the quoted material from the EPA report dealing with sea level rise comes from this chapter.
30. J. G. Titus, "Greenhouse Effect, Sea Level Rise, and Coastal Zone Management," *Coastal Zone Management Journal* 14 (1986): 3, 147–71.
31. S. P. Leatherman, "Coastal Geomorphic Responses to Sea Level Rise: Galveston Bay, Texas," in *Greenhouse Effect and Sea Level Rise: A Challenge for this Generation,* eds. M. C. Barth and J. G. Titus (New York: Van Nostrand, Reinhold, 1984).
32. Smith and Tirpak, *Potential Effects,* vol. 2, 9–38.
33. J. Oerlemans, "A Projection of Future Sea Level," *Climatic Change* (in press).
34. T. M. L. Wigley and S. C. B. Raper, "Thermal Expansion of Sea Water Associated with Global Warming," *Nature* 330 (1987): 127–31.
35. M. F. Meier, "Contribution of Small Glaciers to Global Sea Level," *Science* 226 (1984): 1418–21.
36. Oerlemans's estimates for sea level decrease from east Antarctic ice sheet buildup were based on precipitation changes calculated by S. Manabe and R. J. Stouffer in their general circulation model ("Sensitivity of a Global Climate Model to an Increase in CO_2 Concentration in the Atmosphere," *Journal of Geophysical Research* 85 [1980]: 5529–54). However, GCMs do not well represent precipitation processes in Antarctica, where most of the accumulation is due not to snowfall from clouds but to ice crystals forming near the surface, often in clear skies, as well as frost deposition to the surface. These mechanisms depend on the extreme temperature inversion, where a thin layer of air (only about 30 metres, or 100 feet, thick) near the surface is sometimes as much as 30° C (54° F) colder than the air above it—which is not resolved by most GCMs. Clearly, this area needs more study in order to determine whether the processes whereby snowfall occurs on Antarctica would change with climatic warming, processes presently not well modeled in general circulation models.
37. W. F. Budd, B. J. McInnes, D. Jenssen, and I. N. Smith, "Modelling the Response of the West Antarctic Ice Sheet to a Climatic Warming," in *Glaciers, Ice Sheets and Sea*

Level: Effects of a CO_2-Induced Climatic Change, eds. C. J. van der Veen and J. Oerlemans (Washington, D.C.: National Academy Press, 1987), 172–77.

38. M. L. Parry, T. R. Carter, and N. T. Konijn, eds., *The Impact of Climate Variations on Agriculture* (Dordrecht, the Netherlands: Kluwer Academic Publishers, 1988).
39. K. Meyer-Abich, "Socioeconomic Impacts of CO_2-Induced Climatic Changes and the Comparative Chances of Alternative Political Responses: Prevention, Compensation and Adaptation," *Climatic Change* 2 (1980): 373–85.
40. P. E. Waggoner, "Agriculture and a Climate Changed by More Carbon Dioxide," in *Changing Climate,* Carbon Dioxide Assessment Committee, National Research Council (Washington, D.C.: National Academy Press, 1983) 383–418.
41. W. E. Easterling, M. L. Parry, and P. R. Crosson, "Adapting future agriculture to changes in climate," in *Greenhouse Warming: Abatement and Adaptation,* eds. N. J. Rosenberg, W. E. Easterling, III, P. Crosson, and J. Darmstadter (Washington, DC: Resources for the Future, 1989) 91–104.
42. P. Martin, N. J. Rosenberg, and M. S. McKenney, "Sensitivity of Evapotranspiration in a Wheat Field, a Forest, and a Grassland to Changes in Climate and Direct Effects of Carbon Dioxide," *Climatic Change,* Vol. 14, 1989, pp. 117–151.
43. A number of these studies, including one by Cynthia Rosenzweig at the Goddard Institute for Space Studies and a number of others, are given in Smith and Tirpak, *Potential Effects,* vol. 2, chap. 10.
44. D. F. Peterson and A. A. Keller, "Irrigation," in *Climate Change,* Waggoner.
45. D. J. Dudek, Statement before the House Subcommittee on Water and Power Resources Concerning Implications of Global Warming for Natural Resources, September 27, 1988, Subcommittee on Water and Power Resources of the Committee on Interior and Insular Affairs, U.S. House of Representatives, 100th Cong., 2nd sess., 1988, Serial No. 100–58, *Oversight Hearing on Implications of Global Warming for Natural Resources* (Washington, D.C.: U.S. Government Printing Office, 1989).
46. This and other quotes from Sinha and colleagues, who are with the Indian Agriculture Research Institute, New Delhi, were taken from a conference paper of theirs (S. K. Sinha, N. H. Rao, and M. S. Swaminathan, "Food Security in the Changing Global Climate," in Conference Proceedings, The Changing Atmosphere: Implications for Global Security, Toronto, Ontario, Canada, June 27–30, 1988, WMO, Geneva.
47. The quotation is attributed to G. Bruce, of the Canadian International Development Agency, in *The Changing Atmosphere,* S. K. Sinha.
48. This interdisciplinary project is known as COHMAP, the Cooperative Holocene Mapping Project, which combines ecologists, meteorologists, geologists, and other specialists interested in the earth's past climate history who reconstructed the details of climate change over the past 20,000 years. They have a large volume of published results, but presented a condensed version in an article in *Science* magazine. See COHMAP Members, "Climatic Changes."
49. T. Webb III, P. J. Bartlein, and J. E. Kutzbach, "Climatic Change in Eastern North America During the Past 18,000 Years; Comparisons of Pollen Data with Model Results," in *North America and Adjacent Oceans During the Last Glaciation,* vol. K–3, eds. W. F. Ruddiman and H. E. Wright, Jr. (Boulder, Colo.: Geological Society of America, 1987), 447–62.
50. Emanuel, Shugart, and Stevenson, "Climatic Change"; Pastor and Post, "Response of Northern Forests"; Botkin and Nisbet, "Effects of Climate Change." A. M. Solomon, "Transient Response of Forests to CO_2-Induced Climate Change: Simulation Modeling Experiments in Eastern North America," *Oecologia* 68 (1986): 567–579.
51. Smith and Tirpak, *Potential Effects,* vol. 2, chap. 11, "The Potential Impact of Rapid Climatic Change on Forests in the United States," 11–34.
52. R. L. Peters and J. D. S. Darling, "The Greenhouse Effect and Nature Reserves," *BioScience* 35 (1985): 707–17.
53. L. Roberts, "Is There Life After Climate Change?" *Science* 242 (1988): 1010–12. Quoted material on p. 1012.
54. P. R. Ehrlich and A. H. Ehrlich, *Extinction: The Causes and Consequences of the Disappearance of Species* (New York: Random House, 1981); N. Meyers, *The Sinking Ark* (New York: Pergammon Press, 1979); J. Gradwohl and R. Greenberg, *Saving the Tropical Forests* (Washington, D.C.: Island Press, 1988).

55. R. H. MacArthur and E. O. Wilson, *The Theory of Island Biogeography* (Princeton, N.J.: Princeton University Press, 1967).
56. P. R. Ehrlich, D. D. Murphy, and B. A. Wilcox, "Islands in the Desert," *Natural History* (October 1988): 59–64. Quotation on p. 64.
57. This and other health issues in the EPA report are given in Smith and Tirpak, *Potential Effects,* vol. 2, chap. 14, "Human Health."
58. A. Dobson, "Effects of Global Warming on Parasites and Pathogens in Ecological Communities," in *Consequences of the Greenhouse Effect,* Peters.
59. See Smith and Tirpak, *Potential Effects,* vol. 2, chap. 15, "Urban Infrastructure."
60. R. Sedjo and A. Solomon, "Climate and Forests," in *Greenhouse Warming, Abatement and Adaptation,* V. Rosenberg, W. E. Easterling III, P. R. Crosson and J. Darmstadter (eds.), (Washington D.C., Resources for the Future, 1989): 105–119.

Chapter 7

1. Subcommittee on the Environment and the Atmosphere of the Committee on Science and Technology, U.S. House of Representatives, 94th Congress, 2nd sess., 1976, vol. 78. *The National Climate Program Act,* Hearings, May 18–20, 25–27, 1976 (Washington, D.C.: U.S. Government Printing Office).
2. H. A. Ingram, H. J. Cortner, and M. K. Landy, "The Political Agenda," chap. 18 in *Climate Change,* Waggoner.
3. My schedule conflict was a previous agreement to discuss and debate the strategic defence initiative and "nuclear winter" with the controversial physicist Edward Teller in front of a group of students in Flagstaff, Arizona. Interestingly, when Teller arrived and the organizers joined us for lunch, I found it difficult to engage him in discussions about "star wars" or any other defence issues because all he wanted to talk to me about was the greenhouse effect and how to deal with it.
4. J. E. Hansen, "The Greenhouse Effect: Impacts on Current Global Temperature and Regional Heat Waves," prepared statement (U.S. Congress, Senate, Committee on Energy and Natural Resources, *The Greenhouse Effect and Global Climate Change,* Hearing, June 23, 1988, 100th Cong., 1st sess., 1988, pt. 2, 42–49).
5. Indeed, if I may indulge in a fond recollection, Jim helped me in a very early and controversial paper I published with my institute mentor, S. I. Rasool. Jim suggested an effective yet accurate and simple method for making a radiative-transfer calculation, which I then used with Rasool – a method originally brought to widespread scientific attention by Carl Sagan and James Pollack (C. Sagan and J. B. Pollack, "Anisotrophic Nonconservative Scattering and the Clouds of Venus," *Journal of Geophysical Research* 72 [1967]: 469).
6. The record that Hansen showed to Congress can be found in Hansen and Lebedeff, "Global Surface Air Temperatures." The concern about the correlation of global temperature anomalies over many years is discussed in T. R. Karl, "Multi-year Fluctuations of Temperature and Precipitation: The Gray Area of Climate Change," *Climatic Change* 12 (1988): 179–98.
7. Karl and Jones, "Urban Bias."
8. W. Sullivan, "Goals for U.S. Urged on Weather Control," *New York Times* (December 29, 1972): 50.
9. S. H. Schneider, "What Makes a Good Science Story?" panel discussion, in *Scientists and Journalists: Reporting Science as News,* eds. S. M. Friedman, S. Dunwoody, and C. L. Rogers (New York: Free Press, 1986), 103–15.
10. One good story that appeared was by A. H. Malcolm ("In Ashes of Burned Forests, a Rare Chance to Study Nature's Recovery," *New York Times* [September 27, 1988]: C-1).
11. Klinger, "Successional Change."
12. A. Solow, "Pseudo Scientific Hot Air; The Data on Climate Are Inconclusive," *New York Times* (December 28, 1988): A-15. In connection with the problem of scientists with narrow perspectives who communicate to broad audiences, physicist and policy analyst John Ahearne commented: "When an expert does not communicate effectively, it usually stems from inability or unwillingness. Failure to communicate well also can be connected, however, with an overestimation or overvaluation of one's own

expertise. People who are well informed about science and engineering in general but not about the specific policy questions in dispute should not be called experts, but often they believe they are experts. In contrast to those who are aware of all the complexities bearing on the issues at hand, the less informed often take a paternalistic or maternalistic attitude toward the general public. Sometimes they express the belief that controversy would disappear if only the public were better educated—if only, that is, the public became as well informed as they believe themselves to be. Sometimes they act as though the solution to conflict is simply for the public to trust them and what they claim." (J. F. Ahearne, "Addressing Public Concerns in Science," *Physics Today* [September 1988]): 36–42.)

13. S. Singer, "Fact and Fancy on Greenhouse Earth," *Wall Street Journal* (August 30, 1988); P. Michaels, "The Greenhouse Climate of Fear," *Washington Post* (January 8, 1989).
14. S. H. Schneider, "Global Warming: Scientific Reality or Political Hype?" (U.S. Congress, House, Committee on Energy and Commerce, Subcommittee on Energy and Power, Hearing, February 21, 1989, *Global Warming*, 101st Cong., 1st sess., in press).
15. J. Hansen, letter to *New York Times* (January 11, 1989); Trenberth, Branstator, and Arkin, "Origins."
16. Kerr, "Global Warming Is Real."
17. P. Shabecoff, "Global Warmth in '88 Is Found To Set a Record," *New York Times* (February 4, 1989): 1. Phil Jones told me by letter that Shabecoff did indeed correctly report Jones and Wigley's latest results.
18. Schneider, "Science and Policy." See also endnote 17, chapter 4.
19. S. H. Schneider, "The Greenhouse Effect and the U.S. Summer of 1988: Cause and Effect or a Media Event?—An Editorial," *Climatic Change* 13 (1988): 113–16.
20. For an excellent account of the early phase of the public and private ozone debate (before the ozone hole was discovered), see L. Dotto and H. Schiff, *The Ozone War* (Garden City, N.Y.: Doubleday, 1978).
21. For a recent survey of atmospheric chemistry, see McElroy and Salawitch, "Changing Composition." For a discussion of health and environmental effects see J. G. Titus, ed., *Effects of Changes in Stratospheric Ozone and Global Climate, Volume I: Overview* (Washington, D.C.: United Nations Environment Programme and U.S. Environmental Protection Agency, 1986).
22. R. C. Worrest, "The Effect of Solar UV-B Radiation on Aquatic Systems: An Overview," in *Effects of Changes*, Titus, 175–91.
23. M. J. Molina and F. S. Rowland, "Stratospheric Sink for Chlorofluoromethanes: Chlorine Atom Catalysed Destruction of Ozone," *Nature* 249 (1974): 810–12.
24. Dotto and Schiff, *Ozone War.*
25. Sir B. J. Mason, in *Proceedings of the World Climate Conference,* World Climate Conference (Geneva: World Meteorological Organization, 1979); Sir B. J. Mason, "Has the Weather Gone Mad?" *New Republic Magazine* (July 30, 1977): 23.
26. This *Time* magazine conference resulted in the 1988 *Time* cover story in which the earth was named "Planet of the Year" (*Time,* January 2, 1989). A transcript made by *Time* of the conference's climate change working group is the source of a number of quotes throughout this book.
27. J. C. Farman, B. G. Gardiner, and J. D. Shanklin, "Large Losses of Total Ozone in Antarctica Reveal Seasonal ClO_x/NO_x Interaction," *Nature* 315 (1985): 207–10.
28. This quote from Richard Benedick and the many others that appear in the remainder of this chapter were taken from a manuscript he prepared for the Conservation Foundation on the history of the problem: R. Benedick, *The Ozone Protocol: A New Global Diplomacy* (Washington, D.C.: Georgetown University Institute for the Study of Diplomacy and The Conservation Foundation, in press).
29. R. Kerr, "Stratospheric Ozone Is Decreasing," *Science* 239 (1988): 1489–91; K. E. Trenberth, "Executive Summary of the Ozone Trends Panel Report," *Environment* 30 (July/August 1988): 25–26.
30. K. von Moltke, "International Agreement to Stabilize Climate: Lessons from the Montreal Protocol—An Editorial," *Climatic Change,* Vol. 14, No. 3, June 1989: 211–212. In early November 1989, an attempt was made in Noordwijk, Holland to work toward a framework convention on global warming. Although weakened by U.S., Japan and U.K. objections to specific CO_2 emission cuts, the ministerial conference on Atmospheric

Pollution and Climate Change did build on the 1988 Toronto momentum and stated: "Stabilizing the atmospheric concentration of greenhouse gases is an imperative goal." Of course, putting quantitative commitments for specific emission cuts in writing will be a much harder agreement to achieve.

Chapter 8

1. For example, Jones was a panelist at the national television broadcast of the main event of Greenhouse '88. Jones responded to a questioner who criticized the lack of mass transit in Australia in what I am told is a typical style, by quipping that Australians would never have decent mass transit since they were hopelessly locked into their "auto eroticism." An account of the Greenhouse '88 meeting can be found in "Opening Night," *In Future* 11 (December 1988).
2. Pearman, *Greenhouse.*
3. There were several articles addressing the greenhouse issue, one among them in the Melbourne, Australia, *Herald* (November 4, 1988): 4.
4. S. E. Schwartz, "Acid Deposition: Unraveling a Regional Phenomenon," *Science* 243 (1989): 753–63.
5. Although this came from a preliminary draft and the numbers quoted here should be considered just illustrative of the magnitude, these authors should be congratulated for even attempting this important kind of analysis: J. A. Edmonds, W. B. Ashton, H. C. Cheng, and M. Steinberg, *An Analysis of U.S. CO_2 Emissions Reduction Potential in the Period to 2010,* draft prepared for U.S. Department of Energy, September 24, 1988. Edmonds and Ashton are from the Pacific Northwest Laboratory in Washington, D.C., and Cheng and Steinberg are from the Brookhaven National Laboratory in Upton, New York. The information on CO_2 removal comes from the authors' Table 17.
6. C. Marchetti, "On Geoengineering and the CO_2 Problem," *Climatic Change* 1 (1977): 59–68. Another possibility is taking hydrogen from natural gas, which may exist in greater abundance deep in the earth than most believe: T. Gold, "Ancient carbon sources of atmospheric methane," *Nature* 335 (1988): 404.
7. M. I. Hoffert, Y-C. Wey, A. J. Calligari, and W. S. Broecker, "Atmospheric Response to Deep-Sea Injections of Fossil-fuel Carbon Dioxide," *Climatic Change* 2 (1979): 53–68.
8. Budyko noted that warming effects of human activities "can result in drastic changes of global climate which are of great practical importance. In this connection, the development of methods of climate modification for the purpose of climate control would be essential. One such method can be based on maintaining certain aerosol concentrations of the stratosphere" (Budyko, *Climatic Changes*). See also my critical discussion of this and other climate control schemes in Schneider with Mesirow, *Genesis Strategy,* 215–46. Of course, over the very long term, nature will drop us into an ice age. By that time thousands of years in the future, I suspect we will have the knowledge and I hope the wisdom to use safe countermeasures to stabilize climate (e.g., some form of CFC).
9. W. W. Kellogg and S. H. Schneider, "Climate Stabilization: For Better or Worse?" *Science* 186 (1974): 1163–72.
10. Klinger, "Successional Change."
11. Schneider and Londer, *Coevolution.* See especially chap. 11.
12. For a nice collection of articles on adaptation to environmental variations see Glantz, *Societal Responses.* Also, for an interesting discussion of security and environmental change, see P. Gleick, "Global Climatic Changes and Geopolitics: Pressures on Developed and Developing Countries," in *Climate and Geo-Sciences,* Berger, Schneider, and Duplessy.
13. All the quotes in this chapter from the *Time* magazine environment conference were taken from the transcript of the climate change working group, which met on Friday, November 11, 1988. Its participants included Sir Crispin Tickell, co-chair; Charles P. Alexander, co-chair; Jack A. Eddy; Michael H. Glantz; Albert Gore; F. Kenneth Hare; Brice Lalonde; Michael B. Lemonick; Paulo Nougeira-Neto; Vladimir Sakharov; Stephen H. Schneider; Gus Speth; Dick Thompson; Linda Williams; Maggie McComas; Thomas Sancton; and Sue Rafferty.

14. Not everyone agrees that we should replace some coal with natural gas. For example, Electric Power Research Institute President Emeritus Chauncey Starr recently articulated a typical view in the power industry: "Coal will continue to be widely used, especially in the developing world, because it is abundant and inexpensive. It is a pipe dream to think that poor nations will forgo the use of cheap energy resources to avoid some as yet unproven change in climate. Moreover, natural gas is too valuable as a petrochemical feedstock. I think it is almost a scandalous waste of natural resources to burn it to make electricity." (Quoted in B. Williams, "Greenhouse, Acid Rain Worries Buoy Prospects for U.S. Gas, Clean Coal," *Oil and Gas Journal* [August 29, 1988].) However, ten years earlier most industry representatives argued that if we didn't have substantial energy growth then we couldn't have economic growth. They were wrong, as post-OPEC price hikes set off market signals that led to energy efficiency investments and a steadily growing economy, with virtually no energy growth. For an opposite opinion to Starr, see T. Gold, "Ancient carbon sources of atmospheric methane," *Nature* 335 (1988): 404; or J. H. Ausubel, A. Grubler, and N. Nakicenovic, "Carbon Dioxide Emissions in a Methane Economy," *Climatic Change* 12 (1988): 245–264. In addition, some coal advocates and others have argued that methane system leakage of only a few percent would offset natural gas advantage over coal because CH_4 traps thirty times more infrared radiation on a per molecule basis than coal. However, as noted, in footnote 28 of Chapter 5, if one properly accounts for the much longer lifetime of CO_2 concentration anomalies in the atmosphere, then some 10% leakage from natural gas systems would be needed for methane to catch up to coal as a greenhouse gas, although it would still be environmentally superior from the acid point of view. Most industry analysts think that gas system leakage is at the few percent level, not the 10% level. However, methane does use up OH, one of nature's important cleansers. Methane also is broken up into carbon dioxide and water vapour in the stratosphere, the latter of which could contribute further to global heat trapping. Thus, methane emissions may yet prove to have additional problems.
15. S. Seidel and D. Keyes, *Can We Delay a Greenhouse Warming?* Strategic Studies Staff, Office of Policy and Resources Management (Washington, D.C.: U.S. Environmental Protection Agency, 1983).
16. P. R. Ehrlich, A. H. Ehrlich, and J. P. Holdren, *Ecoscience: Population, Resources, Environment* (San Francisco: Freeman Press, 1977); A. B. Lovins, L. H. Lovins, F. Krause, and W. Bach, *Least-Cost Energy: Solving the CO_2 Problem* (Andover, Mass.: Brick House, 1981); W. U. Chandler, H. S. Geller, and N. R. Ledbetter, *Energy Efficiency: A New Agenda* (Springfield, Va.: GW Press, 1988); J. Goldemberg, T. B. Johansson, A. K. N. Reddy, and R. H. Williams, *Energy for Development* (Washington, D.C.: World Resources Institute, 1987); I. N. Mintzer, *A Matter of Degrees: The Potential for Controlling the Greenhouse Effect* (Washington, D.C.: World Resources Institute, 1987).
17. Chandler, Geller, and Ledbetter, *Energy Efficiency*. See Table 3, p. 43.
18. For a general discussion of environment/development issues, see L. R. Brown et al., *State of the World* (New York: Norton, 1989) or the World Commission on Environment and Development, *Our Common Future*. Oxford University Press: New York, 1987.
19. A similar planetary bargain was proposed by J. P. Holdren, "Energy and Prosperity," *Bulletin of the Atomic Scientists* (January 1975): 26–28.
20. Schneider and Londer, *Coevolution*, 464–65.
21. W. W. Kellogg and M. Mead, *The Atmosphere: Endangered and Endangering*, No. NIH 77-1065 (Washington, D.C.: U.S. Department of Health, Education, and Welfare, 1975).
22. Conference Statement (*The Changing Atmosphere: Implications for Global Security*, Toronto, Ontario, Canada, June 27–30, 1988, Environment Canada [Toronto, 1988]).
23. Edmonds et al., *Analysis of U.S. CO_2*.
24. J. Goldemberg, T. B. Johansson, A. K. N. Reddy, and R. H. Williams, *Energy for a Sustainable World* (Washington, D.C.: World Resources Institute, 1987).
25. See text of a speech given by Prime Minister Margaret Thatcher at a Royal Society Dinner on Tuesday, September 27, 1988.
26. Senators Timothy Wirth of Colorado and John Heinz of Pennsylvania sponsored a study of environmental problems and solutions led by Harvard assistant professor of public policy, Dr. Robert N. Stavins. Their study explored the issue of extrapolating tradeable rights to global bargains. T. E. Wirth and J. Heinz, *Project 88, Harnessing*

Market Forces to Protect Our Environment: Initiatives for the New President (Washington, D.C.: December 1988; copies can be obtained by writing to either senator).

27. T. C. Schelling, "Climatic Change: Implications for Welfare and Policy," chap. 9 in *Changing Climate,* National Research Council, 449–82. Quotation from p. 481.

EPILOGUE

1. George C. Marshall Institute, *Scientific Perspectives on the Greenhouse Problem* (Washington D.C.: George C. Marshall Institute, 1989): 37 pp.
2. S. H. Schneider, 1989 personal correspondence to Alan Hecht as follows:

September 1, 1989

Dr. Alan Hecht
Office of International Activities
Environmental Protection Agency

Dear Alan:

Congratulations on your new job. I predict it will be stimulating both for you and the people you interact with. As you requested, here is my brief analysis of the recent Marshall Institute report (Scientific Perspectives on the Greenhouse Problem, 1989, George C. Marshall Institute, Washington, D.C.)

The Marshall Institute report is really three reports in one. The first part is a fairly standard review of the greenhouse science problem. It is typical of that in most National Academy of Science reports, and reflects the caveats normally given in the presentations of most of the climatic modelers that talk or write on this issue. There is an impression in the Marshall report that somehow these caveats have been underrepresented by some members of the community, but this is not strongly argued. In any case, up to page 14, I have few disputes with the narrative, other than its constant focus on uncertainty and its omission of the primary reasons that climate modelers are concerned for the future: validation of the models' performance against actual climatic changes. There is insufficient mention of model validation in the Marshall report. For example, the authors do not mention the excellent performance of most General Circulation Models in simulating the very large seasonal cycle of surface temperature, or the successes in simulation of neighboring planets, Mars and Venus, each of which has radically different greenhouse properties than earth, nor is there mention of the daily variability validation studies that have recently been published. While these validations do not remove the uncertainties focused on in the first 14 pages, the absence of the mention of the substantial degree of model validation of climate sensitivity suggests a bias.

On page 15 begin what I believe to be the major problems with this report. First of all, the authors note, citing my *Science* article (S. H. Schneider, *Science 243,* 771, 1989), that the global temperature response to CO_2 doubling is in the range of 1° to 5°C. They correctly point out that a temperature drop of roughly 1°C relative to today's temperatures occurred in the Little Ice Age. That episode occurred between about 1400 and 1850 in various parts of the world at various times and differing degrees. Thus, a natural cooling of 1°C would, the Marshall authors argue, completely compensate for the lower end of any greenhouse warming if such a Little Ice Age-like event recurred in the next century. Indeed, this is correct, but it fails to mention that the upper range of the CO_2 warming limit, (i.e., 5°C), would totally swamp any natural fluctuations of the type that have occurred in the past 10,000 years, the time since the end of the last ice age. Moreover, natural centuries-long fluctuations on the order of 1° *warming* (e.g., the so-called Medieval Optimum around 1,000 years ago) are just as likely to occur, and could *add* to any anthropogenic greenhouse warming in the next century. Indeed, what is absent in the remainder of the Marshall document is any statement, even intuitive ones, about the relative probabilities of the greenhouse warming being at the low end of the range and comparable natural cooling occurring relative to the probabilities that more middle or upper severity warming scenarios could occur. (e.g., see *Current Issues in Atmospheric Change,* National Academy of Sciences, 1987 for statements of the high probability of warming greater than 1°C in the 21st century.)

In section four the authors delve into the very complicated and long debated controversial issues of potential solar effects on climate. They cite the fact that sunspots have been relatively absent for certain century long periods, which have typical recurrence intervals of hundreds of years. The coincidence between the last such minimum, the "Maunder Minimum" between 1650 and 1700, and the "Little Ice Age" is cited. But what is not cited by the Marshall authors is the fact that before and after this 50 year long minimum the so-called Little Ice Age was in full force in many parts of the world. Furthermore, the coolest periods during this time did not necessarily occur in all places at the same time nor was it fully coincident with Maunder Minimum. In addition, a 10,000 year look at tree ring evidence shows many other such minima, but does not reveal consistent correlation of these sunspots minima with glacial advance periods, and therefore, no firm correlation can be established. This is why this area of research is still controversial, a point not adequately stressed in the Marshall document. Finally, the magnitude of natural, several century long global climate fluctuations during the past 10,000 year inter-glacial period is on the order of 1°C, which is at the low end of the 1° to 5°C range the Marshall authors concede as the consensus estimate for the next century.

Therefore, the higher probabilities are that whatever nature does on the sun or earth to cause natural fluctuations is likely to be swamped by manmade influence of greenhouse gases sometime between now and the first few decades of the 21st century. These relative probabilities get no prominent attention. Rather the lower probability scenarios are what the Marshall Institute authors focus on. Moreover, they cite recent evidence that some stars have been observed to have energy output changes on the order of a few tenths of a percent to explain recent global warming trends. By no means can this be cited to explain with high confidence the observed approximately 0.5°C warming trend of the past century. It is equally likely, as not mentioned by the Marshall Institution report, that a solar constant *decrease* from our sun could have been in progress during the past 100 years, and therefore, any greenhouse warming that has taken place could have been masked by such an event. Since the solar constant has not been well observed, except in the past decade, both scenarios are equally likely. Furthermore, if only a few tenths of a percent change in solar energy were responsible for the 0.5°C century long trend in climate over the past century, then this would suggest a planet that is relatively sensitive to small energy inputs. The Marshall Institute simply can't have it both ways: they can't argue on the one hand that small changes in solar energy output can cause large temperature changes, but that comparable changes in the energy input from greenhouse gases will not also produce comparable large signals. Either the system is sensitive to large scale radiative forcing or it is not, another factor not mentioned by the Marshall authors. Although they argue that a 33-year running average of sunspot numbers superficially resembles the temperature record of the past 100 years, they do wisely admit that this could well be a coincidence.

The Marshall authors go on to claim that "scientists' concerns for offering sound advice on the greenhouse problem have tended to rely on the observed temperature increase of 0.5°C since 1880 as their best evidence that the greenhouse effect is already here and that steps should be taken now to cope with its full development in the next century." (Marshall Institute, page 28.)

However, only very few people have made such a claim. Most scientists I know argue that it is not the performance of the planet in the past century, which only in the last decade is at the margin of the noise level of natural climate fluctuations, that motivates their concerns for the next century. Rather, our concerns are grounded in the very well-validated understanding of how radiative trapping by important trace gases like carbon dioxide, methane and chlorofluorocarbons can heat the surface. Further concern is based on the (less well) validated modeling sensitivity studies which suggest that something between 1° and 5°C warming is the most probable prediction for the next century, (e.g., NAS, 1987). Indeed, it will take another decade or two of observations, with a presumed continuing of the record heat of the 1980s, to establish to a high degree of confidence that the greenhouse forcing of the past century is finally unambiguously detected. But, if we follow the Marshall Institute's advice and wait for such certainty before implementing actions we will then have to adapt to a much larger dose of change than if we attempt to slow it down now. That decision, of course, is a *value judgement,* not a decision that can be made based solely on any scientific method.

Herein is my principal objection to the Marshall report, and in fact, an objection so serious as to make me doubt the capability of the authors to provide a balanced scientific view. They claim "current forecasts of the manmade greenhouse effect do not appear to be sufficiently accurate to be used as a basis for sound national policy decisions." (Marshall Institute, page 23.) Furthermore, they argue that "it is our judgement that if a prudent investment is made in computing power, observing programs and added manpower, answers that have a usable degree of reliability can be provided to policy makers within three to five years." Thus, they recommend no current policy response to the present debate, other than more research. What I find so objectionable in these statements, particularly the former one, is that it is not a scientific judgement, but their value judgement, which they do not explicitly claim as their personal views of how to respond to the range of uncertainties. The second statement is technically inaccurate, I believe, in that virtually all knowledgeable scientists would never claim that we will have a strong consensus of atmospheric researchers as to the reliability of the regional distribution of time evolving climate changes in three to five years, regardless of the level of research effort. If left to its own devices, the climate community is likely to take twenty-five years to provide such information. If the kinds of "prudent investments" that the Marshall Institute authors do wisely call for were implemented, then it is my opinion, and one I believe is widely shared in the scientific community, that we could *accelerate* substantially the rate of building such consensus. But, a decade or so would still be needed before which high resolution, coupled with models of atmosphere, ocean, sea ice, land hydrology, ecosystems, and chemistry could be adequately run, tested and validated to provide credible regional forecasts. Of course, it is better to produce such forecasts in ten years than in twenty-five and the costs are small relative to the potential risks. Thus, I agree with the advice to invest in such research. But to claim answers will be clear within three to five years and therefore, policy should wait for them, is in my opinion a scientifically erroneous judgment as well as a value judgment with which I personally disagree strongly. People may legitimately disagree about values, but scientists must always make this explicit, and the absence of this is what has me so disturbed about the Marshall Report.

Finally, the authors provide no cost/benefit analysis of the risks or benefits of alternative actions now versus the risks and benefits of delay. In the absence of such studies, or even the citation to the few such studies that exist, the Marshall report's primary conclusion to delay policy actions has little merit.

Alan, I hope this is useful to you and that you can use it to good purpose.

Warm regards and best wishes in your new position.

Sincerely,
Stephen H. Schneider
Head
Interdisciplinary Climate Systems

3. National Research Council, *Carbon Dioxide and Climate: A Scientific Assessment* (Washington, D.C.: National Academy Press, 1979): 11 pp; National Research Council, *Changing Climate: Report of the Carbon Dioxide Assessment Committee* (Washington D.C.: National Academy Press, 1983): 496 pp; National Academy of Sciences, *Current Issues in Atmospheric Change,* (Washington D.C.: National Academy Press, 1987): 39 pp., are examples of National Academy reports in which the well-travelled 1.5°C–4.5°C warming range is cited. A current study by the National Research Council's Committee on Science, Engineering and Public Policy is likely to reaffirm that range, in full awareness of the current debate.
4. Intergovernmental Panel on Climate Change (IPCC), *Scientific Assessment of Climate Change,* 2nd Draft, March 1990 (Geneva: World Meteorological Organization).
5. R. S. Lindzen, "Greenhouse Warming: Science Versus Consensus" (unpublished—but widely circulated—manuscript, dated October 10, 1989). A less *ad hominem* ver-

sion was published as: R. S. Lindzen, "Some Coolness Concerning Global Warming," *Bulletin of the American Meteorological Society* 77 (1990): 288–299. In the first version Lindzen asserted that the NAS consensus overestimated global warming by a factor of four to eight. The published version some months later suggested a more moderate factor of two to five overestimation.

6. W. T. Brookes, "The Global Warming Panic," *Forbes*, (25 December 1989): 96–102.
7. W. T. Brookes, *op. cit.*
8. W. T. Brookes, "Greenhouse Showdown or Show Trial?," *Washington Times (January* 24, 1990): F2.
9. J. F. B. Mitchell, C. A. Senior and W. J. Ingram, "CO_2 and Climate: A Missing Feedback?" *Nature* 341 (1989): 132–134.
10. J. Hansen, D. Johnson, A. Lacis, S. Lebedeff, P. Lee, D. Rind and G. Russell, "Climate Impact of Increasing Atmospheric Carbon Dioxide," *Science* 213 (1981): 957–966; R. L. Gilliland and S. H. Schneider, "Volcanic, CO_2 and Solar Forcing of Northern and Southern Hemisphere Surface Air Temperatures," *Nature* 310 (1984): 38–41. See also, T. M. L. Wigley and S. C. B. Raper, "Natural variability of the Climate System and Detection of the Greenhouse Effect," *Nature 344* (1990): 324–327.
11. H. W. Ellsaeser, "The Climatic Effect of CO_2: A Different View," *Atmospheric Environment* 18 (1984): 431–434.
12. M. Grubb, "The Greenhouse Effect: Negotiating Targets," Energy and Environmental Program (London: Royal Institute of International Affairs, 1989): 56 pp.
13. S. Engelberg, "Sununu Says He Revised Speech on Warming," *New York Times* (February 5, 1990): A15.
14. W. D. Nordhaus, "To Slow or Not To Slow: The Economics of the Greenhouse Effect," manuscript submitted to the Panel on Policy Implications of Greenhouse Warming; Committee on Science, Engineering and Public Policy (Washington D.C.: National Research Council, 1990).
15. To understand why one prominent economic model is very questionable in projecting economic costs of CO_2 emissions cuts (usually stated as a loss of gross national product) we need to explain briefly how it works. In essence, the econometric model used to project future GNP performance as a function of energy prices is calibrated by statistical methods (called regression equations) that match GNP change over time with energy prices over time. For example, following the OPEC oil price rise of 1973, GNP responded by decreasing for several years. Then, as economies adjusted to the higher prices and energy efficiency was encouraged by the price shock, GNP grew once again even though total energy consumption changed very little for the next half dozen years. Clearly, the immediate transient response of the GNP to energy price changes will be different from the medium term adjustment, which should again be different over the very long term. Yet, an econometric model used to project GNP response to CO_2 controls (which were dealt with as an equivalent energy tax) was run over 110 years into the future based on curve matching that included data on short term, transient conditions. Seen in this light it is not surprising that such models would predict large GNP losses 100 years into the future when energy prices go up—it is built into the model by regression equations that include transient situations and then apply them over the very long term. It is also no wonder such models were severely criticized at the U.S. National Academy of Sciences debate. More on this debate can be found in Williams, R. H., 1990 (Draft). "Will Constraining Fossil Fuel Carbon Dioxide Emissions Really Cost So Much?" A Critique of: Manne, A. S. and R. Richels, 1990 (Draft), "Global CO_2 Emissions Reduction—the Impacts of Rising Energy Costs," February 1990.
16. A. Raval and V. Ramanathan, "Observational Determination of the Greenhouse Effect," *Nature* 341 (1989): 758.
17. J. F. B. Mitchell et al., *Nature.*
18. In 1990 two NASA satellite experts [Spencer, R. W. and J. R. Christy, "Precise Monitoring of Global Temperature Trends from Satellites," *Science* 247 (1990): 1558–1562] stated the following:

> Passive microwave radiometry from satellites provides more precise atmospheric temperature information than that obtained from the relatively sparse

> distribution of thermometers over the earth's surface. Accurate global atmospheric temperature estimates are needed for detection of possible greenhouse warming, evaluation of computer models of climate change, and for understanding important factors in the climate system. Analysis of the first ten years (1979 to 1988) of satellite measurements of lower atmospheric temperature changes reveals a monthly precision of 0.01°C, large temperature variability on time scales from weeks to several years, but no obvious trend for the ten-year period. The warmest years, in descending order, were 1987, 1988, 1983, and 1980. The years 1984, 1985, and 1986 were the coolest.

The usually warm early 1980s were followed by a few cooler years that were coincident with the explosive volcanic eruption of El Chichon. This was followed a few years later by the warmest years in the instrumental record, 1987 and 1988 (see Fig. 4, p. 85). Whether the cool year-volcano correlation was coincidence or cause and effect is not certain. What is important is that the instrumental record for the 1980s correlated well with the satellite remote-sensing technique for mid-tropospheric temperature of Spencer and Christy. Yet, most of the media stories following the publication of this paper ignored the correlation between the thermometer and satellite records, and instead extrapolated out of context a sentence of the authors to the effect that no global warming trend was evident in the ten years of their analysis. Of course, no responsible scientist would ever claim that a global warming signal could be detected above background noise from ten years of data! Nevertheless, a frequent media interpretation of this study was that there was "no global warming trend"—clearly a serious misinterpretation of the facts given the one-hundred-year record (on Fig. 4).

Index